T0245450

CAMBRIDGE LIBRARY COLLECTION

Books of enduring scholarly value

Life Sciences

Until the nineteenth century, the various subjects now known as the life
sciences were regarded either as arcane studies which had little impact
on ordinary daily life, or as a genteel hobby for the leisured classes. The
increasing academic rigour and systematisation brought to the study of
botany, zoology and other disciplines, and their adoption in university
curricula, are reflected in the books reissued in this series.

My Life

Alfred Russel Wallace (1823–1913) was a British naturalist, explorer,
geographer and biologist, best remembered as the co-discoverer, with
Darwin, of natural selection. His extensive fieldwork and advocacy of the
theory of evolution led to him being considered one of the nineteenth
century's foremost biologists. He was later moved by a variety of personal
experiences to examine the concept of spirituality, but his exploration into
the potential for compatibility between spiritualism and natural selection
alienated him from the scientific community. He was also a social activist,
highly critical of unjust social and economic systems in nineteenth-century
Britain, and one of the first prominent scientists to express concern over
the environmental impact of human activity. This autobiography was first
published in 1905. Volume 2 deals with his many eminent acquaintances,
including Darwin and Huxley, his lecture tour in America, and his
involvement with spiritualism and with social activism.

Cambridge University Press has long been a pioneer in the reissuing of out-of-print titles from its own backlist, producing digital reprints of books that are still sought after by scholars and students but could not be reprinted economically using traditional technology. The Cambridge Library Collection extends this activity to a wider range of books which are still of importance to researchers and professionals, either for the source material they contain, or as landmarks in the history of their academic discipline.

Drawing from the world-renowned collections in the Cambridge University Library, and guided by the advice of experts in each subject area, Cambridge University Press is using state-of-the-art scanning machines in its own Printing House to capture the content of each book selected for inclusion. The files are processed to give a consistently clear, crisp image, and the books finished to the high quality standard for which the Press is recognised around the world. The latest print-on-demand technology ensures that the books will remain available indefinitely, and that orders for single or multiple copies can quickly be supplied.

The Cambridge Library Collection will bring back to life books of enduring scholarly value (including out-of-copyright works originally issued by other publishers) across a wide range of disciplines in the humanities and social sciences and in science and technology.

My Life

A Record of Events and Opinions

VOLUME 2

ALFRED RUSSEL WALLACE

CAMBRIDGE UNIVERSITY PRESS

Cambridge, New York, Melbourne, Madrid, Cape Town,
Singapore, São Paolo, Delhi, Tokyo, Mexico City

Published in the United States of America by Cambridge University Press, New York

www.cambridge.org
Information on this title: www.cambridge.org/9781108029599

This edition first published 1905
This digitally printed version 2011

ISBN 978-1-108-02959-9 Paperback

MY LIFE

"OLD ORCHARD," BROADSTONE.
(Built in 1902.)

MY LIFE

A RECORD OF EVENTS AND OPINIONS

BY

ALFRED RUSSEL WALLACE

AUTHOR OF

"MAN'S PLACE IN THE UNIVERSE," "THE MALAY ARCHIPELAGO," "DARWINISM,"
"GEOGRAPHICAL DISTRIBUTION OF ANIMALS," "NATURAL
SELECTION AND TROPICAL NATURE," ETC.

*WITH FACSIMILE LETTERS, ILLUSTRATIONS
AND PORTRAITS*

TWO VOLUME

VOLUME II

LONDON: CHAPMAN & HALL, LD.

1905

CONTENTS

CONTENTS

ILLUSTRATIONS

viii ILLUSTRATIONS

NOTE.

The portraits of the Author illustrating this work show him at the following ages :—

AGE 25. BEFORE GOING TO THE AMAZON.

„ 30. BEFORE THE MALAYAN JOURNEY.

„ 46. AT PUBLICATION OF "MALAY ARCHIPELAGO."

„ 55. AT PUBLICATION OF "TROPICAL NATURE."

„ 80. FRONTISPIECE.

The Frontispiece to "Darwinism" (age 66) completes the Series.

MY LIFE

A RECORD OF EVENTS AND OPINIONS

CHAPTER XXV

MY FRIENDS AND ACQUAINTANCES—DARWIN

SOON after I returned home, in the summer of 1862, Mr. Darwin invited me to come to Down for a night, where I had the great pleasure of seeing him in his quiet home, and in the midst of his family. A year or two later I spent a week-end with him in company with Bates, Jenner Weir, and a few other naturalists ; but my most frequent interviews with him were when he spent a few weeks with his brother, Dr. Erasmus Darwin, in Queen Anne Street, which he usually did every year when he was well enough, in order to see his friends and collect information for his various works. On these occasions I usually lunched with him and his brother, and sometimes one other visitor, and had a little talk on some of the matters specially interesting him. He also sometimes called on me in St. Mark's Crescent for a quiet talk or to see some of my collections.

My first letter from him dealing with scientific matters was in August, 1862, and our correspondence was very extensive during the period occupied in writing or correcting his earlier books on evolution, down to the publication of "The Expression of the Emotions in Man and Animals," in 1872, and afterwards, at longer intervals, to within less than a year of his death. A considerable selection of our correspondence has been published in the "Life and Letters" (1887), and especially in "More Letters" (1903); while several of the more interesting of these were contained in the one-volume life, entitled "Charles Darwin," which appeared in 1892. As many of my readers, however, may not have these works to refer to, I will here give a few of his letters to myself which have not yet been published, together with some of my own, and also occasional extracts from some of Darwin's that have already appeared, in order to make clear the nature of our discussions, and also, perhaps, to throw a little light upon our respective characters.

In a letter entirely without date, but which was evidently written in 1863, he gives me some information for which I had asked about reviews of the "Origin of Species."

"Down, Bromley, Kent (1863).

"MY DEAR MR. WALLACE,

"I write one line to thank you for your note, and to say that the B. of Oxford wrote the *Quarterly R.* (paid £60), aided by Owen. In the *Edinburgh,* Owen no doubt praised himself. Mr. Maw's review in *Zoologist* is one of the best, and staggered me in parts, for I did not see the sophistry of (those) parts. I could lend you any which you might wish to see, but you would soon be tired. Hopkins in *Fraser* and Pictet are two of the best.

"I am glad you like the little orchid book; but it has not been worth the ten months it has cost me : it was a hobby-horse, and so beguiled me.

"How puzzled you must be to know what to begin at! You will do grand work, I do not doubt. My health is, and always will be, very poor: I am that miserable animal, a regular valetudinarian.

"Yours very sincerely,

"C. DARWIN."

In March, 1864, he wrote me from Malvern Wells that he had been very ill at home, having fits of vomiting every day for two months, and been able to do nothing. These attacks

were brought on by the least mental excitement, which often
rendered it impossible for him to see his friends, and which
appear to have lasted at intervals throughout his life. This
must always be remembered when we consider the enormous
amount of work he was able to do ; but, fortunately, the quiet
interest of carrying out observations or experiments lasting
for months, and often for years, seem to have been beneficial.
On the other hand, writing his books and correcting the
MSS. and the proofs in the very careful manner he always
practised were most wearying and distasteful to him.

On February 23, 1867, he wrote to me asking if I could
solve a difficulty for him.   He says : " On Monday evening I
called on Bates, and put a difficulty before him which he could
not answer, and, as on some former similar occasion, his first
suggestion was, ' You had better ask Wallace.'   My difficulty
is, Why are caterpillars sometimes so beautifully and artistic-
ally coloured ?   Seeing that many are coloured to escape
dangers, I can hardly attribute their bright colour in other
cases to mere physical conditions.   Bates says the most
gaudy caterpillar he ever saw in Amazonia was conspicuous
at the distance of yards, from its black and red colours,
whilst feeding on large, green leaves.   If any one objected to
male butterflies having been made beautiful by sexual
selection, and asked why they should not have been made
beautiful as well as their caterpillars, what would you answer ?
I could not answer, but should maintain my ground.   Will
you think over this, and some time, either by letter or when
we meet, tell me what you think ? "

On reading this letter, I almost at once saw what seemed
to be a very easy and probable explanation of the facts.   I
had then just been preparing for publication (in the *Westminster Review*) my rather elaborate paper on " Mimicry and
Protective Colouring," and the numerous cases in which
specially showy and slow-flying butterflies were known to
have a peculiar odour and taste which protected them from
the attacks of insect-eating birds and other animals, led me at
once to suppose that the gaudily-coloured caterpillars must

have a similar protection. I had just ascertained from Mr. Jenner Weir that one of our common white moths (*Spilosoma menthrastri*) would not be eaten by most of the small birds in his aviary, nor by young turkeys. Now, as a *white* moth is as conspicuous in the *dusk* as a *coloured* caterpillar in the *daylight,* this case seemed to me so much on a par with the other that I felt almost sure my explanation would turn out correct. I at once wrote to Mr. Darwin to this effect, and his reply, dated February 26, is as follows :—

"MY DEAR WALLACE,
          "Bates was quite right ; you are the man to apply to in a difficulty. I never heard anything more ingenious than your suggestion, and I hope you may be able to prove it true. That is a splendid fact about the white moths ; it warms one's very blood to see a theory thus almost proved to be true."

The following week I brought the subject to the notice of the Fellows of the Entomological Society at their evening meeting (March 4), requesting that any of them who had the opportunity would make observations or experiments during the summer in accordance with Mr. Darwin's suggestion. I also wrote a letter to *The Field* newspaper, which, as it explains my hypothesis in simple language, I here give entire :—

"CATERPILLARS AND BIRDS.
  "SIR,
          "May I be permitted to ask the co-operation of your readers in making some observations during the coming spring and summer which are of great interest to Mr. Darwin and myself ? I will first state what observations are wanted, and then explain briefly why they are wanted. A number of our smaller birds devour quantities of caterpillars, but there is reason to suspect that they do not eat all alike. Now we want direct evidence as to which species they eat and which

they reject. This may be obtained in two ways. Those who
keep insectivorous birds, such as thrushes, robins, or any of
the warblers (or any other that will eat caterpillars), may
offer them all the kinds they can obtain, and carefully note
(1) which they eat, (2) which they refuse to touch, and (3)
which they seize but reject. If the name of the caterpillar
cannot be ascertained, a short description of its more prominent characters will do very well, such as whether it is
hairy or smooth, and what are its chief colours, especially
distinguishing such as are green or brown from such as are of
bright and conspicuous colours, as yellow, red, or black. The
food plant of the caterpillar should also be stated when known.
Those who do not keep birds, but have a garden much fre-
quented by birds, may put all the caterpillars they can find
in a soup plate or other vessel, which must be placed in a
larger vessel of water, so that the creatures cannot escape,
and then after a few hours note which have been taken and
which left. If the vessel could be placed where it might be
watched from a window, so that the kind of birds which took
them could also be noted, the experiment would be still more
complete. A third set of observations might be made on
young fowls, turkeys, guinea-fowls, pheasants, etc., in exactly
the same manner.

"Now the purport of these observations is to ascertain the
law which had determined the coloration of caterpillars.
The analogy of many other insects leads us to believe that
all those which are green or brown, or of such speckled or
mottled tints as to resemble closely the leaf or bark of the
plant on which they feed, or the substance on which they
usually repose, are thus to some degree protected from the
attacks of birds and other enemies. We should expect,
therefore, that all which are thus protected would be greedily
eaten by birds whenever they can find them. But there are
other caterpillars which seem coloured on purpose to be conspicuous, and it is very important to know whether they have
another kind of protection, altogether independent of disguise,
such as a disagreeable odour and taste. If they are thus protected, so that the majority of birds will never eat them, we

can understand that to get the full benefit of this protection they should be easily recognized, should have some outward character by which birds would soon learn to know them and thus let them alone; because if birds could not tell the eatable from the uneatable till they had seized and tasted them, the protection would be of no avail, a growing caterpillar being so delicate that a wound is certain death. If, therefore, the eatable caterpillars derive a partial protection from their obscure and imitative colouring, then we can understand that it would be an advantage to the uneatable kinds to be well distinguished from them by bright and conspicuous colours.

"I may add that this question has an important bearing on the whole theory of the origin of the colours of animals, and especially of insects. I hope many of your readers may be thereby induced to make such observations as I have indicated, and if they will kindly send me their notes at the end of the summer, or earlier, I will undertake to compare and tabulate the whole, and to make known the results, whether they confirm or refute the theory here indicated.

"ALFRED R. WALLACE.

"9, St. Mark's Crescent, Regent's Park, N.W.,
    "March, 1867."

This letter brought me only one reply, from a gentleman in Cumberland, who informed me that the common "gooseberry" caterpillar, which is the lava of the magpie moth (*Abraxas grossulariata*), is refused by young pheasants, partridges, and wild ducks, as well as sparrows and finches, and that all birds to whom he offered it rejected it with evident dread and abhorrence. But in 1869 two entomologists, Mr. Jenner Weir and Mr. A. G. Butler, gave an account of their two seasons' experiments and observations with several of our most gaily-coloured caterpillars, and with a considerable variety of birds, and also with lizards, frogs, and spiders, confirming my explanation in a most remarkable manner. An account of these experiments is given in the second and all later editions of my book on "Natural Selection;" but it

is more fully treated in my "Darwinism," chap. ix., under
the heading "Warning Colours among Insects," and it has thus
led to the establishment of a general principle which is very
widely applicable, and serves to explain a not inconsiderable
proportion of the colours and markings in the animal world.
It is, of course, only a wider application of the same fundamental fact by which Bates had already explained the purpose
of "mimicry" among insects, and it is a matter of surprise to
me that neither Bates himself nor Darwin had seen the probability of the occurrence of inedibility in the larvæ as well as
in the perfect insects.

In the year 1870 Mr. A. W. Bennett read a paper before
Section D. of the British Association at Liverpool, entitled
"The Theory of Natural Selection from a Mathematical
Point of View," and this paper was printed in full in *Nature*
of November 10, 1870. To this I replied on November 17,
and my reply so pleased Mr. Darwin that he at once wrote to
me as follows :—

"Down, November 22.

"MY DEAR WALLACE,

"I must ease myself by writing a few words to say
how much I and all in this house admire your article in
Nature. You are certainly an unparalleled master in lucidly
stating a case and in arguing. Nothing ever was better
done than your argument about the term Origin of Species,
and about much being gained if we know nothing about
precise cause of each variation."

At the end of the letter he says something about the
progress of his great work, "The Descent of Man."

"I have finished 1st vol. and am half-way through proofs
of 2nd vol. of my confounded book, which half kills me by
fatigue, and which I fear will quite kill me in your good
estimation.

"If you have leisure, I should much like a little news of
you and your doings and your family,

"Ever yours very sincerely,

"CH. DARWIN."

The above remark, "kill me in your good estimation," refers to his views on the mental and moral nature of man being very different from mine, this being the first important question as to which our views had diverged. But I never had the slightest feeling of the kind he supposed, looking upon the difference as one which did not at all affect our general agreement, and also as being one on which no one could dogmatize, there being much to be said on both sides. The last paragraph shows the extreme interest he took in the personal affairs of all his friends.

As my article of which he thought so highly is buried in an early volume of *Nature*, I will here reproduce the rather long paragraph which so specially interested him. It is as follows:—

" The first objection brought forward (and which had been already advanced by the Duke of Argyll) is, that the very title of Mr. Darwin's celebrated work is a misnomer, and that the real ' origin of species ' is that spontaneous tendency to variation which has not yet been accounted for. Mr. Bennett further remarks that, throughout my volume of ' Essays,' I appear to be unconscious that the theory I advocate does not go to the root of the matter. It is true that I am 'unconscious' of anything of the kind, for I maintain, and am prepared to prove, that the theory, if true, does go to the very root of the question of the ' origin of species.' The objection, which from its being so often made, and now again brought forward, is evidently thought to be an important one, is founded on a misapprehension of the right meaning of words. It ignores the fact that the word 'species' denotes something more than 'variety' or 'individual.' A species is an organic form (or group) which, for periods of great and indefinite length, as compared with the duration of human life, fluctuates only within narrow limits. But the 'spontaneous tendency to variation' is altogether antagonistic to such comparative stability, and would, if unchecked, entirely destroy all 'species.' Abolish, if possible, selection and survival of the fittest, so that every spontaneous variation should survive in equal proportion with all others, and the result must inevitably be an endless variety of *unstable forms*

no one of which would answer to what we mean by the word
'species.' No other cause but selection has yet been discovered capable of perpetuating and giving stability to some
forms, and causing the disappearance of others, and therefore
Mr. Darwin's book, if there is any truth in it at all, has a
logical claim to its title. It shows how 'species,' or stable
forms, are produced out of unstable spontaneous variations,
which is certainly to trace their ' origin.' The distinction of
'species' and 'individual' is equally important. A horse, or
a number of horses, as such, do not constitute a 'species.' It
is the comparative *permanence* of the form as distinguished
from the ass, quagga, zebra, tapir, camel, etc., that makes
them one. Were there a mass of intermediate forms connecting all these animals by fine gradations, and hardly a
dozen individuals alike—as would probably be the case had
selection not acted—there might be a few horses, but there
would be no such thing as a *species* of horse. That could only
be produced by some power capable of eliminating intermediate forms as they arose, and preserving all of the true horse
type ; and such a power was first shown to exist by Mr.
Darwin. The origin of varieties and individuals is one thing,
the origin of species another."

It is a remarkable thing that this very simple preliminary
misunderstanding of the very meaning of the term "species"
continued to appear year after year in most of the criticisms
of the theory of natural selection. It was put forward both
by mere literary critics and also by naturalists, and was in
many cases adduced as a discovery which completely over-
threw the whole of Darwin's work. So frequent was it that
twenty years later, when writing my "Darwinism," I found it
necessary to devote the first chapter to a thorough explanation of this point, under the heading, "What are ' Species,' and
what is meant by their ' Origin'?" and I think I may feel
confident that to those who have read that work this particular
purely imaginary difficulty will no longer exist.

Soon after the "Descent of Man" appeared, I wrote to
Darwin, giving my impressions of the first volume, to which
he replied (January 30, 1871). This letter is given in the

"Life and Letters" (iii. p. 134), but I will quote two short passages expressing his kind feelings towards myself. He begins, "Your note has given me very great pleasure, chiefly because I was so anxious not to treat you with the least disrespect, and it is so difficult to speak fairly when differing from any one. If I had offended you, it would have grieved me more than you will readily believe." And the conclusion is, "Forgive me for scribbling at such length. You have put me quite in good spirits; I did so dread having been unintentionally unfair towards your views. I hope earnestly the second volume will escape as well. I care now very little what others say. As for our not agreeing, really, in such complex subjects, it is almost impossible for two men who arrive independently at their conclusions to agree fully; it would be unnatural for them to do so."

I reviewed "The Descent," in *The Academy*, early in March, and Darwin wrote to me on the 16th, expressing his gratification at its whole tone and matter, and then, referring to the differences between us, making what was then a good point against me—that my objections to sexual selection having produced certain results in man, had not much force if, as he believed, I admitted that the plumes of the birds of paradise had been thus gained. At that time, though I had begun to doubt, I had not definitely rejected the whole of that part of "sexual selection" depending on female preference for certain colours and ornaments.

On July 9, 1871, he wrote me a long letter, chiefly about Mr. Mivart's criticisms and accusations in his book on "The Genesis of Species," and again in a severe article in the *Quarterly Review*. These he proposed replying to in a new edition of the "Origin," but the incident worried him a good deal. In a postscript he says, "I quite agree with what you say, that Mivart fully intends to be honourable, but he seems to me to have the mind of a most able lawyer retained to plead against us, and especially against me. God knows whether my strength and spirit will last out to write a chapter *versus* Mivart and others; I do so hate controversy, and feel I shall do it so badly."

Again, on July 12, he writes: "I feel very doubtful how far I shall succeed in answering Mivart. It is so difficult to answer objections to doubtful points and make the discussion readable. The worst of it is, that I cannot possibly hnnt through all my references for isolated points—it would take me three weeks of intolerably hard work. I wish I had your power of arguing clearly. At present I feel sick of everything, and if I could occupy my time and forget my daily discomforts, or rather miseries, I would never publish another word. But I shall cheer up, I dare say, soon, having only just got over a bad attack. Farewell. God knows why I bother you about myself.

"I can say nothing more about missing links than I have said. I should rely much on pre-Silurian times ; but then comes Sir W. Thompson like an odious spectre. Farewell."

I give these extracts because they serve to explain why Darwin did not publish the systematic series of volumes dealing with the whole of the subjects treated in the "Origin." With his almost constant and most depressing ill-health, the real wonder is that he did so much. We can, therefore, fully understand why, when he had published the "Descent of Man," in 1871, and the second editions of that work and of the "Animals and Plants," in 1875, with the intervening "Expression of Emotions," in 1872, he should devote himself almost entirely to the long series of observations and experiments upon living plants, which constituted his relaxation and delight, and resulted in that series of volumes which are of the greatest value and interest to all students of the marvels and mysteries of vegetable life. And when, in 1881, he published his last volume upon "Worms," giving the result of observations and experiments carried on for forty-four years, he enjoyed the great satisfaction of its being a wonderful success, while it was received by the reviewers with unanimous praise and applause.

During this latter period of his life I had but little correspondence with him, as I had no knowledge whatever of the subjects he was then working on. But he still continued to write to me occasionally, either referring kindly to my own work or sending me facts or suggestions which he thought would be of interest to me. I will here give only some extracts from a few of the latest of the letters I received from him.

On November 3, 1880, he wrote me the following very kind letter upon my "Island Life," on which I had asked for his criticism :—

"I have now read your book, and it has interested me deeply. It is quite excellent, and seems to me the best book which you have ever published ; but this may be merely because I have read it last. As I went on I made a few notes, chiefly where I differed slightly from you ; but God knows whether they are worth your reading. You will be disappointed with many of them ; but it will show that I had the will, though I did not know the way to do what you wanted.

"I have said nothing on the infinitely many passages and views, which I admired and which were new to me. My notes are badly expressed, but I thought that you would excuse my taking any pains with my style. I wish my confounded handwriting was better. I had a note the other day from Hooker, and I can see that he is much pleased with the dedication."

With this came seven foolscap pages of notes, many giving facts from his extensive reading which I had not seen. There were also a good many doubts and suggestions on the very difficult questions in the discussion of the causes of the glacial epochs. Chapter xxiii., discussing the Arctic element in south temperate floras, was the part he most objected to, saying, " This is rather too speculative for my old noddle. I must think that you overrate the importance of new surfaces on mountains and dispersal from mountain to mountain. I still believe in Alpine plants having lived on the lowlands and in the southern tropical regions having been cooled during glacial periods, and thus only can I understand character of floras on the isolated African mountains. It appears to me that you are not justified

in arguing from dispersal to oceanic islands to mountains.
Not only in latter cases currents of sea are absent, but what
is there to make birds fly direct from one Alpine summit to
another? There is left only storms of wind, and if it is
probable or possible that seeds may thus be carried for great
distances, I do not believe that there is at present any
evidence of their being thus carried more than a few
miles."

This is the most connected piece of criticism in the notes,
and I therefore give it verbatim. My general reply is printed
in " More Letters," vol. iii. p. 22. Of course I carefully considered all Darwin's suggestions and facts in later editions
of my book, and made use of several of them. The last,
as above quoted, I shall refer to again when considering the
few important matters as to which I arrived at different
conclusions from Darwin. But I will first give another
letter, two months later, in which he recurs to the same
subject.

"Down, January 2, 1881.

" MY DEAR WALLACE,

"The case which you give is a very striking one,
and I had overlooked it in *Nature*;[1] but I remain as great a
heretic as ever. Any supposition seems to me more probable
than that the seeds of plants should have been blown from
the mountains of Abyssinia, or other central mountains of
Africa, to the mountains of Madagascar. It seems to me
almost infinitely more probable that Madagascar extended
far to the south during the glacial period, and that the
S. hemisphere was, according to Croll, then more temperate ;
and that the whole of Africa was then peopled with some
temperate forms, which crossed chiefly by agency of birds
and sea-currents, and some few by the wind, from the shores
of Africa to Madagascar subsequently ascending to the
mountains.

" How lamentable it is that two men should take such

---

[1] *Nature*, December 9, 1880. The substance of this article by Mr. Baker, of
Kew, is given in " More Letters," vol. iii. p. 25, in a footnote.

widely different views, with the same facts before them ;
but this seems to be almost regularly our case, and much do
I regret it. I am fairly well, but always feel half dead with
fatigue. I heard but an indifferent account of your health
some time ago, but trust that you are now somewhat
stronger.

"Believe me, my dear Wallace,
"Yours very sincerely,
"CH. DARWIN."

It is really quite pathetic how much he felt difference of
opinion from his friends. I, of course, should have liked to
have been able to convert him to my views, but I did not
feel it so much as he seemed to do. In letters to Sir Joseph
Hooker (in February and August, 1881) he again states his
view as against mine very strongly ("More Letters," iii.
pp. 25 and 27); and this, so far as I know, is the last reference
he made to the subject. The last letter I received from him
was entirely on literary and political subjects, and, as usual,
very kind and friendly. As it makes no reference to our
controversies, and touches on questions never introduced before
in our correspondence, I think it will be interesting to give it
entire.

"Down, July 12, 1881.

"MY DEAR WALLACE,

"I have been heartily glad to get your note and
hear some news of you. I will certainly order 'Progress and
Poverty,' for the subject is a most interesting one. But I
read many years ago some books on political economy, and
they produced a disastrous effect on my mind, viz., utterly
to distrust my own judgment on the subject, and to doubt
much every one else's judgment! So I feel pretty sure that
Mr. George's book will only make my mind worse confounded
than it is at present. I also have just finished a book which
has interested me greatly, but whether it would interest any
one else I know not. It is the 'Creed of Science,' by W.
Graham, A.M. Who or what he is I know not, but he

discusses many great subjects, such as the existence of God, immortality, the moral sense, the progress of society, etc. I think some of his propositions rest on very uncertain foundations, and I could get no clear idea of his notions about God. Notwithstanding this and other blemishes, the book has interested me *extremely*. Perhaps I have been to some extent deluded, as he manifestly ranks too high what I have done.

" I am delighted to hear that you spend so much time out-of-doors and in your garden. From Newman's old book (I forget title) about the country near Godalming, it must be charming.

" We have just returned home after spending five weeks on Ullswater. The scenery is quite charming, but I cannot walk, and everything tries me, even seeing scenery, talking with any one, or reading much. What I shall do with my few remaining years of life I can hardly tell. I have everything to make me happy and contented, but life has become very wearysome to me. I heard lately from Miss Buckley in relation to Lyell's Life, and she mentioned that you were thinking of Switzerland, which I should think and hope that you would enjoy much.

" I see that you are going to write on the most difficult political question, the land. Something ought to be done, but what is the rub. I hope that you will (not) turn renegade to natural history ; but I suppose that politics are very tempting.

" With all good wishes for yourself and family,
        " Believe me, my dear Wallace,
            " Yours very sincerely,
                " CHARLES DARWIN."

This letter is, to me, perhaps the most interesting I ever received from Darwin, since it shows that it was only the engrossing interests of his scientific and literary work, performed under the drawback of almost constant ill-health, that prevented him from taking a more active part in the discussion of those social and political questions that so

deeply affect the lives and happiness of the great bulk of the
people. It is a great satisfaction that his last letter to me,
written within nine months of his death, and terminating a
correspondence which had extended over a quarter of a
century, should be so cordial, so sympathetic, and broad-
minded.

In 1870 he had written to me, "I hope it is a satisfaction
to you to reflect—and very few things in my life have been
more satisfactory to me—that we have never felt any jealousy
towards each other, though in some sense rivals. I believe
I can say this of myself with truth, and I am absolutely sure
that it is true of you." The above long letter will show that
this friendly feeling was retained by him to the last, and to
have thus inspired and retained it, notwithstanding our many
differences of opinion, I feel to be one of the greatest honours
of my life. I have myself given an estimate of Darwin's work
in my "Debt of Science to Darwin," published in my "Natural
Selection and Tropical Nature," in 1891. But I cannot here
refrain from quoting a passage from Huxley's striking
obituary notice in *Nature*, summing up his work in a single
short paragraph: "None have fought better, and none
have been more fortunate than Charles Darwin. He found
a great truth, trodden underfoot, reviled by bigots, and
ridiculed by all the world; he lived long enough to see it,
chiefly by his own efforts, irrefragably established in science,
inseparably incorporated with the common thoughts of men,
and only hated and feared by those who would revile but
dare not. What shall a man desire more than this?"

*The Chief Differences of Opinion between Darwin and
myself.*—As this subject is often referred to by objectors to
the theory of natural selection, and it is sometimes stated
that I have myself given up the most essential parts of that
theory, I think it will be advisable to give a short statement
of what those differences really are, and how they affect the
theory in question. Our only important differences were on
four subjects, which may be considered separately.

1. *The Origin of Man as an Intellectual and Moral*

Being.—On this great problem the belief and teaching of Darwin was, that man's whole nature—physical, mental, intellectual, and moral—was developed from the lower animals by means of the same laws of variation and survival ; and, as a consequence of this belief, that there was no difference in *kind* between man's nature and animal nature, but only one of degree. My view, on the other hand, was, and is, that there is a difference in kind, intellectually and morally, between man and other animals ; and that while his body was undoubtedly developed by the continuous modification of some ancestral animal form, some different agency, analogous to that which first produced organic *life*, and then originated *consciousness*, came into play in order to develop the higher intellectual and spiritual nature of man. This view was first intimated in the last sentence of my paper on the "Development of Human Races under Natural Selection," in 1864, and more fully treated in the last chapter of my "Essays," in 1870.

These views caused much distress of mind to Darwin, but, as I have shown, they do not in the least affect the general doctrine of natural selection. It might be as well urged that because man has produced the pouter-pigeon, the bull-dog, and the dray-horse, none of which could have been produced by natural selection alone, therefore the agency of natural selection is weakened or disproved. Neither, I urge, is it weakened or disproved if my theory of the origin of man is the true one.

2. *Sexual Selection through Female Choice.*—Darwin's theory of sexual selection consists of two quite distinct parts—the combats of males so common among polygamous mammals and birds, and the choice of more musical or more ornamental male birds by the females. The first is an observed *fact*, and the development of weapons such as horns, canine teeth, spurs, etc., is a result of natural selection acting through such combats. The second is an *inference* from the observed facts of the display of the male plumage or ornaments ; but the statement that ornaments have been developed by the female's choice of the most beautiful male

because he is the most beautiful, is an inference supported by singularly little evidence. The first kind of sexual selection I hold as strongly and as thoroughly as Darwin himself; the latter I at first accepted, following Darwin's conclusions from what appeared to be strong evidence explicable in no other way; but I soon came to doubt the possibility of such an explanation, at first from considering the fact that in butterflies sexual differences are as strongly marked as in birds, and it was to me impossible to accept female choice in their case, while, as the whole question of colour came to be better understood, I saw equally valid reasons for its total rejection even in birds and mammalia.

But here my view really extends the influence of natural selection, because I show in how many unsuspected ways colour and marking is of use to its possessor. I first stated my objections to "female choice" in my review of the "Descent of Man" (1871), and more fully developed it in my "Tropical Nature" (1878), while in my "Darwinism" (1889), I again discussed the whole subject, giving the results of more mature consideration. I had, however, already discussed the matter at some length with Darwin, and in a letter of September 18, 1869, I gave him my general argument as follows :—

"I have a general and a special argument to submit.

"1. Female birds and insects are usually exposed to more danger than the male, and in the case of insects their existence is necessary for a longer period. They therefore require, in some way or other, an increased amount of protection.

"2. If the male and female were distinct species, with different habits and organizations, you would, I think, admit that a difference of colour, serving to make that one less conspicuous which evidently required more protection than the other, had been acquired by natural selection.

"3. But you admit that variations appearing in one sex are (sometimes) transmitted to that sex only. There is, therefore, nothing to prevent natural selection acting on the two sexes as if they were two species.

"4. Your objection that the same protection would, *to a*

certain extent, be useful to the male seems to me quite unsound, and directly opposed to your own doctrine so con-
vincingly urged in the 'Origin,' that natural selection *never
improves an animal beyond its needs.* Admitting, therefore,
abundant variation of colour in both sexes, it is impossible
that the male can be brought by natural selection to
resemble the female (unless such variations are always
transmitted), because the difference in their colours is for
the purpose of making up for their different organization
and habits, and natural selection cannot give to the male
more protection than he requires, which is less than in the
female.

" 5. The striking fact that in all *protected* groups the females
usually resemble the males (or are equally brightly coloured)
shows that the usual tendency is to transmit colour to both
sexes when it is not injurious to either.

" Now for the special argument.

" 6. In the very weak-flying Leptalis *both* sexes mimic
Heliconidæ. But in the much stronger-flying Papilio, Pieris,
and Diadema, it is the female only that mimics the protected
group, and in these cases the females often acquire brighter
and more conspicuous colours than the male.

" 7. No case is known of a male Papilio, Pieris, or Diadema,
alone, mimicking a protected species ; yet *colour* is more
frequent in males, and *variations* are always ready for the
purpose of sexual or other forms of selection.

" 8. The fair inference seems to be that each *species*, and
also each *sex*, can only be modified by selection just as far
as is absolutely necessary—not a step further. A male,
being by structure and habits less exposed to danger, and
therefore *requiring* less protection than the female, cannot
have an equal amount of protection given to it by natural
selection ; but the female *must* have some extra protection to
balance her greater exposure to danger, and she rapidly
acquires it in one way or another.

" 9. The objection as to male fish, which *seem* to require
protection, yet have sometimes bright colours, seems to me
of no more weight than is the existence of some unprotected

species of white Leptalis as a disproof of Bates' theory of
mimicry,—or that only a few species of butterfly resemble
leaves,—or that the habits and instincts that protect one
animal are absent in allied species. These are all illustrations of the many and varied ways in which nature works to
give the exact amount of protection it needs to each species."

3. *Arctic Plants in the Southern Hemisphere, and on
Isolated Mountain-tops within the Tropics.*—Having paid great
attention to the whole question of the distribution of organisms, I was obliged to reject Mr. Darwin's explanation of the
above phenomena by a cooling of the tropical lowlands of the
whole earth during the glacial period to such an extent as to
allow large numbers of north-temperate and Arctic plants to
spread across the continents to the southern hemisphere, and,
as the cold passed away, to ascend to the summits of isolated
tropical mountains. The study of the floras of oceanic
islands having led me to the conclusion that the greater part
of their flora was derived by aërial transmission of seeds,
either by birds or by gales and storms, I extended this view
to the transmission along mountain ranges, and from
mountain-top to mountain-top, as being most accordant with
the facts at our disposal. I explained my views at some
length in " Island Life," and later, with additional facts, in
" Darwinism."

The difficulties in the way of Darwin's view are twofold.
First, that a lowering of temperature of inter-tropical lowlands to the required extent would inevitably have destroyed
much of the overwhelming luxuriances and variety of plant,
insect, and bird life that characterize those regions. This
has so impressed myself, Bates, and others familiar with the
tropics as to render the idea wholly inconceivable ; and the
only reason why Darwin did not feel this appears to be that
he really knew nothing personally of the tropics beyond a
few days at Bahia and Rio, and could have had no conception
of its wonderfully rich and highly specialized fauna and flora.
In the second place, even if a sufficient lowering of temperature had occurred during the ice-age, it would not account for

the facts, which involve, as Sir Joseph Hooker remarks, "a continuous current of vegetation from north to south," going much further back than the glacial period, because it has led to the transmission not of existing species only, but of distinct representative species, and even distinct genera, showing that the process must have been going on long before the cold period. The reason why Darwin was unaffected by these various difficulties may perhaps be found in the circumstance that he had held his views for so many years almost un challenged. In a letter to Sir Charles Lyell, in 1866, he says, "I feel a strong conviction that soon every one will believe that the whole world was cooler during the glacial period. Remember Hooker's wonderful case recently discovered of the identity of so many temperate plants on the summit of Fernando Po, and on the mountains of Abyssinia. I look at it as certain that these plants crossed the whole of Africa, from east to west, during the same period. I wish I had published a long chapter, written in full, and almost ready for the press, on this subject which I wrote ten years ago. It was impossible in the 'Origin' to give a fair abstract" ("More Letters," vol. i. p. 476). Having thus held his views for twenty-five years, they had become so firmly impressed upon his mind that he was unable at once to give them up, however strong might be the arguments against them. This particular difference, however, is not one which in any way affects the theory of natural selection.

4. *Pangenesis, and the Heredity of Acquired Characters.* —Darwin always believed in the inheritance of acquired characters, such as the effects of use and disuse of organs and of climate, food, etc., on the individual, as did almost every naturalist, and his theory of pangenesis was invented to explain this among other affects of heredity. I therefore accepted pangenesis at first, because I have always felt it a relief (as did Darwin) to have *some* hypothesis, however provisional and improbable, that would serve to explain the facts ; and I told him that "I shall never be able to give it up till a better one supplies its place." I never imagined

that it could be directly disproved, but Mr. F. Galton's experiments of transfusing a large quantity of the blood of rabbits into other individuals of quite different breeds, and afterwards finding that the progeny was not in the slightest degree altered, did seem to me to be very nearly a disproof, although Darwin did not accept it as such. But when, at a much later period, Dr. Weismann showed that there is actually no valid evidence for the transmission of such characters, and when he further set forth a mass of evidence in support of his theory of the continuity of the germ-plasm, the "better theory" was found, and I finally gave up pangenesis as untenable. But this new theory really simplifies and strengthens the fundamental doctrine of natural selection.

It will thus appear that none of my differences of opinion from Darwin imply any real divergence as to the overwhelming importance of the great principle of natural selection, while in several directions I believe that I have extended and strengthened it. The principle of "utility," which is one of its chief foundation-stones, I have always advocated unreservedly; while in extending this principle to almost every kind and degree of coloration, and in maintaining the power of natural selection to increase the infertility of hybrid unions, I have considerably extended its range. Hence it is that some of my critics declare that I am more Darwinian than Darwin himself, and in this, I admit, they are not far wrong.

CHAPTER XXVI

MY FRIENDS AND ACQUAINTANCES—SPENCER, HUXLEY, MIVART, ETC.

SOON after my return home, in 1862 or 1863, Bates and I, having both read "First Principles" and been immensely impressed by it, went together to call on Herbert Spencer, I think by appointment. Our thoughts were full of the great unsolved problem of the origin of life—a problem which Darwin's "Origin of Species" left in as much obscurity as ever—and we looked to Spencer as the one man living who could give us some clue to it. His wonderful exposition of the fundamental laws and conditions, actions and inter-actions of the material universe seemed to penetrate so deeply into that "nature of things" after which the early philosophers searched in vain and whose blind gropings are so finely expressed in the grand poem of Lucretius, that we both hoped he could throw some light on that great problem of problems. I forget the details of the interview, but I think Bates was chief spokesman, and expressed our immense admiration of his work, and that as young students of nature we wished to have the honour of his acquaintance. He was very pleasant, spoke appreciatively of what we had both done for the practical exposition of evolution, and hoped we would continue to work at the subject. But when we ventured to touch upon the great problem, and whether he had arrived at even one of the first steps towards its solution, our hopes were dashed at once. That, he said, was too funda-mental a problem to even think of solving at present. We did not yet know enough of matter in its essential constitution nor of the various forces of nature ; and all he could say was that everything pointed to its having been a development out of matter—a phase of that continuous process of evolution by which the whole universe had been brought to its present condition. So we had to wait and work contentedly at minor problems. And now, after forty years, though Spencer and Darwin and Weismann have thrown floods of light on the phenomena of life, its essential nature and its origin remain as great a mystery as ever. Whatever light we do possess is from a source which Spencer and Darwin neglected or ignored.

In 1865, when Spencer was, I believe, one of the editors of *The Reader*, he asked me to write an article on the treatment of savage races, with special reference to some cases of the barbarity of settlers in Australia that had recently been published. This I did, and the article appeared in the issue of June 17. Ten years later, on November 13, 1875, he wrote to ask me where and when this article had appeared, adding, " I ask the question because I contemplate giving Dr. Bridges a castigation for the unwarranted sneer at the close of his article in the *Fortnightly*." I may add that I have reprinted my article (with some additions referring to recent facts) in my " Studies Scientific and Social," vol. ii. p. 107.

The first letter I received from Spencer was when I sent him my paper on " The Origin of Human Races under the Law of Natural Selection." He said that he had read it with great interest, and added, " Its leading idea is, I think, undoubtedly true," concluding with a hope that I would pursue the inquiry.

Soon afterwards he invited me to dine with him in Bayswater, where he lived for many years in a boarding-house with rather a commonplace set of people—retired Indian officers and others ; and I afterwards visited him there several times. I was amused when some popular error was solemnly put forth at dinner as the explanation of some phenomenon ; Spencer would coolly tell them that it was quite incorrect, and then proceed to explain *why* it was so, and on principles of evolution could not be otherwise. In the evening, after we had had a little private conversation, we would go into the drawing-room where there was music, and Spencer would sometimes play on his flute. On remarking to him one day that I wondered he could live among such unintellectual people, he said that he had purposely chosen such a home in order to avoid the mental excitement of too much interesting conversation; that he suffered greatly from insomnia, and that he found that when his evenings were spent in commonplace conversation, hearing the news of the day or taking part in a little music, he had a better chance of sleeping.

In the autumn of 1867 I read the Duke of Argyll's "Reign of Law," and though I found much that was erroneous and weak in argument, I thought his discussion of the mode of flight in birds, founded largely on personal observation, was very good; in fact, the best I had seen. Spencer had also read this, and differed from me, thinking that important parts of the duke's theory of flight was not true, because they would not apply equally to bats; and we had quite a discussion on the subject. The next day, after thinking the matter over, I wrote him a long letter of eight pages, trying to show that the *general* principles of flight in birds, bats, and insects were the same; but that in birds there were additional *special* adaptations that render their flight more perfect, and their power of motion through the air, under adverse conditions, more varied and more complete. The duke, dealing with birds only, had dwelt most on these special adaptations (chiefly, if I remember, the beautiful overlapping and movements of the separate feathers increasing resistance during the downward, and decreasing it during the upward stroke) which did not exist in bats or in insects. I also showed that although this adaptation was absent in the wings of insects, the *general* form and movements of the wings were similar and produced similar, but not identical results. In his reply he admitted the accuracy of my description of the flight of insects, but made the following remark in furtherance of his former objection as regards the duke's account of the flight of birds : " If you will move an outstretched wing backwards and forwards with equal velocity, I think you will find that the difference of resistance is nothing like commensurate with the difference of size between the muscles that raise the wings and the muscles that depress them." The reason of this great difference could not be accurately explained at that time, but a few years later, Marey, by his ingenious experiments and photographs, showed that while the whole upward motion of the wing is very gradual, the downward stroke, though equally gradual at the beginning and the end, is two or three times as rapid in the middle, thus giving the great upward and onward impulse, necessitating the extremely large muscles noted by Spencer. An excellent short account of the whole mechanism of the flight of birds, with many of Marey's diagrams and illustrations, is given in Professor A. Newton's " Dictionary of Birds," in the article " Flight," and is the clearest exposition of the subject I have yet seen.

In 1872, in my presidential address to the Entomological Society, I endeavoured to expound Herbert Spencer's theory of the origin of insects, on the view that they are fundamentally *compound animals*, each segment representing one of the original independent organisms. This theory is expounded at some length in the second volume of his " Principles of Biology " (chapter iv., " The Morphological Composition of Animals "), but had apparently been almost unnoticed by English entomologists. On sending him a copy of the address, he wrote to me as follows : " It is gratifying to me to find that your extended knowledge does not lead you to scepticism respecting the speculation of mine which you quote, but rather enables you to cite further facts in justification of it. Possibly your exposition will lead some of those, in whose lines of investigation the question lies, to give deliberate attention to it."

This communication gave me much pleasure, because the subject was one quite out of my own domain, and though I had taken a good deal of trouble to understand his views and to represent them accurately, and had also adduced a few additional facts in support of it, yet the subject was so novel and so complex that I was rather afraid I might have made

XXVI] HERBERT SPENCER

some blunders in my abstract of it. I was much relieved, therefore, to find that my account of his views was satisfactory to him.

In 1874, when writing "The Principles of Sociology," Herbert Spencer asked me to look over the proofs of the first six chapters, and give him the benefit of my criticisms, "alike as naturalist, anthropologist, and ˙traveller." I found very little indeed requiring emendation, but I sent him a couple of pages of notes with suggestions on points of detail, which, I believe, were of some use to him.

During the year 1881 I had several letters from him, dealing with subjects of general interest. In consequence of an article I wrote on "How to Nationalize the Land," especially showing how to avoid the supposed insuperable objection of State management, a "Land Nationalization Society" was formed, of which I was chosen president. As I had been induced to study the question by Herbert Spencer's early volume on "Social Statics," I sent him a copy of our programme and asked if he would join us. His reply is very instructive, as showing how nearly he agreed with us at that time, and also how slight were the difficulties he suggested as the most important.

The letter is as follows :—

"38, Queen's Gardens, Bayswater, W.,
"April 25, 1881.

"DEAR MR. WALLACE,
        "As you may suppose, I fully sympathize in the general aims of your proposed Land Nationalization Society ; but for sundry reasons I hesitate to commit myself, at the present stage of the question, to a programme so definite as that which you send me. It seems to me that before formulating˙the idea in a specific shape, it is needful to generate a body of public opinion on the general issue, and that it must be some time before there can be produced such recognition of the general principle involved as is needful before definite plans can be set forth to any purpose.

"It seems to me that the thing to be done at present is to arouse public attention to (1) the abstract inequity of the present condition of things ; (2) to show that even now there is in our law a tacit denial of absolute private ownership, since the State reserves the power of resuming possession of land on making compensation ; (3) that this tacitly admitted ownership ought to be overtly asserted ; (4) and that having been overtly asserted, the landowner should be distinctly placed in the position of a tenant of the State on something like the terms proposed in your scheme : namely, that while the land itself should be regarded as public property, such value as has been given to it should vest in the existing so-called owner.

"The question is surrounded with such difficulties that I fear anything like a specific scheme for resumption by the State will tend, by the objections made, to prevent recognition of a general truth which might otherwise be admitted. For example, in definitely making the proposed distinction between 'inherent value as dependent on natural conditions, etc.,' and the 'increased value given by the owner,' there is raised the questions—How are the two to be distinguished? How far back are we to go in taking account of the labour and money expended in giving fertility ? In respect of newly enclosed tracts, some estimation may be made ; but in respect of the greater part, long reduced to cultivation, I do not see how the valuations, differing in all cases, are to be made.

"I name this as one point ; and there are many others in respect of which I do not see my way. It appears to me that at present we are far off from the time at which action may advantageously be taken.

"Truly yours,

"HERBERT SPENCER."

On this I may remark that, during the twenty-five years that has elapsed, the Land Nationalization Society has been continuously at work, doing the very things that our critic seemed to think ought to be done *before* we formed the society. We have now "generated a body of public opinion"

in our favour, which could hardly have been effected without
the work of a society, and we have long since satisfied most
thinking men that the special difficulty as to the valuation
of the owners' improvements is a purely imaginary one, since
it is continually done.   But the remarkable thing is, that only
ten years later, in his volume on " Justice," the writer of this
letter should have so far changed his opinions as to arrive
ultimately at the conclusion thus stated : " A fuller consideration of the matter has led me to the conclusion that individual
ownership, subject to State suzerainty, should be maintained."
Those who care to understand what were the supposed facts
leading to this most impotent conclusion, will find them
stated and exposed in vol. ii., chap. xviii. of my " Studies." [1]
They were first given in an address to the Land Nationalization Society in 1892.

A few months later he wrote me again on the land
question, in reply to my recommendation of Henry George's
book " Progress and Poverty," and this letter, as exhibiting
his ideas on human progress generally, and also his somewhat
hasty judgments on particular writers, seems well worthy of
preservation, and I therefore give it verbatim.

" 38, Queen's Gardens, Bayswater, July 6, 1881.
" DEAR MR. WALLACE,
        " I have already seen the work you name—' Pro-
gress and Poverty ; ' having had a copy, or rather two copies,
sent me.   I gathered, from what little I glanced at, that I
should fundamentally disagree with the writer, and have not
read more.
    " I demur entirely to the supposition, which is implied in
the book, that, by any possible social arrangements whatever,
the distress which humanity has had to suffer in the course of
civilization could have been prevented.   The whole process,

---

[1] H. Spencer's treatment of the land question in this work is criticized
and controverted in great detail by Henry George in " A Perplexed Philosopher,"
published in 1893.   Neither H. Spencer nor any of his disciples have refuted
these destructive criticisms.

with all its horrors and tyrannies, and slaveries and abomi-
nations of all kinds, has been an inevitable one accompanying
the survival and spread of the strongest, and the consolidation
of small tribes into large societies ; and among other things,
the lapse of land into private ownership has been, like the
lapse of individuals into slavery, at one period of the process
altogether indispensable. I do not in the least believe that
from the primitive system of communistic ownership to a high
and finished system of State ownership, such as we may
look for in the future, there could be any transition without
passing through such stages as we have seen, and which
exist now.

"Argument aside, however, I should be disinclined to
commit myself to any scheme of immediate action, which, as
I have indicated to you, I believe, at present, premature.
For myself, I feel that I have to consider not only what I
may do on special questions, but also how the action I take
on special questions may affect my general influence ; and
I am disinclined to give more handles against me than are
needful. Already, as you will see by the enclosed circular, I
am doing in the way of positive action more than may be
altogether prudent.

"Sincerely yours,
"HERBERT SPENCER."

I do not remember, and I do not think that Henry George
either stated or implied that the course of civilization "might
have been different" from what it has been. His whole work
was devoted to showing the injustice and the evils of private
property in land, just as Herbert Spencer himself had done
in "Social Statics ;" and both works are alike beneficial, inasmuch as they demonstrate these facts and serve as incentives
and guides for our future attempts to remedy them. If Mr.
Spencer had not hastily laid aside the book, owing to this
prepossession against it, even he might have been benefited
by the thorough examination of the whole subject which
Mr. George gave, while he could hardly have failed to
admire its admirable and forcible exposition of the problem

and his often eloquent delineations of its results. I remember
that some years earlier, when I asked Herbert Spencer what
he thought of Buckle's " History of Civilization," which I took
for granted that he had read, his reply was somewhat similar
to that here given in the case of Henry George—that on
looking into the book he saw that its fundamental assumption
was erroneous, and therefore he did not care to read it. I
believe he referred to Buckle's view of the immense influence
of the aspects of nature in influencing human character, which,
even if much exaggerated, cannot be said to be wholly untrue,
and certainly does not destroy the value of a work of such
research, eloquence, and illumination as the " History of
Civilization."

The next letter of much interest I have from Herbert
Spencer is, when acknowledging receipt of a copy of my little
book, entitled " Bad Times," on November 21, 1885. In it
he says, " Much of what I read I quite agreed with, especially
the chapters on 'Foreign Loans' and 'War Expenditure.' . . .
There is one factor which seems to me not an improbable
one, which neither you nor any others have taken account of.
During the past generation, one of the causes of the great
exaltation of prosperity has been the development of the
railway system, which while it had the effect of opening up
sources of supply and means of distribution, had also the
effect during a long period of greatly exalting certain industries concerned in construction. There was consequently
a somewhat abnormal degree of prosperity, which lasted long
enough to furnish a standard of good times, and to be mistaken for the normal condition. Now that this unusual and
temporary cause of prosperity has in considerable measure
diminished, we are feeling the effect."

This was no doubt true, and in the case of America I
had adduced the railway mania in the United States, from
1869 to 1873, and our own over-production of shipping
while we were supplying the whole world with rails and
engines, as causes of the subsequent depression in both
countries.

The last three letters I received from Herbert Spencer were in 1894 and 1895, all on the subject of what he termed "the absurdity of Lord Salisbury's representation of the process of natural selection" in his British Association address at Oxford, wishing me to write to the *Times*, pointing out his errors, which were influencing many persons and writers in the press, and suggesting certain points I should especially deal with. He concluded, "It behoves you of all men to take up the gauntlet he has thus thrown down." I replied, declining the task, on the ground that I did not think Lord Salisbury's influence in a matter of science of much importance, and that I thought my time better employed in writing such articles on social and political, as well as general scientific questions which then interested me. To this he replied that he did not at all agree with me, and that "articles in the papers show that Lord Salisbury's argument is received with triumph, and unless it is disposed of, it will lead to a public reaction against the doctrine of evolution at large."

As I still declined to go into this controversy, having dealt with the whole matter in my "Darwinism," and still being sceptical as to any great effects being produced by the address in question, he wrote me a month later as follows: "As I cannot get you to deal with Lord Salisbury, I have decided to do it myself, having been finally exasperated into doing it by this honour paid to his address in France—the presentation of a translation to the French Academy. The impression produced upon some millions of people in England cannot be allowed to be thus further confirmed without protest." He then asked me for some references, which I sent him, and his criticism of Lord Salisbury duly appeared, and was thoroughly well done, so that I had no reason to regret not having undertaken it myself. This was the latest letter I received from him ; but during his last illness my wife, being in Brighton, called to make inquiries after his health, and left our cards, and I received a kindly expressed card in reply, written by his amanuensis, but signed with his own initials. It is dated November 28, 1903, ten days before his death.

Among his intimate friends, Herbert Spencer was always interesting from the often unexpected way in which he would apply the principles of evolution to the commonest topics of conversation, and he was always ready to take part in any social amusement. He once or twice honoured me by coming to informal meetings of friends at my little house in St. Mark's Crescent, and I also met him at Sir John Lubbock's very pleasant week-end visits, and also at Huxley's, in St. John's Wood. Once I remember dining informally with Huxley, the only other guests being Tyndall and Herbert Spencer. The latter appeared in a dress-coat, whereupon Huxley and Tyndall chaffed him, as setting a bad example, and of being untrue to his principles, quoting his Essay on " Manners and Fashion," but all with the most good-humoured banter. Spencer took it in good part, and defended himself well, declaring that the coat was a relic of his early unregenerate days, and where could he wear it out if not at the houses of his best friends? " Besides," he concluded, " you will please to observe that I *am* true to principle in that I do *not* wear a white tie ! "

Those who are acquainted only with the volumes of Herbert Spencer's " Synthetic Philosophy " can have no idea of the lightness, the energy, and the bright satire of some of his more popular writings. Such are many of his earlier Essays, and in his volume on " The Study of Sociology " we find abundant examples of these qualities. In conclusion, I may remark that, although I differ greatly from him on certain important matters, both of natural and social science, and have never hesitated to state my reasons for those differences with whatever force of fact and argument I could bring to bear upon them, I yet look upon these as but spots on the sun of his great intellectual powers, and feel it to be an honour to have been his contemporary, and, to a limited extent, his friend and coadjutor.

With the remainder of my scientific friends I had, for the most part, only social intercourse, with no correspondence of general interest. Those I saw most of during my residence in London, and with whom I became most intimate, were Huxley, Tyndall, Sir John Lubbock, Dr. W. B. Carpenter, Sir William Crookes, Sir Joseph Hooker, Mr. Francis Galton, Professor Alfred Newton, Dr. P. L. Slater, Mr. St. George Mivart, Sir William Flower, Sir Norman Lockyer, Professor R. Meldola, and many others whose names are only known to specialists. All these I met very frequently at scientific meetings, or at some of their houses at which I was occasionally a guest. To all of them I have been more or less indebted for valuable information or useful suggestions in the course of my work, and to these I must also add Professor E. B. Poulton, of Oxford; F. W. H. Myers, Professors W. F. Barrett and Percival Wright, of Dublin, with Professors Patrick Geddes, of Edinburgh, and J. A. Thomson, of Aberdeen. For the last quarter of a century I have lived so completely in the country that I have ceased to have personal intercourse with most of them; and of those still among us, I can only say here that I hope and believe they all continue to be my very good friends. In future chapters I may have to refer to some of them again, in connection with special conditions of my life. Here I will only give a few indications as to my personal relations with a few of them.

Of all those I have mentioned I became, I think, most intimate with Huxley. At an early date after my return home he asked me to his house in Marlborough Place, where I soon became very friendly with his children, then all quite young, all very animated, and not at all shy. Mrs. Huxley was also exceedingly kind and pleasant, and the whole domestic tone of the house was such as to make me quite at my ease, which is what happens to me with only a few persons. I used often to go there on Sunday afternoons, or to spend the evening, while I was several times asked to dine to meet persons of similar pursuits to my own. One of those occasions that I particularly remember was to meet Dr. Miklucho Maklay, a Russian anthropologist, who was going to New Guinea, and as I was the only Englishman who had lived some months alone in that country, Huxley thought we should be interested with each other.

Maklay was a small, wiry man, somewhat younger than myself; he spoke English well, and told us all about what he was going to do. His idea was that you could really learn nothing about natives unless you lived with them and became almost one of themselves; above all, you must win their confidence, and must therefore begin by trusting them absolutely. He proposed to go in a Russian warship, and be left for a year at some part of the north coast where Europeans were wholly unknown, with one servant, but without visible arms. This was, I think, in the winter of 1870-71. Both Huxley and myself thought this plan exceedingly risky, but he determined to try it; and he succeeded, but through the exercise of an amount of coolness and courage which very few men indeed possess. He returned to Russia to complete his preparations, and in September, 1871, was landed in Astrolabe Bay with two servants, one a Swede and the other a Polynesian. The ship's carpenter built him a small hut, fourteen feet by seven feet, and then the ship sailed away and left him totally unprotected. As soon as it was seen that the ship was completely out of sight, large numbers of natives, armed with knives, bows, and spears, gathered round his hut and soon began to make warlike demonstrations, which went on more or less for some days. They would shoot arrows close to his head or body, or draw their bow to the full with the arrow directed to his chest, and then loose the string with a twang, while holding back the arrow; but he sat still and smiled, knowing, I suppose, that if they really meant to kill him that was hardly the way they would do it, and that in any case he could not possibly escape them. At other times they would run at him with their spears, or press the spear-point against his teeth till he was forced to open his mouth. But finding that he was brave, that he did not try to escape them, and also finding that he was a " medicine " man, could heal their wounds and cure the sick, they gradually came to consider him as a friend and even as a supernatural being. Soon one servant died, and the other was almost constantly ill, so that the doctor had plenty to do; but he lived with these people for fifteen months, learnt their language, studied them minutely, and explored much of the surrounding country. I know of no more daring feat by any traveller. A short account of this exploration is given in *Nature*, vol. ix. p. 328.

I used often to call in at Jermyn Street if I had any question to ask Huxley, and he was always ready to give me all the information in his power; while I am pretty sure I owe partly, if not largely, to his influence the grant of the royal medal of the Royal Society, and perhaps also of the Darwin medal. Once only there was a partial disturbance of our friendly relations, of the exact cause of which I have no record or recollection. I had published some paper in which, I believe, I had stated some view which he had originated without mentioning his name, and in such a way as to leave the impression that I put it forth as original. This I had no notion of doing; but I think it was an idea which had become quite familiar to me, and that I had quite forgotten *who* originated it. I fancy some one must have called Huxley's attention to it, and when I next met him, I think just as he was leaving Jermyn Street to go home, he was much put out, and said something intimating that after what I had said in this paper, he wondered at my speaking to him again. I forget what more was said, but on going home I looked at the article, and found that I had used some expression that *might* be interpreted as a slight to him. I immediately wrote a letter of explanation and regret, and I here give his reply, which greatly relieved me, and our relations at once resumed their usual friendly character.

"MY DEAR WALLACE,

"Very many thanks for your kind letter.

"I am exceedingly callous to the proceedings of my enemies, but (I suppose by way of compensation) I am very sensitive to those of valued friends, and I certainly felt rather sore when I read your paper. But I dare say I should have 'consumed my own smoke' in that matter as I do in most, if I had not been very tired, very hungry, very cold, and consequently very irritable, when I met you yesterday. Pray forgive me if I was too plain spoken,

"And believe me, as always,

"Yours very faithfully,

"T. H. HUXLEY.

"Jermyn Street, February 14, 1870."

In a letter he wrote me, in 1881, on another matter he refers to my former intimacy with his children. "Your little friends are grown to be big friends. Two are married, and one has made me a grandfather. Leonard, my eldest boy, is six feet high, and at Balliol; even the smallest of the mites you knew is taller than her mother. All within reach unite with me in kindest regards and remembrances."

In 1891 I had read two books by Mr. Arthur J. Bell, a Devonshire gentleman who had devoted himself to the study of the modern physical sciences in their relation to the deepest problems of our nature and destiny. The first was entitled, "Whence comes Man, from 'Nature' or from 'God'?" The second, published two years after the first, as a sequel to it, was called, "Why does Man Exist?" I was greatly struck with the power of reasoning, the clearness of style, and the broad grasp of the whole subject displayed by the author, and having written to him to say how much I had enjoyed his books, he called upon me at Parkstone, and in the course of conversation he expressed a great desire that Huxley should be induced to read them—at all events the second, which, though a sequel to the first, is quite independent of it.[1] I therefore wrote to Huxley, telling him the author

[1] As I am sure that there are many persons who have never heard of these books who would greatly enjoy them, I will here quote the subject-matter of the second as stated in the last page of the first work, as follows:—

"Before replying to the question with which we started—the question, 'Whence comes Man, from "Nature" or from "God"?'—we must, I think, state what man is.

"As it seems to me, man is the highest development of the 'Power' called 'Life'—a Power added, at a comparatively late period of geological time, to Powers already existing.

"To the question, then, 'Whence comes man; does he come from Nature or from God?' we must, I think, reply that not only man, but Nature also, owe

would be pleased to send him the books if he would like to have them, and in that case would be glad if he would give his opinion of the work. His reply, dated November 23, 1891, is a characteristic example of his style, and as it is also the last letter of his I possess, I here reproduce it.

"Hodeshea, Eastbourne.

"MY DEAR WALLACE,

"The instinct of self-preservation leads me, as a rule, to decline to read and still more to give an opinion about books that are sent to me. But, then, they do not usually come with such a recommendation as yours, and if your friend Mr. Bell is kind enough to send me a copy of his book, I will not only read it, but pay him the highest compliment in my power, by doing my best to pick holes in it! I 'can't say no fairer.'

their existence to the Infinite Eternal Being—God, who 'created' all things." Then follows the striking passage which he reprints as the "Argument" of the second work, "Why does Man Exist?"

"ARGUMENT.

"Supposing these answers to be accepted, other questions suggest themselves. We want to know why man exists. We want to know why God 'created' him. Did God desire that man should be good? Is there any reason why he should be good? If there be, then why does evil exist? And there arises also the further question, that, supposing there be a good reason why man *should* be good, is goodness possible to him? If his character be made *for* him, not *by* him, how *can* he be good if his character, which he did not make himself, be not good? Does his existence terminate at death? Does he come into the world only for the sake of what he therein does—suffers—enjoys? or is his existence continued after death? Is that existence, if it be continued after death, to be desired or to be dreaded? Is the having been born a misfortune or a blessing? What is the character of God? Is He a Being to be feared—to be hated—or to be loved? What are man's relations to his fellow-man? What are man's relations to God—that awful Being whose power over us seems to be absolute? And that last, most terrible of questions, Is man's existence owing to God's malevolence—to His indifference—or to His love?"

Here are surely subjects enough for a volume of 420 pages, and Mr. Bell discusses them all thoroughly and honestly, with wonderful knowledge and sagacity, with sound logic, and in clear, forcible, and often brilliant language. And he arrives at a grand—a magnificent conclusion—a conclusion that comes as near to a satisfactory solution of these seemingly insoluble problems as with our limited faculties we can attain to.

" I get along very well under condition of keeping quiet here, and I am happy to say that my wife, who joins with me in kind remembrances, has greatly improved in health since we settled here.

" Ever yours very faithfully,
" T. H. HUXLEY."

Although Huxley was as kind and genial a friend and companion as Darwin himself, and that I was quite at ease with him in his family circle, or in after-dinner talk with a few of his intimates (and although he was two years younger than myself), yet I never got over a feeling of awe and inferiority when discussing any problem in evolution or allied subjects—an inferiority which I did not feel either with Darwin or Sir Charles Lyell. This was due, I think, to the fact that the enormous amount of Huxley's knowledge was of a kind of which I possessed only an irreducible minimum, and of which I often felt the want. In the general anatomy and physiology of the whole animal kingdom, living and extinct, Huxley was a master, the equal—perhaps the superior—of the greatest authorities on these subjects in the scientific world ; whereas I had never had an hour's instruction in either of them, had never seen a dissection of any kind, and never had any inclination to practise the art myself. Whenever I had to touch upon these subjects, or to use them to enforce my arguments, I had to get both my facts and my arguments at second hand, and appeal to authority both for facts and conclusions from them. And because I was thus ignorant, and because I had a positive distaste for all forms of anatomical and physiological experiment, I perhaps over-estimated this branch of knowledge and looked up to those who possessed it in a pre-eminent degree as altogether above myself.

With Darwin and Lyell, on the other hand, although both possessed stores of knowledge far beyond my own, yet I did possess *some* knowledge of the same kind, and felt myself in a position to make use of their facts and those of all other students in the same fields of research quite as well as the majority of those who had observed and recorded them. I had, however, very early in life noticed, that men with immense *knowledge* did not always know how to draw just conclusions from that knowledge, and that I myself was quite able to detect their errors of reasoning. I also found that when, in my early solitary studies in physics or mechanics, I came upon some conclusion which seemed to me, for want of clear statement in the books at my command, contrary to what it ought to be, yet when, later, the matter was clearly explained, I at once saw where my error lay and had no further difficulty. I will here mention one of these smaller stumbling-blocks, which I know are to this day quite impassable by large numbers of persons who are interested in physical science. It is the fact that degrees of latitude *increase* in length from the equator to the pole, the only explanation usually given being that this is due to the compression at the poles, or, in other words, of the polar diameter being less than the equatorial. Now nine persons out of ten (probably more) who know what a " degree" is, and have an elementary knowledge of geometry, and perhaps a much more than elementary knowledge of several other sciences, could not explain offhand why this is so ; while many of them, meeting with the statement for the first time and trying to understand it, would come to the conclusion that it was a mistake—perhaps a printer's error, and that degrees really *decrease* towards the pole. For they know that a circle is divided into 360 parts, each being a degree, and if you draw a circle round the earth, passing through the two poles with a radius of half the equatorial diameter, and divide it into 360 equal parts, each of those parts will be a degree. But the earth's radius at the poles will be about 132 miles *less* than at the equator ; therefore the degrees will be proportionately *less*, not more as stated. I possess a pamphlet addressed to the President of the Royal Astronomical Society by a Mr. Gumpel, pointing this out, and asking them to correct so important an error. But I presume he was only laughed at, as what Professor de Morgan called a " paradoxer," and the Americans a " crank," and I dare say

the poor man lived and died in the conviction that astronomers were ashamed to confess their error. Now the
essential point, rarely explained in popular books, is, that if
the earth were of exactly the same shape it is now, but did
not turn on its axis, then degrees of latitude at and near the
pole would really be *shorter* than those at and near the
equator ; but the bulging out at the equator is caused by
the *rotation*, owing to centrifugal force diminishing the force
of gravity there, and causing the true sphere, which gravity
would produce in a non-rotating fluid or flexible mass, to be
changed into a spheroid of greater diameter at the equator,
where the rotating motion is swifter, and therefore the
centrifugal force greater. The *surface* will therefore become
a surface of equilibrium, due to the two forces everywhere
acting upon it, and the direction of a plumb-line will be also
determined by the same two forces, and will necessarily be
at right angles to that *surface*. It follows that as the curvature along a meridian is more rapid near the equator than
that of a sphere of the mean diameter of the earth, and less
rapid or flatter near the poles, therefore two or more plumb-
lines near the equator will meet at a point *nearer* than the
geometrical centre of the earth, while those near the poles
will meet at a point *beyond* the geometrical centre, and therefore the degrees near the latter, being measured on a circle
of longer radius, will be longer than those near the equator.
It appears, then, that the problem is not a geometrical one,
as the mere statement of the fact seems to make it, but one
of mechanics and the laws of motion, and what we really
measure is the amount of curvature on different parts of the
earth's surface, *not* an equal angle measured from its centre,
which is what the term "degree" usually and properly
means. From this point of view the astronomers *are* all
wrong, since they use the term "degree" of latitude in a
technical sense, which is not its geometrical meaning, and
they very rarely explain this to their readers. Degrees of
latitude are dynamical, not geometrical quantities.

This rather long digression may be considered to be out
of place, but it is given in order to illustrate the steps by

which I gradually acquired confidence in my own judgment, so that in dealing with any body of facts bearing upon a question in dispute, if I clearly understood the nature of the facts and gave the necessary attention to them, I would always draw my own inferences from them, even though I had men of far greater and more varied knowledge against me. Thus I have never hesitated to differ from Lyell, Darwin, and even Spencer, and, so far as I can judge, in all the cases in which I have so differed, the weight of scientific opinion is gradually turning in my direction. In reasoning power upon the general phenomena of nature or of society, I feel able to hold my own with them ; my inferiority consists in my limited knowledge, and perhaps also in my smaller power of concentration for long periods of time.

With Huxley also I felt quite on an equality when dealing with problems arising out of facts equally well known to both of us ; but wherever the structure or functions of animals were concerned, he had the command of a body of facts so extensive and so complex that no one who had not devoted years to their practical study could safely attempt to make use of them. I therefore never ventured to infringe in any way on his special departments of study, though I occasionally made use of some of the results which he so lucidly explained.

One of my near neighbours while I lived in London was Dr. W. B. Carpenter, the well-known physiologist and microscopist, and a voluminous writer on various branches of natural science. I often called on him in the evening, when I usually found him at work with his microscope, and he always took pleasure in showing me some special structure or some obscure organism, and explaining the nature of what I saw. The great controversy was then at its height as to the alleged animal nature of a substance found in the Laurentian formation of Canada, supposed to be the oldest of all the stratified rocks. Dr. Carpenter maintained that it was a low form of Foraminifera, a group of which he had made a special study. This supposed organism had been named by Sir William

Dawson, the geologist, *Eozoon Canadense*, and he was supported in his view by Dr. Carpenter, to whom he sent the finest procurable specimens. By making sections in various directions, and by the knowledge he possessed of the minute structure of living and fossil Foraminifers, he arrived at his conclusions ; while other observers declared that this supposed primitive organism was entirely of mineral origin, and that all the apparent details of organic structure were deceptive. Dr. Carpenter showed me these specimens, and pointed out the details of structure on which he relied, but having no knowledge of the actual structures with which he compared them, I could myself see nothing sufficiently definite to settle such an important question. The discussion went on very fiercely for years, but the general opinion now is that all the appearances are due to forms of crystallization in these very ancient metamorphic rocks. Dr. Carpenter was also at work on the anatomy and physiology of the Crinoidea or sea-lilies, on which he published some important papers, and these, too, he would dilate upon and explain, though not much to my enlightenment.

We often walked across the Regent's Park into town together, and we were very friendly, though never really intimate ; and a few years later we entered on a rather acute controversy upon mesmerism and clairvoyance, to which I shall refer later on.

Among the more prominent naturalists, one of my chief friends, and the one whose society I most enjoyed, was Professor St. George Mivart, who for some years lived not far from us in London. He was a rather singular compound mentally, inasmuch as he was a sincere but thoroughly liberal Catholic, and an anti-Darwinian evolutionist. But his friendly geniality, his refined manners, his interesting conversation and fund of anecdote of the most varied kind, rendered him a charming companion. His most intimate friends seemed to be priests, one or two of whom were almost always among the guests, and often the only ones, when I dined with him. And they, too, were excellent company, full of humour and

anecdote of the most varied kind, though also ready for
serious talk or discussion ; but in either case, with none of
the reserve or somewhat rigid decorum of the majority of our
clergy.   Mivart visited a good deal in the country houses of
the aristocracy and country gentlemen, and he used often to
tell me things that happened in some of them, or that were
spoken of as common knowledge, which I could not have
believed on less direct authority, and which went to prove
that some of the worst features of society morals, such as are
occasionally revealed in the divorce courts, are by no means
uncommon.

Mivart thoroughly enjoyed a good dinner (as did I myself)
and was rather fond of illustrative stories on gastronomic
subjects.   One that has remained in my memory for its
almost pathetic humour was of two friends recalling old times
together.   "Do you remember," said one, "that splendid
dinner we had at Grantham, and how we did enjoy it?"   "I
do indeed," said his friend, "and it has been a constant
regret to me ever since that I did not have a second helping
of that magnificent haunch of mutton!"

He would also sometimes tell of the incredible doings of
some of the fashionable *roues* among the wealthy, and if I
doubted the possibility of such things being true, would
appeal to the priests, who would assure me that such things,
and worse, did really occur.

Mivart was a very severe and often an unfair critic of
Darwin, and I never concealed my opinion that he was not
justified in going so far as he did.   I also criticized some of
his own writings, but he took it all very good-naturedly, and
we always remained excellent friends.   Besides natural history
we had other tastes in common.   He enjoyed country life,
and for some time had a small country house in the wilds of
Sussex, about midway between Forest Row and Hayward's
Heath, where we sometimes spent a few days ; and some
years later he built a house on the Duke of Norfolk's estate
near Albury, where he had to make a new garden and began
to take an interest in horticulture.   He was also greatly
interested in psychical research and spiritualistic phenomena ;

but this I shall refer to again when I come to my own experiences and inquiries on this intensely interesting subject.

Even more completely than Darwin, Mivart was almost a self-taught biologist. He was educated and trained for the bar, but never practised, his father being a wealthy man. When about five and twenty he began to take an interest in anatomy, and determined to study it systematically; and he one day told me that when he announced his intention, his father remarked, " Well, you never have earned a penny yet, and I suppose you never will." This rather put him on his mettle, and shortly afterwards he wrote an article for some periodical, and on receiving a liberal honorarium he produced the cheque, jokingly telling his father that he had earned it to prove that his prediction was a wrong one. This is a curious parallel to Darwin's statement that when he left school he was considered by his masters and by his father as "a very ordinary boy, rather below the common standard in intellect."

Considering the period of life at which Mivart first turned his attention either to science or literature, the amount of knowledge of comparative anatomy he acquired, largely from dissections and study carried on at home, was very great, and placed him in the first rank among the many great anatomists of his time. This is the opinion of the very competent writer of his obituary notice in *Nature* (vol. lxi. p. 569). His writings on biological subjects were almost as extensive as those of Darwin himself, and his total literary work, largely metaphysical and generally of high merit, was very much larger. In the excellent obituary notice already referred to full justice is done both to the wide knowledge, the intellectual ability, and the charming personality of one whose friendship I continue to look back upon with pleasure and satisfaction.

I will conclude this chapter with a few words about the meetings of the British Association at which I was present. In 1862 I was invited by my kind friend, Professor Alfred Newton, to be his guest at Magdalen College during the meeting, in company with a party of scientific friends, chiefly ornithologists. This was both my first visit to Cambridge and to the Association, and under such pleasant conditions I thoroughly enjoyed both. Besides the number of eminent men of science I had the opportunity of hearing or seeing, I had the pleasure of spending an evening with Charles Kingsley in his own house, and enjoying his stimulating conversation. There was also a slight recrudescence of the evolution controversy in the rather painful dispute between Professor Richard Owen and Huxley, supported by Flower, on certain alleged differences between the brains of man and apes.

I so much enjoyed the meeting, both in its scientific and social aspects, that I attended the next eleven meetings, and generally took part in some of the discussions, besides occasionally reading short papers. One of the most enjoyable meetings socially was that at Exeter, where I and a large party of scientific men were hospitably entertained at a country mansion eight or ten miles from the city, into which we were driven and brought back every day. Among the guests there was Professor Rankin, who entertained us by singing some of his own descriptive or witty compositions, especially the " Song of the Engine Driver," and that inimitable Irish descriptive song on " The City of Mullingar." On this occasion there appeared one of the most humorous parodies of the work of the association that has ever been written, called " Exeter Change for the British Lions." It was in the form of a small magazine, giving reports of the meetings, with absurd papers, witty verses, and clever parodies of the leading members, all worthy of Hood himself in his most humorous vein. One of the best of the parodies is the following, as all will admit who are familiar with the style of the supposed author.

ON THE ALCOHOLIC COMPOUND TERMED PUNCH.

BY JOHN T—ND—LL, LL.D., F.R.S.

Chastened and invigorated by the discipline of physical research, the philosopher fearlessly climbs the never-trodden peaks of pure thought, whence he surveys without dizziness the shadowy domain which lies beyond the horizon of ordinary observation. The empirical art of punch-brewing is co-extensive with civilization. But the molecular commotion which agitates the palate of the punch-drinker and awakes in his brain an indescribable feeling of satisfaction could only be apprehended by one whose mind had been previously exercised on the parallel bars of accoustics and optics.

Taste is due to vibratory motion. A peppermint lozenge, for example, dissolving in the mouth, may be likened to a vast collection of minute tuning-forks vibrating synchronously. Pulses are imparted to the nervous filaments of the tongue and palate, and are translated by the internal sense into peppermint. What was molecular agitation is now taste.

With punch properly compounded, we obtain saporous vibrations of various degrees of rapidity, but so related that their simultaneous action on the organ of taste produces an agreeable harmony. The saccharine, acid, and ethylic trills are rhythmical, and a glass of punch is truly the analogue of the sonnet. The instinct of man has detected many such harmonies which have yet to be investigated. For example : what palate is insensible to the harmonious effect of roast hare and currant-jelly? But where is the philosopher who can lay his hand upon his heart and say he has determined the relation of the saporous vibrations of the jelly to those of the hare? My own researches on this point have deepened my natural humility, and I now eat my currant-jelly with the simple faith of a little child.

Experiment has proved that the juice of three or four lemons, and three-quarters of a pound of loaf-sugar dissolved in about three pints of boiling water, give saporous waves which strike the palate at such intervals that the thrilling acidity of the lemon-juice and the cloying sweetness of the sugar are no longer distinguishable. We have, in fact, a harmony of saporific notes. The pitch, however, is too low, and to heighten it, we infuse in the boiling water the fragrant yellow rind of one lemon. Here we might pause, if the soul of man craved no higher result than lemonade. But to attain the culminating saporosity of punch, we must dash into the bowl, at least, a pint of rum and nearly the same volume of brandy. The molecules of alcohol, sugar, and citric acid collide, and an entirely new series of vibrations are produced—tremors to which the dullest palate is attuned.

In punch, then, we have rhythm within rhythm, and all that philosophy can do is to take kindly to its subtle harmonies. It will depend in some measure upon previous habits, whether the punch, when mixed, will be taken in excess or in moderation. It may become a dangerous ally of gravity and bring a sentient being to the gutter. But, on the other hand, it may become the potent inner stimulus of a noble outward life.

I was also honoured by being admitted to the fraternity of the " Red Lions," who fed together during each meeting of the association and expressed applause by gentle roars and wagging of (coat) tails. On these occasions all kinds of jokes were permissible, and speeches were made and songs sung by the scientific humourists assembled. At Edinburgh in 1871, Lord Neaves, a well-known wit and song-writer, was a guest, and gave us some of his own compositions, especially that on " The Origin of Species a là Darwin "—which he recited standing up and with very fine humour. The following verses are samples :—

> " A *very* tall Pig with a *very* long nose
> Sends forth a proboscis right down to his toes,
> And then by the name of an Elephant goes,
>   Which Nobody can Deny !

> " An Ape with a pliable thumb and big brain,
> When the gift of the gab he had managed to gain,
> As Lord of Creation established his reign,
>   Which Nobody can Deny ! "

And so on for twelve verses, and encouraging roars and great final tail-wagging.

The most deplorable event in my experience of the association was the choice of the late Duke of Buccleuch as President for 1867, at Dundee ; proposed, as I understood, by Sir Roderick Murchison and weakly agreed to by his colleagues. The President's Address has, in every other case, been considered a very serious affair, requiring the labour of some months to compose, in order to render it worthy of an audience consisting practically of the best scientific intellect of our country. But the president on this occasion evidently considered it a condescension on his part to be there at all. He began by telling us that he had never written a speech in his life, and never intended to ; that he knew very little about science, though no doubt it was very useful in its way. Of course it helped us to find coal, " and that kind of thing," to support our manufactures; chemistry, too, very useful, dyeing, manure, and many other things—and thus he went on, with a lot of commonplaces hardly up to the level of an audience of tenant-farmers, for, I suppose, nearly an hour; and then there were complimentary speeches! The address —or rather *an* address—was, of course, printed, but I never read it, as I felt sure it would be so altered and almost wholly remodelled that it would not at all resemble the poor stuff we had been compelled to hear.

At Glasgow, in 1876, I was President of the Biological Section, and our meeting was rendered rather lively by the announcement of a paper by Professor W. F. Barrett on experiments in thought-reading. The reading of this was opposed by Dr. W. B. Carpenter and others, but as it had been accepted by the section, it was read. Then followed a rather heated discussion; but there were several supporters of the paper, among whom was Lord Rayleigh, and the public evidently took the greatest interest in the subject, the hall being crowded. After having studied the matter some years longer, Professor Barrett, with the assistance of the late Frederick Myers, Professor Sidgwick, Edmund Gurney, and a few other friends, founded the Society for Psychical Research, which has collected a very large amount of evidence and is still actively at work.

I and my wife were entertained at Glasgow by Mr. and Mrs. Mirlees, and at one of their dinner-parties we enjoyed the company of William Pengelly, of Torquay, the well-known explorer of Kent's Cavern, whose acquaintance I had made some years before while spending a few days at Torquay with my friend and publisher, Mr. A. Macmillan. He sat on one side of our hostess, and I and my wife on the other, and during the whole dinner he kept up such a flow of amusing and witty conversation that the entire party (a large one) looked at us with envy. He was certainly among the most genial and witty men I have ever met, and could make even dry scientific subjects attractive by his humorous

way of narrating them. It was a rather curious coincidence
that on this occasion, when "psychical research" had first
been introduced to the British Association, I learnt from Mr.
Pengelly that he had himself had one of the most amazing
psychical experiences on record, which I may perhaps find
an opportunity of narrating when I give an account of my
own investigation of these subjects.

After this year I felt that I had pretty well exhausted
the interests of the association meetings, and preferred to
take my autumn holiday, with my wife and two children,
either by the sea or among the mountains, where we could
quietly enjoy the beauties of nature in aspects somewhat
new to us; the only exception I afterwards made being the
jubilee meeting at York, and even here the chief attractions
were the beautiful Alpine gardens of Mr. Backhouse, the
excursion to Rievaulx Abbey, and a visit afterwards to my
friend Dr. Spruce in his retirement at Welburn, near Castle
Howard.

CHAPTER XXVII

MY FRIENDS AND ACQUAINTANCES : SIR JAMES BROOKE,
PROFESSOR ROLLESTON, MR. AUG. MONGREDIEN, SIR
RICHARD OWEN, DR. RICHARD SPRUCE

ABOUT a year or two after I had returned home, Sir James
Brooke had also returned to England, and had retired to
a small estate at the foot of Dartmoor, where he lived in
a comfortable cottage-farmhouse amid the wild scenery in
which he delighted.   I had met him once or twice in London,
and, I think in the summer of 1863 or 1864, he invited me
to spend a week with him in Devonshire, to meet his former
private secretary and my old friend in Sarawak, Mr. (now Sir
Spencer) St. John.   We had a very pleasant time, strolling
about the district or taking rides over Dartmoor ; while at
meals we had old-time events to talk over, with discussions
of all kinds of political and social problems in the evening.
At the same time Lady Burdett-Coutts, with her friend Mrs.
Brown, were staying near, and often drove over and took us
all for some more distant excursions.

This meeting and my friendship with Sir James Brooke
led to my receiving several invitations to dine in Stratton
Street, where my friend George Silk was also a frequent
guest ; but my unfortunate habit of speaking my thoughts too
plainly broke off the acquaintance.   The rajah's nephew,
Captain Brooke, who had been formerly designated as Sir
James's successor under the Malay title of Tuan Muda (young
lord), had done or written something (I forget what) to which
Sir James objected, and a disagreement ensued, which resulted in the captain being deposed from the heirship, and

his younger brother Charles, the present rajah, being nomi-
nated instead. As I was equally friendly and intimate with
both parties and heard both sides, I thought the captain had
been rather hardly treated, and one day, when the subject was
mentioned at Stratton Street, I ventured to say so. This
evidently displeased Lady Burdett-Coutts, and I was never
invited again—a matter which did not at all disturb me, as the
people I met there were not very interesting to me. When
Sir James Brooke heard of my indiscretion, he wrote to me
very kindly, saying that he knew that I was the captain's
friend and had a perfect right to take his part, and that my
doing so did not in the least offend him and would make
no difference in our relations, and I continued to receive
friendly letters from him till he went to Borneo for the last
time, in 1866. Soon after his return he died at his Devonshire home, in June, 1868. I have given my estimate of his
character and of his beneficent work at Sarawak in my "Malay
Archipelago."

One of my early friends, though I did not see a great deal
of him, was Professor George Rolleston, whose death in the
prime of life (in 1885) was a great loss to the biological
sciences. I possess, however, only one letter from him,
accompanying some remarks by a friend of his, Dr. Kay,
principal of a theological college in Calcutta, on my article
in *The Reader* on " How to Civilize Savages," in which I had
criticized missionary work, and, by implication, popular ideas
of the value of Christianity. The MSS. sent has been lost,
but I happen to have a rough copy of my reply, and as it
argues the missionary question more fully than was thought
necessary in the article (included with additions in my
" Studies "), I think it may be well to print it here.

" 9, St. Mark's Crescent, Regent's Park,
" September 23, 1865.
" DEAR ROLLESTON,
          " Your friend has very fairly stated my argument,
yet does not seem to me to touch the point of it in his

answer.   For instance, he says, 'the principal doctrines of
Christianity were held at the beginning as now.'   True, but
what was that beginning? and *where* did the doctrines and
dogmas of Christianity spring up?   It was in the very focus
of all the highest and most ancient civilizations of the world
—the Jewish, the Egyptian, the Assyrian, the Greek, and the
Roman.   These peoples had already gone through the long
process of mental development which the savage has not even
begun.   The doctrines (of Christianity) grew among them,
as they do *not* grow among savages, because they were adapted
to the mental state in the one case, but are not in the other.

" What savage nations have (as he asserts) been raised out
of their degradation by Christianity?   The Abyssinians are
a good case to show that Christianity *alone* does nothing.
The circumstances have not been favourable to the growth of
civilization in Abyssinia, and therefore, though they have had
Christianity as long as we have (or longer), they are scarcely
equal morally to many pagan and certainly inferior to some
Mohammedan nations.   This is a crucial instance.

" He says the Britons did not arrive at any 'great moral
elevation' under the Romans.   But will he point out any
savages who have arrived at a 'great moral elevation' in the
same time under Christianity?   I know of none.   No doubt
there has been often a superficial improvement, as in some
of the South Sea islands; but it is an open question how
much of that is due to the purely moral influence of a higher
and more civilized race.

" Of course, if you claim all virtue as Christian virtue, and
impute all want of goodness to want of true Christianity, you
may prove the value of any religion.   The Mohammedan
argues exactly the same (see Lady Duff Gordon's ' Letters
from Egypt').   Your friend would no doubt impute whatever
scraps of goodness there may exist in myself to the Christianity
in which I was educated; but I know and feel (though it
would no doubt shock him to hear) that I acted from lower
motives than I do now, and that I was really inferior
morally as a Christian than I am now as, what he would
call, an infidel.

" I look upon the doctrine of future rewards and punish-
ments as a motive to action to be radically bad, and as bad
for savages as for civilized men. I look upon it, above all, as
a bad preparation for a future state. I believe that the *only*
way to teach and to civilize, whether children or savages, is
through the influence of love and sympathy ; and the great
thing to teach them is to have the most absolute respect for
the rights of others, and to accustom them to receive pleasure
from the happiness of others. After this education of habit,
they should be taught the great laws of the universe and of
the human mind, and the precepts of morality must be placed
on their only sure foundation—the conviction that they
alone can guide mankind to the truest and most widespread
happiness.

" I cannot see that the teaching of all this can be furthered
by the dogmas of any religion, and I do not believe that
those dogmas really have any effect in advancing morality in
one case out of a thousand.

" My article, by-the-bye, was considerably pruned, and I,
of course, think spoilt by the editor.

" Yours very sincerely,

" ALFRED R. WALLACE."

In the year 1869 it was proposed to establish a scientific
weekly paper to serve as a record of progress for workers, to
furnish reviews of scientific books by specialists dealing with
them on their merits alone, to give reports of the meetings
of societies, and popular yet accurate accounts of all remarkable new facts or theories of general interest. I took
part in the meetings at which the subject was discussed,
and undertook to contribute occasionally to its pages, and
for the next quarter of a century almost every volume of
Nature, as the new periodical was called, contains either
reviews, letters, or articles from my pen. In the fifth issue
(December 2, 1869) there was an article on science reform,
giving an account of the report of a committee of the British
Association on a question suggested by a paper read by
Lieut.-Colonel Strange, entitled, " On the Necessity for State

Intervention to secure the Progress of Physical Science."
The committee, almost all professors or officially employed
men of science, reported that State aid *was* required, and the
article in *Nature* supported the view.   Believing that this
was not only injudicious, but wrong, I thought it advisable to
state my reasons for opposing it, and sent a rather long letter
to the editor.   It was published on January 13, 1870, but in
order to counteract its supposed dangerous tendency a
leading article accompanied it, headed, " Government Aid
to Science," strongly controverting my views, somewhat mis-
representing them, and omitting to deal with the main ethical
question which I raised.   As my letter is buried in the first
volume of a periodical which few of my readers will possess,
and as I hold the same views still, and consider their advocacy
to be now more important than ever, I here reproduce my
letter.

" GOVERNMENT AID TO SCIENCE.

"The public mind seems now to be going wild on the
subject of education ; the Government is obliged to give way
to the clamour, and men of science seem inclined to seize the
opportunity to get, if possible, some share of the public
money.   Art education is already to a considerable extent
supplied by the State, technical education (which I presume
means education in 'the arts') is vigorously pressed upon the
Government, and science also is now urging her claims to a
modicum of State patronage and support.

"Now, I protest most earnestly against the application of
public money to any of the above-specified purposes, as being
radically vicious in principle, and as being, *in the present state
of society*, a positive wrong.   In order to clear the ground, let
me state that, for the purpose of the present argument, I
admit the right and duty of the State to educate its citizens.
I uphold national, but I object absolutely to all sectional or
class education ; and all the above-named schemes are simply
forms of class education.   The broad principle I go upon is
this—that the State has no moral right to apply funds raised
by the taxation of all its members to any purpose which is

not directly available for the benefit of all. As it has no
right to give class preferences in legislation, so it has no right
to give class preferences in the expenditure of public money.
If we follow this principle, national education is not forbidden,
whether given in schools supported by the State, or in
museums, or galleries, or gardens fairly distributed over the
whole kingdom, and so regulated as to be equally available
for the instruction or amusement of all classes of the community. But here a line must be drawn. The schools, the
museums, the galleries, the gardens must all alike be *popular*
—that is, adapted for and capable of being fully used and
enjoyed by the people at large—and must be developed by
means of public money to such an extent only as is needful
for the highest attainable *popular* instruction and benefit.
All beyond this should be left to private munificence, to
societies, or to the classes benefited, to supply.

" In art, all that is needed only for the special instruction of
artists or for the delight of amateurs, should be provided by
artists or amateurs. To expend public money on third-rate
prints or pictures, or on an intrinsically worthless book, both
of immense money value on account of their rarity, and as
such of great interest to a small class of literary and art
amateurs, and to them only, I conceive to be absolutely
wrong. So, in science, to provide museums such as will at
once elevate, instruct, and entertain all who visit them may
be a worthy and just expenditure of public money ; but to
spend many times as much as is necessary for this purpose in
forming enormous collections of all the rarities that can be
obtained, however obscure and generally uninteresting they
may be, and however limited the class who can value or
appreciate them, is, as plainly, an unjust expenditure. It
will perhaps surprise some of your readers to find a naturalist
advocating such doctrines as these ; but though I love nature
much, I love justice more, and would not wish that any man
should be compelled to contribute towards the support of an
institution of no interest to the great mass of my countrymen,
however interesting to myself.

" For the same reason, I maintain that all schools of art or of science, or for technical education, should be supported by the parties who are directly interested in them or benefited by them. If designs are not forthcoming for the English manufacturer, and he is thus unable to compete with foreigners, who should provide schools of design but the manufacturers and the pupils who are the parties directly interested? It seems to me as entirely beyond the proper sphere of the State to interfere in this matter, as it would be to teach English bootmakers or English cooks at the public expense in order that they may be able to compete with French *artistes* in these departments. In both cases such interference amounts to protection and class legislation, and I have yet to learn that these can be justified by the urgent necessity of our producing shawls and calicoes, or hardware and crockery, as elegantly designed as those of our neighbours. And if our men of science want more complete laboratories, or finer telescopes, or more costly apparatus of any kind, who but our scientific associations and the large and wealthy class now interested in science should supply the want? They have hitherto done so nobly, and I should myself feel that it was better that the march of scientific discovery should be a little less rapid (and of late years the pace has not been bad) than that science should descend one step from her lofty independence and sue *in formâ pauperis* to the already overburthened taxpayer. In like manner, if our mechanics are not so well able as they might be to improve the various arts they are engaged in, surely the parties who ought to provide the special education required are the great employers of labour, who by their assistance are daily building up colossal fortunes ; and also that great and wealthy class which is, professionally or otherwise, interested in the constructive or decorative arts.

" I maintain further, not only that money spent by Government for the purposes here indicated is wrongly spent, but also that it is, in a great measure, money wasted. The best collectors (whether in art or science) are usually private amateurs ; the best workers are usually home-workers or the *employés* of scientific associations, not of Governments. Could any Government institution have produced results so much superior to those of our Royal Institution, with its Davy, Faraday, and Tyndall, as to justify the infringement of a great principle ? Would the grand series of scientific and mechanical inventions of this century have been more thoroughly or more fruitfully worked out if Government had taken science and invention under its special patronage in the year 1800, and had subjected them to a process of forcing (in a kind of Laputa College) from that day to this ? No one can really believe we should have got on any better under such a *régime*, while it is certain that much power would have been wasted in the attempt to develop inventions and discoveries before the age was ripe for them, and which would therefore have inevitably languished and been laid aside without producing any great results. Experience shows that free competition ensures a greater supply of the materials and a greater demand for the products of science and art, and is thus a greater stimulus to true and healthy progress than any Government patronage. Let it but become an established rule that all institutions solely for the advancement of science and art must be supported by private munificence, and we may be sure that such institutions would be quite as well kept up as they are now, and I believe much better. If they were not, it would only prove more clearly how unjust it is to take money from the public purse to pay for that which science and art lovers would very much like to have, but are not willing themselves to pay for.

"The very common line of argument, which attempts to prove the widespread uses and high educating influence of art and of science, is entirely beside the question. Every product of the human intellect is more or less valuable ; but it does not therefore follow that it is just to provide any special product for those who want it at the expense of those who either do not want or are not in a condition to make use of it. Good architecture, for instance, is a very good thing, and one we are much in want of ; but it will hardly be maintained that architects should be taught their profession at the public expense. The history of old china, of old clothes, or of postage-stamps are each of great interest to more or less extensive sections of the community, and much may be said in each case to prove the value of the study; but surely no honest representative of the nation would vote, say, the moderate sum of a million sterling for three museums to exhibit these objects, with a full staff of beadles, curators, and professors at an equally moderate expenditure of £10,000 annually, with perhaps a like sum for the purchase of specimens. But if we once admit the right of the Government to support institutions for the benefit of any class of students and amateurs, however large and respectable, we adopt a principle which will lead us to offer but a weak resistance to the claims of less and less extensive interests whenever they happen to become the fashion.

" If it be asked (as it will be) what we are to do with existing institutions supported by Government, I am prepared to answer. Taking the typical examples of the National Gallery and the British Museum, I would propose that these institutions should be reorganized, so as to make them in the highest degree instructive and entertaining to the mass of the people; that no public money should be spent on the purchase of specimens, but what they already contain should be so thoroughly cared for and utilized as to render these establishments the safest, the best, and the most worthy receptacles for the treasures accumulated by wealthy amateurs and students, who would then be ready to bestow them on the nation to a greater extent than they do at present. From the duplicates which would thus accumulate in these institutions the other great centres of population in the kingdom should be proportionately supplied, and from the Metropolitan centres trained officers should be sent to organize and superintend local institutions, such a proportion of their salaries being paid by Government as fairly to equalize the expenditure of public money over the whole kingdom, and thus not infringe that great principle of equality and justice which, I maintain, should be our guide in all such cases.

" ALFRED R. WALLACE."

I received one solitary letter from a scientific man supporting my views—Mr. G. R. Crotch, of the University Library, Cambridge, a very good naturalist and reasoner. But the process of forcing on expenditure for scientific purposes has gone on increasing : the *Challenger* expedition, with its enormously costly publication of results in thirty-seven large quarto volumes, of not the least interest to any but specialists in biology and physics ; the new buildings at South Kensington for the Science and Art Department ; the enormous and unending increase of new buildings for the housing of all the output of the modern book trade, and of the hundreds and thousands of daily and weekly newspapers, and the monthly magazines and endless trade and art and specialist periodicals—huge mountains of rubbish that each succeeding year will render more utterly impossible of examination by any human being who may live in the next century. In connection with South Kensington, the suggestion has been put forward that a million of money is required to properly house the various scientific departments there ; while, most recent of all, there has been an influential request for an anthropometrical survey and sickness registration of the whole population, at a cost comparable with that of the geological survey ! the grounds being that it is the only way to ascertain if there *is* any physical deterioration of the people, and thus enable the Government to stave off any fundamental remedial measures by the excuse of want of further information !

Among the many pleasant episodes of my life was my connection with Mr. Augustus Mongredien, a member of the Corn Exchange, a writer on free trade, and author of a book published by Murray in 1870—"Trees and Shrubs for English Gardens." When I got my chalk-pit at Grays in 1871, built a house there, and began to take a great interest in gardening, I bought this book, and in consequence wrote to the author. Soon afterwards he invited me to visit him at Heatherside, on the Bagshot sands, where he had formed a nursery of several hundred acres, planted with a great variety of trees and shrubs then just coming to maturity. He then formed it into a joint-stock company, and persuaded me, along with Mr. Fortune, the well-known traveller and plant-collector in China and Japan, and several other persons connected with horticulture, to become directors. After two or three years (there being a mortgage on the property) the company had to be dissolved, and Mr. Mongredien lost all he had invested in it. During the time it lasted, however, I and my wife often spent from Saturday to Monday at Heatherside with Mr. Mongredien, his wife, and two daughters; and among the friends we occasionally met there was Professor (afterwards Sir Richard) Owen, the great anatomist, and one of the most charming of companions. Mr. Mongredien himself was a highly educated and most energetic man, and a great converser. He knew most European languages well, including modern Greek, and was a good classical scholar. He was also well read in general literature, devotedly fond of plants and of nature generally, and somewhat of a *bon vivant;* and when I add that his wife was agreeable, and his daughters intellectual, it will be seen that we had all the elements to make our visits delightful. I had had some correspondence with Professor Owen many years before about the specimens of orang-utang I sent home from Borneo, and I had occasionally met him at scientific societies or at the British Museum; but here I saw him in his social aspect, telling us curious little anecdotes about animals, or quoting the older poets for the gratification of the young ladies. He was also very fond of gardening, and we spent much of our time in long walks about the grounds, where there were quantities of the finest species of conifers from about ten to thirty feet high and in perfect health, and showing all the exquisite beauties of their special type of vegetation in form, foliage, and colour more completely than when at a greater age. These visits gave me a knowledge and love of trees and shrubs, which has been a constant pleasure to me in the three gardens I have since had to make, from the very beginning.

Among the dearest of my friends, the one towards whom I felt more like a brother than to any other person, was Dr.

Richard Spruce, one of the most cultivated and most charming of men, as well as one of the most enthusiastic and observing of botanists. As he lived in Yorkshire after 1867, I only saw him at rather long intervals, but I generally took the opportunity of lecture engagements in the north to pay him a few days' visit. Our correspondence also was scanty, as he was a great invalid and could not write much, and I only preserved such letters as touched upon subjects connected with my own work. I will, however, give a few extracts from these, both to illustrate the character of a little-known man of science, and also because some of the matters touched on are of general scientific interest.

I sent Spruce a copy of my little volume of Essays on "Natural Selection," in 1870, and after reading it he sent it on to *his* friend, W. Wilson, of Warrington, a British botanist, and, like my friend, an enthusiast in mosses. His reply Spruce sent to me, and it is rather amusing, as showing the feelings of the older school of naturalists towards the new heresy of Darwinism.

"MY DEAR FRIEND,

"You will think me a wayward chiel when you hear my confession that to-day, feeling very squeamish mentally, I happened to bethink myself of Wallace's book, and ventured to open it with great misgivings about my coming into *rapport* with one whom you introduced to me as the champion of Darwinian philosophy. With fear and trembling I paused on the threshold of the book, just to see what I should have to grapple with. The 'Contents,' therefore, engaged such attention as I could command, and after examining, or rather glancing, at the contents of the first seven chapters without much emotion of either attractive or repulsive character, skipped over to chapter x., the last of the series, not greatly excited at either pole of the intellect, until I came to 'Matter is Force; all Force is probably Will-force.' 'Oho!' said I, 'now we come to something of interest and connected with my friend Rev. T. P. Kirkman's rather unskilfully written pamphlet on this very subject—we shall have everything in shape and properly argued by the clear-minded Wallace, no doubt.'

"My inquisitiveness, however, did not prevent my beginning at the beginning of the chapter, and I now write before I have come to the question of force and matter. I am delighted and most agreeably surprised to discover that Wallace, whom I least expected to agree with me, confirms what I said to you in a previous letter about Darwin's theory being *one truth* in conjunction with another (and perhaps higher) truth ; not the *only truth* in reference to created entities.

" Well, if Wallace has nothing more contrarient than the contents of this chapter are likely to present to me, I shall not fear to read the rest of the book despondent of coming into complete harmony with him, neither need you fear that I shall remain sceptical on those points where already I am willing to receive them in hypothesis for all really useful or practical purposes in reference to classification. I have as yet to assure myself that chapter x. is not a delusive phantasmal addition written or dreamed by myself, and which I shall soon find, on waking, to be unreal and imaginary.

" As it is, all my apprehensions of a *soporific*, such as I found Darwin's book to be, are dispelled. The book is a very readable one, at any rate, and no one need go to sleep over it . . . (a long passage here on origin of sense of justice).

" Many, many thanks for the loan of this book. Even the little I have read would demand a most grateful return, and I would not have missed it for a good deal. I now anticipate an intellectual feast over the whole of the book, and shall carry it with me joyfully and hopefully to Southport.

" I am glad to learn that Terrington Carr is not entirely obsolete and abolished. I do hope to see it again with my own eyes, and to gather the sphagnum.

"Ever affectionately and truly yours,

" W. WILSON."

It is curious that this chapter x., which was so grievous a falling-off to Darwin that he scored it with "No! No!"

and could hardly believe I wrote it, should have been the means of attracting one good botanist to read it with attention, and thus probably to make a convert.

A letter from Spruce, dated Welburn, Yorkshire, December 28, 1873, gives some interesting matter on a botanical subject on which I had consulted him.

"My article on the modifications in plant-structure produced by the agency of ants was never printed. After I had been told that the MSS. was in the printer's hands, it was returned to me with the request that I would strike out of it two or three short passages, amounting altogether to hardly a page of the *Linnæan Journal.* I declined to do this, for the obnoxious passages summarized my views on the permanent effects produced on certain species of plants by the unceasing operations of ants, extending doubtless through thousands of ages ; and these views were founded on observations continued during eight consecutive years. The bare *reading* of the paper, at the Linnæan, seems to have left a very erroneous impression on some of the auditors. Somebody—I believe it was at a meeting of your own Entomological Society—has credited me with the theory that plants take to climbing to get out of the way of the ants! As I read this absurd statement I thought that none of the plants I had commented on had a climbing habit ; but on looking over the list of two or three hundred species, I find there is a single one that climbs.

"When you go to the British Museum or to the Kew Herbarium, ask to look at the genus *Tococa* or *Myrmidone,* in Melastomaceæ, and you will see examples of the curious sacs on the leaves which are inhabited by ants. Similar sacs are found on the leaves also of certain Chrysobalaneæ, Rubiaceæ, etc., and analogous ones on the branches of cordias and other plants. I believe that in many cases these sacs have become inherited structures—as much as the spurs of orchids and columbines, and thousands of other asymetrical structures, all of which I suppose to have originated in some long continued *external* agency.

"I know that I ought to have gone carefully over all my

specimens again, and to have had drawings prepared to illustrate my memoir. It is the inability to do this which has kept me from writing on many subjects which engaged my attention during the course of my travels. . . .

"The ants cannot be said to be useful to the plants, any more than fleas and lice are to animals. They make their habitation in the melastomas, etc., and suck the juice of the sweet berries; and the plants have to accommodate to their parasites as they best may. But even an excrescence may be turned into a 'thing of beauty,' as witness the galls of the wild rose.

"That diseased structures *may* become inherited—even in the human subject—there is plenty of evidence to prove. Some curious instances are given in Dr. Elam's 'Physician's Problems.'"

At this period Dr. Spruce was, of course, not aware of the very strong evidence against the inheritance of acquired characters of any kind, nor had he the advantage of Kerner's wonderful series of observations on the nature of protective plant-structures against enemies of various kinds—" unbidden guests." Nor was he aware of Belt's remarkable explanation of the use to the plant of one of the most remarkable of these ant-structures—the bull's-horn thorns of a species of acacia. He shows that the ants encouraged by these structures to inhabit the plants are stinging species, are very pugnacious, and thus protect the foliage both from browsing mammals, from other insects, and even from the large leaf-cutting ants.[1] In a later letter, however, Dr. Spruce adopts *utility* to the plant as a general principle.

In a letter, dated Coneysthorpe, Malton, Yorkshire, July 28, 1876, he writes as follows :—

"I can hardly say that I have ever speculated on the purport of the odours of leaves, but I have (at your instance) rummaged in my notes and my memory, for such evidence as I possess on that head, and will lay it before you.

" Every structure, every secretion, of a plant is (before all)

---

[1] " The Naturalist in Nicaragua," pp. 218, 223.

beneficial to the plant itself. That is, I suppose, an incontrovertible axiom. Odoriferous glands, especially if imbedded
in the leaf, act as a protection against leaf-cutting ants, and
(to some extent) also against catterpillars. I can remember
no instance of seeing insects attracted to a plant, to aid in its
fertilization, or for any other purpose, by their presence. The
glands on which some insects feed are (so far as I know)
always exposed, either in the shape of cups on the petioles,
involucres &c., or of hairs with dilated and hollow bases,
and of sessile or stalked cysts, on the leaves, petioles,
pedicels &c.; and the secretion is either tasteless or slightly
sweet, but inodorous—to our senses at least.

"Trees with aromatic leaves abound in the plains of
equatorial America. Those which have the aromatic (and
often resinous) secretion imbedded in distinct cysts include
all Myrtacea, Myrsineæ, Sanydeæ, and many Euphorbiaceæ,
Compositæ etc. The leaves of very few of these are, when
growing, ever touched by leaf-cutting ants. In the few cases,
however, where the secretion is slightly but pleasantly bitter,
and wholesome, as in the Orange, the leaves are quite to their
taste. At a farm house on the Trombetas[1] I was shown
orange-trees which had been entirely denuded in a single
night by Saúba ants. Various expedients are resorted to by
the inhabitants of Saúba-infested lands to protect their fruit-
trees, such as a small moat, kept constantly filled with water,
around each tree; or wrapping the base of the trunk with
cotton kept soaked with andiroba oil, etc.

[Note.—Leaf-cutters in the vicinity of man work chiefly
by night, taught doubtless by painful experience of his vicious
propensity to interfere with their operations. But in the
depths of the forest I have often caught them at work, some
up a tree cutting off leaves and even slender young branches,
others on the ground sawing them up and carrying them off.
When at San Carlos,[2] I one day went into the forest to
gather a *Securidaca* (woody Polygaleous twiner) I had seen
coming into flower a few weeks before. I found it in full

---

[1] A northern tributary of the Amazon above Santarem.
[2] The first village in Venezuela on the Upper Rio Negro.—A. R. W.

flower, but the little tree on which it grew—a Phyllanthus,
with slightly milky and quite innocuous juice, had been taken
possession of by a horde of ants, and I had to wait until they
had stripped it of every leaf before I could pull down my
Securidaca, which they had left quite untouched. It was
probably preserved by its drastic properties from sharing the
fate of the Phyllanthus.]

"Many odoriferous leaves seem destitute of special oil-
glands, and their essential oil probably exists in nearly
every cell, along with the chlorophyll as I have found it in
several aromatic Hepatics. Many Laurineæ and Burseraceæ
(Amyrideæ of Lindley) are in this case. The latter are
eminently resiniferous, and yield the best native pitch (the
brea branca) of the Amazon valley. I have never seen their
leaves mutilated by ants, and I think never by catterpillars.
Oil-glands indeed exist in many plants where they are either
so deeply imbedded or so minute as only to be detected by
close scrutiny. Their presence was denied in the Nutmegs
(see Lindley, etc.)[1] until I found them in the American
species, and one species has them so conspicuous that I have
called it *Myristica punctata*.

"In nearly all these plants, however, when the essential
oil has been wholly or in part dissipated by drying, the
leaf-cutters find the leaves apt material for their purpose—
whatever that may be.[2] They once fell on some of my dried
specimens, and first cut up a Croton—a genus I had never
seen them touch in the living state. It reminded me of our
cows in England, which cautiously avoid the fresh foliage of
Buttercups, but eat it readily when made into hay. The
acrid principle in these and many other plants, odorous and
inodorous, is known to be highly volatile.

"Where aromatic plants most abound is in the dry—often
nearly treeless—mountainous parts of southern Europe and
Western Asia, especially in the sierras of Spain. When I

[1] In Lindley's " Vegetable Kingdom " (3rd ed.) he gives among the characters
of the *Order* Myristicaceæ, "Leaves not dotted."—A. R. W.
[2] The ants store these leaves in extensive underground cavities, where fungi
grow on them on which the ants feed (see Bates and Belt).—A. R. W.

was with Dufour at St. Sever, in April, 1846, he received
a large parcel of plants recently gathered in the Sierra
Guadarama by Prof. Graells, of Madrid. A very large proportion were aromatic, and many of them Labiates.

"I cannot make out that plants with scented leaves
abound more in the tropics than in mid-Europe: nor does
there seem to be a larger proportion of them in any zone of
the equatorial Andes than in the Amazonian plain; although,
as hill-plants are often gregarious, and those of hot plains
very rarely so, odoriferous plants may seem more prevalent
in the high Andes than on the Amazon.

"Plants growing nearest eternal snow in the Andes are,
however (so far as I have observed them), all scentless; but
some acquire an aroma in drying, as, for example the thick
roots of the Valerians that abound there.

"Aromatic plants grow in the Andes up to, perhaps,
13,000 feet, and consist chiefly of Composites, Myrtles,
Labiates and Verbenas. I know a hill-side at about 9000
feet, which at this time of year is one mass of odoriferous foliage and flowers, chiefly of a Labiate undershrub
(Gardoquia fasciculata, Bth.). Another slope of far wider
extent is much gayer with varied colour mainly of the blue
flowers of *Dalea Mutisii* H. B. K.—a papilionaceous shrub
allied to the Indigos—and of the red-purple foxglove-like
flowers of *Lamourouxia virgata* H. B. K. (which is parasitic
on the roots of the Dalea) mingled with the yellow flowers of
the Quitenian broom (*Genista Quitensis*, L.), and of many
other herbs and shrubs with flowers of various shades of
colour; but aromatic plants are almost unrepresented
except by scattered bushes of a Salvia and a Eupatorium.
Analogous contrasts are common enough in our own country.

"In those parts of the Peruvian and Quitenian Andes
I have explored, I have not found odoriferous plants more
abundant than in some parts of England and the Pyrenees;
yet they are quite as much so as in the Amazonian plain,
and often belong to the same Natural Orders. Now leaf-
cutting ants are unknown in the Andes; whence I infer that,
although the presence of a pungent smell and taste may be

protective to leaves in hot forests where such ants do exist, it has not been acquired originally to provide the requisite protection.

"I much doubt the correctness of Mr. Belt's theory that the ants which inhabit leaf-sacs protect the leaves from leaf-cutting ants; for the leaves of such plants are almost invariably thin and dry; whereas the Saúba always selects leaves that are more or less coriaceous, and if it really wanted the sacciferous leaves I fancy it would make short work of their frail inhabitants. Besides, there are numbers of Melastomes, allied to Tococa and Myrmidone, which the Saúba never touches, although they have no protective (?) sacs; but it cuts up readily the coriaceous leaves of other Melastomes, such as various Bellucias, Henrietteas, &c.

"RICHARD SPRUCE."

This letter was written in pencil lying on a couch, to which he was confined the greater part of the day during the latter years of his life, and I have much pleasure in printing it here, because it serves to show my friend's acuteness of observation, and the great interest he took, not only in the structure, but in the whole life and nature of the plants he loved so well, and in their relations to the animal world. I have no doubt but that his objections to Belt's theory that the small stinging ants protected the leaves of the trees or shrubs they inhabited from the very powerful and destructive Saúba ants, are quite sound, and that his many years' observations in the Amazonian forests are to be trusted on this point; yet I believe that Belt was right in their being protective, and there are many devourers of leaves that are as destructive as the leaf-cutting ants. Shrubs which always had colonies of stinging ants would probably be avoided by the tapir and by deer, while they would almost certainly check the ravages of caterpillars, locusts, and the large leaf- and stick-insects.

There is another point that this letter illustrates: the wonderful complexity and adaptability of organization of all living things leading to that infinite variety of form and structure, of colour and motion, which constitute the greatest

charm of the study of nature. People continually ask, "If
scented leaves are such a protection, why do not all plants
have them? If so many can do without them they cannot
be of any use." And the same objection is made to all the
other wonderful modes of protection by concealing colours or
patterns, by resembling uneatable or dangerous species, by the
production of spines or various kinds of armour. "Why are
not all protected?" they say. "You admit that the majority
are without these kinds of protection, yet they all continue
to exist. The whole idea is therefore a delusion." And they
think they have thus destroyed a large part of Darwin's theory.
But all this shows that they are either ignorant of, or forget,
the main facts on which that theory is founded—the enormous
rate of possible increase of *all* organisms, the intensity therefore of the struggle to exist, since only the few best adapted
of these enormous numbers can survive to produce offspring;
and also the undoubted fact that species vary enormously in
population, some being common over large areas, some comparatively scarce, others confined to very limited areas, others
again only existing in such small numbers and in such
restricted areas that they are very rarely found. Now, if
some great change of climate comes on slowly, such a mixed
population of species will be affected in different ways and
will require different modifications to become adapted to it.
Some will become extinct, some will be adapted in one way,
some in quite a different way, depending partly on the kind
and amounts of variation that occurs in each species. Some
will therefore become more numerous in individuals, others
less; and when the complete change of climate has been
effected, we should find a new set of species, some differing
very little, others very greatly from the former inhabitants
of the district, but all fairly well adapted to live under the
new conditions. Taking the one case of the protected leaves,
it would be only those which were in some danger of exter-
mination by insect and other enemies that would develop
the various forms of protection by oil-glands, or hairs, or
spines, or by attracting stinging ants; while many which
existed in great numbers and over wide areas, and which

produced abundance of seed annually ready to fill up all vacancies caused by death, could (metaphorically) laugh at all such enemies, and let them devour as they pleased. Such a plant is our own oak tree, which, though infested by galls of many kinds and devoured by numerous caterpillars, is yet not in the least danger of extinction by them, and therefore has developed no special protection against them.

Again, when in any one year much injury is done by caterpillars, that affords such an increase of food to young birds that the insects are almost all destroyed, and in the following year there are comparatively few, giving the trees time to recuperate and attain to their former vigour ; while in the following year the birds have less food and are thus diminished in numbers. This wonderful action and reaction of all living things on each other is beautifully described by Mr. Hudson in the chapter of his " Naturalist in La Plata," entitled " A Wave of Life."

Early in 1879 I read Grant Allen's book on the " Colour Sense " (for the purpose of a review in *Nature*), and wrote to Spruce asking for some information as to the colours of edible fruits in the South American forests. His reply was, as usual, full of interesting and suggestive facts, and I here give it.

" To reply fully to the queries in your last letter would require me to wade through several volumes of my MSS., but I have put together a few *excerpta* which may serve your present purpose, if they only reach you in time.

" I fear I cannot adduce much evidence as to the fruits most sought after by birds and monkeys. I have seen birds feed on various fruits, but on scarcely any that were not food for man—or at least for Indian man—although a few of them might be too austere, or too acid, for my taste. If, as Sterne says, 'dogs syllogize with their noses,' so do birds with their beaks, monkeys and Indians with their teeth : insomuch as relates to the choice of food. In my long voyage on the Cassiquiare, Alto Orinoco, and some of their tributary streams, my Indians met with many fruits new to

them, all of which that looked at all promising, they tried
their teeth on ; and, if the taste suited they ate on without
dread of consequences.   Drupaceous fruits especially were
found almost uniformly wholesome, although the juice of
the bark &c. might be acrid or poisonous.   It is curious
that in the Apocynea—an order notable for its abundant
milky, and usually poisonous juice—the fruits are rarely, or
very slightly, milky, and the succulent fruits (which are
found in about half the species) are almost invariably wholesome.   You know the *Thevetias*, whose large bony triangular
endocarps, strung together, form the rattles which the Uanpé
Indians tie round their ankles in their dances.   The milk of
the bark is a deadly poison—Humboldt says a scratch from
a thumb-nail anointed with it is almost certain death.   At
Marabitanas a well-grown tree of *T. neriifolia* grew near the
Commandante's house.   It bore flowers and ripe fruits—drupes,
with a thin yellowish cuticle, and about as much flesh on
them as on an average plum ; and I noticed that the Commandante's fowls greedily ate up the fleshy part of any fruit
that might chance to fall.   Seeing this, I thought I might
safely eat of them ; so I gathered and ate four.   What little
taste they had was rather pleasant, and no ill effects followed.
I had not then seen (as I saw a few years afterwards) what
a quantity of black pepper and tobacco a fowl can swallow
with impunity, or I might have thought the experiment rather
hazardous.

" Many fruits and seeds are sought by animals of all
kinds for the sake of their farinaceous or oleaginous properties.   The envelope of these, in any part of the world,
is not often gaily coloured, although some pods of Amazonian
Leguminosæ are deep red, and the contained seeds are very
often painted or mottled.   I suppose however it is about
the succulent, sweet or acid fruits—the drupes and berries—
you chiefly enquire.   The great mass of these are certainly
as vividly coloured as any fruits of temperate climes—more
so indeed, in many cases, than the flowers that precede them.
Call to mind the bright reds and yellows of the Peach-palm,
the Mango, the enlarged fleshy pear-like petiole of the

Cashew, &c. &c. Purple or almost black fruits, often with a bloom on them, are found in many genera of Palms; in the delicious little sloe-like fruits called Umirí (species of Humirium); in the *Cocúras*—exquisite grape-like fruits hanging in dense bunches from little trees of the order Artocarpeæ (*Pouruma cecropiæfolia, P. retusa, P. apiculata* &c.). Among the smaller Palms (Bactris and Geonoma) some have bright red, others black fruits. Papaws have the fruit yellow in the species of the plain; in the mountain species greenish, although some of the smaller ones have scarlet fruit. Myrtles (the berried species, all of which have innocuous, although not many sapid fruits) have in the great majority of Amazon species, black-purple fruits; in some they are red and often intensely acid; in others yellow, &c.

"Succulent fruits with a russet or grey coat are not numerous on the Amazon. There, as elsewhere, they owe that peculiarity to the cuticle minutely breaking up and withering, yet still more or less firmly persisting. Of this class are the very fine and large fruits called Cumá in the Tupi language, yielded by two Apocyneous trees of the Rio Negro (*Couma triphylla* and *C. dulcis*) and one of the Orinoko (*C. oblongæ*). The thickish russet rind contains seeds nestling in copious pulp, which eats rather like the fruit of the Medlar or Service, although far sweeter, whence the Portuguese colonists called the tree Sorveira. The bark abounds in thick, sweet and wholesome! milk, which is excellent glue.

" As the Greengage (whose coat is sometimes partly russet-grey) is the finest among European plums, so is the homely-coloured Cumá among all the fruits of the Rio Negro.

" I think I could count on my fingers (if I exclude the melon-tribe) all the edible green drupes and berries of the Rio Negro. The chief of them are the Alligator-pear and some Custard-apples, although some of the latter have a yellow, some a white, and some a red-purple rind."

Then among other home and private matters comes the remark equally appropriate now, " What an awful state the country is getting into! 'War and wasteful expenditure' seems to be the key-note of our Government."

The special points of interest in the above letter are its complete confirmation of the views derived from European plants, as to use of the colours of fruits in indicating those which are edible for birds or arboreal mammals, while the few exceptions as regards colour are of those large and very sweet fruits whose attractions are sufficient without the signal of bright colour. Again, the very frequent occurrence of acrid or poisonous juice or milk in the bark and leaves, protecting the young shoots and trees from herbivorous animals, combined with perfectly innocuous and often agreeable fleshy or juicy fruits in order to assist in their dispersal, so clearly implies a selective agency in two opposite directions in the same species, as almost to amount to the required demonstration of the existence of natural selection.

I cannot forbear calling attention to the extremely careful wording and punctuation of these letters, written from a sick couch, and of which I have not altered a word or a comma. The clearness and accuracy with which the information is conveyed fittingly corresponds with the writer's careful observation of every aspect and detail of plant life. Had his health permitted more continuous work for a few years longer, he would probably have given us a volume upon all the chief aspects and relations of the vegetation of the forests and mountains of equatorial America, which would have been of the greatest scientific and popular interest.

CHAPTER XXVIII

MY FRIENDS AND ACQUAINTANCES—DR. PURLAND,
MR. SAMUEL BUTLER, PROFESSOR HAUGHTON

ONE of the most interesting, amusing, and eccentric men I
became acquainted with during my residence in London, and
with whom I soon became quite intimate, was Dr. T. Purland,
a dentist, living in Mortimer Street, Cavendish Square. He
was a stout, dark, middle-aged man, with somewhat Jewish
features, and of immense energy and vitality—one of those
men whose words pour out in a torrent, and who have always
something wise or witty to say. He had been a great coin-
collector, and had many anecdotes to tell of rarities hit upon
accidentally. He had an unbounded admiration for Greek
coins as works of art, and would dilate upon their beauties as
compared with the poor and inartistic works of our day.
He was something of an Egyptologist, and had many odds
and ends of antiquities, including teeth from mummies and
dentists' instruments found in the old tombs and sarcophagi.
He was a widower with three growing-up children, and had
been obliged to part with all the more valuable parts of his
collection to educate them.

He was a very powerful mesmerist, and helped, with Dr.
Elliotson and others, in establishing the mesmeric hospital
then in existence, and could succeed in sending patients into
the mesmeric trance when other operators failed. He was one
of the few men at that time who had been up in a balloon
(with Green, the celebrated aëronaut, I think), and one evening
at our house in St. Mark's Crescent, when Huxley and Tyndall
were present, he made some remarks which interested
Tyndall, who thereupon asked him many questions as to
his sensations, the general appearance of the earth, clouds, etc.,

to all of which Dr. Purland replied with such promptitude
and intelligence that all our friends were soon gathered
round to hear the discussion, which went on a long time.

Dr. Purland also possessed a most interesting series of
scrap-books, in which he had collected an immense number
of engravings and woodcuts from old magazines, papers, and
books, which, during his life in London, he had picked up at
bookstalls or among his friends. These were beautifully
arranged in a series of uniform quarto volumes, in some of
which he had illustrated his own second marriage by means
of a series of appropriate caricatures, showing the courtship,
the proposal, the ceremony, the wedding breakfast, the
departure, the wedding journey, with numerous incidents to
the return home ; and occasionally among friends he would
go through all these, describing the various incidents in a
most humorous manner, so as to keep us all highly amused.
When he came to any of our evening receptions, he usually
appeared with one of these books under his arm, and it was
always a source of much interest to our guests. Besides
these books, he had a great collection of odd duplicate scraps,
some of which he used to gum on to the envelopes of letters
in place of a seal, or inside to illustrate some matter referred
to in the letter.

I possess about a dozen of his letters—replies to invitations, remarks on reading my early books, or other matters
—all so amusing and so well illustrating the character and
individuality of the man that I will now print some of them,
and give a few in facsimile to show his style of caricature
illustration.

The letter opposite was, I think, the first I had from him,
and I only give it to illustrate two of his peculiarites—his
gastronomical taste indicated by " Beer Month " for October,
and the " piece of plate " represented by *half* a beautiful little
print in blue of an old willow-pattern plate pasted in opposite
the signature.

The next letter is in answer to an invitation to tea. He
had been reading my " Malay Archipelago," and the reference

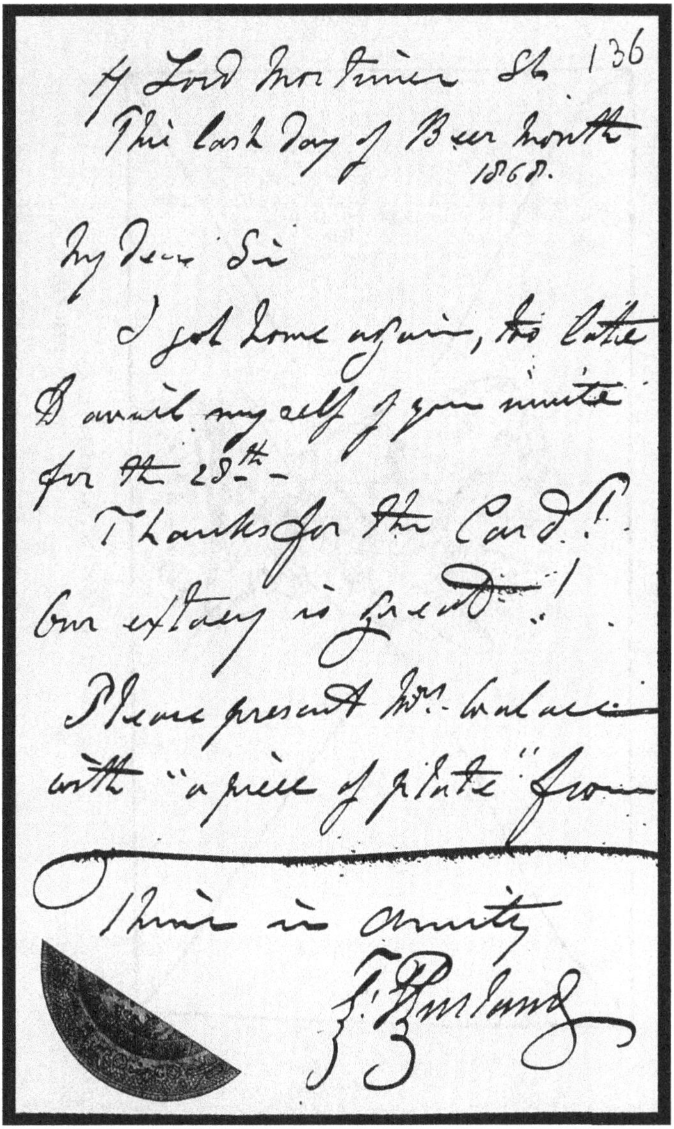

37 Lord Mortimer St. 136
This last Day of Bber month
1868.

My Dear Sir

I got home again, too late
to avail myself of your invite
for the 28th –
Thanks for the Card!!
our extasy is great!!
Please present Mrs. Walker
with "a piece of plate" from

Yours in amity
F. Buckland

[*To face p.* 76, Vol. II.

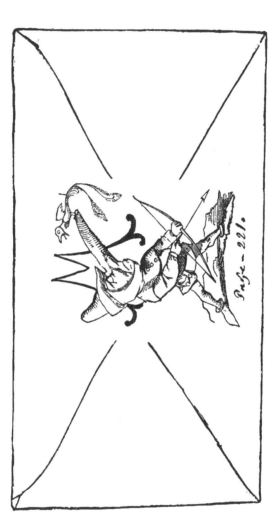

ENVELOPE OF SECOND LETTER.   [*To face p.* 77, VOL. II.

Saturday!

St. markes. W   S Camberwell

Islington N   E 7/3 ow

Which?

Rain?
Mud!
Garotters!!

[*To face p.* 77, VOL. II.

on the envelope (here reproduced) is to the description of the king bird of paradise, and shows how he was able to introduce appropriate cuts from his large stock. The letter itself is in hieroglyphic form, intimating that he had other engagements, indicating himself by his large nose and scrap-book under his arm.

The next I shall give is an account of the sad results of reading one of my books aloud. The heading is a pseudonym for his operating room.

"Fang Castle, June, 1870,
"Therm. 77¼.

"Thanks worthy Signor for the entertainment afforded by your Boke on *Natural Selection*. But good as 'Natural Selection' is, or maybe : I like *Mutual Selection* much better ; and to my thinking it is of much more importance : ex. gr. mutual selection is this—A Lady asks me to become her husband—I ax her to become my wife—that's *Mutual Selection*—aint it 'Natural'? The question of the 'fittest' is a subsequent affair : as is the *Creation by birth*, etc., etc.

"But the pleasure was sadly and suddenly interrupted : I was reading aloud, and got on pretty well through p. 90–91. At 92 Jaws ached terribly! but at p. 94 and 5, even vulcanite could not stand it ; and to my horror my upper set of teeth gave way with a crash! divided between the right lateral and the canine. I was helpless ; and but for an old piece in reserve, my enjoyment of a succulent Roast Pig would have been entirely destroyed : it cost me dear—quite the value of a collection : I must give up reading *scientific (?) names* aloud.

"I picked up a good specimen of Lignum ambulans for a shilling a week ago : and it now forms a prominent feature in our surgery. We are promised a Phyllium in a few days : and a Kallima paralekta. The Rosa Canina is a puzzle at present : I never saw a *Red* Canine tooth! Speaking of teeth—Huxley in his Physiology says Bicuspids *never* have more than two fangs—He knows nothing about it. I have them with three—Molars with 4, 5, and 6! In my lecture case, now before me, there are several : they are not as common as dirt or *earwigs in the country!* but they often turn up.

"I begin the second reading to-night—not aloud—*oh no!*
"With our best Salaam to the Lady, I remain
"Thine in amity,
"THEODOSIUS PURLAND."

The next letter is so wholly and heartily gastronomic that it appeals to me strongly, and reveals the jovial character of the man so amusingly that it must not be omitted.

"Fang Castle, 7, Mortimer St., W.,
"Jan. 9, 1870–1.
"Now you're wuss and wuss!

"Tuesday is the '*University*' of the High-mighty and pious College of Dentists of England, and everywhere else: the 'Collection' of Officers, and when I am to give an acct. of all the four-penny pieces I have received during the year—for, and on behalf of the Jaw-breakers in general, and the Council in particular. We begin at 7—close when we have no more to say; and adjourn to St. James's Hall feeding-Box, for a trial of the Artificials!

"It was lucky I called there this morning. Our Sec. had ordered a *Cold Collation—Cold Veal, Ham* and *Fowls!* COLD DEVILS! You may as well eat a *Hat-box* or *Fire-wood*.

"I have ordered a Hot Supper—*Ducks—Giblet Pies—Plum-pudding*, and such like Comforts—*cold grub indeed, and the Glass at* 26°. So you see, as I cannot well be in two places at once, and where Duty call one must obey, we shall not have the pleasure of Banquetting upon the '*Cold Greens*.'

"As to 'Alcohol'—I do not think I shall venture out—Aunt Loo is going to preside at a *School treat* in the shape of T., Bunns, Plum cake and sundry indigestibles, one a Magic-Lanthorn, which they are to devour. Tom and his Cousin Constance go as well: *So I shall be alone*, as the Gals are at Torquay—capital place for females as it is all Talkée! Talkée!—So, as I said before, I shall be alone—and I contemplate the utter destruction of a KIDNEY PUDDING! Think of that, Master Brook—a *Kidney pudding!* and perhaps a bit of steak or a Sausage or TWO, perhaps THREE: only two of us—the *pudding* and *I!* no weggibles, to take up the room the

Thine in Amity

Theodosia Rutland

[*To face p.* 78, VOL. II.

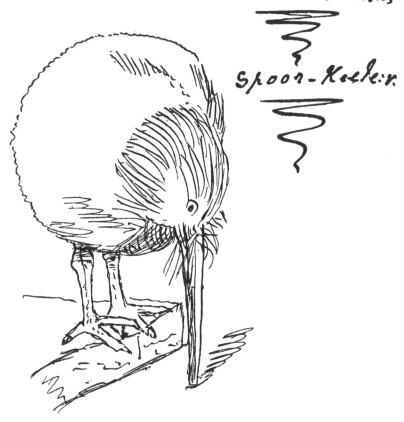

Septembribus — Primus
Senentibus — Omcibus

Spoon — Keeler.

[*This drawing and inscription formed the heading of the third of Dr. Purland's
letters printed on the opposite page.*]

pudding *ought to occupy* ! Oh no! And then the ale—think of the ale—a fresh Cask—Nine Gallons, a shilling a Gallon! goes down your throat like *a wheel-barrow*, washing out the Corners preparatory to a *fresh plate of pudding*—the idea is enchanting, and would, if set to Music, be overpowering! Talk of *quartettes* and *quintettes;* what are they to a *Solo* upon a Kidney pudding? Answer me that! No, you can't; it is unanswerable! So with our blessing upon thee and thine, I remain 'pretty much as usual,'

"Yours,

"T. PURLAND."

I presume the "cold greens" refers to some delicacy (perhaps lobster-salad) I had tempted him with, while "alcohol" in the next line must imply an invitation to a "spiritual" *séance* with some friends, which were very frequent about this period. In like manner, he puts "university" for "anniversary," and "collection" for "election"—all in the exuberance of his spirits, which forbids his writing like other people. But the frank, open, animal enjoyment of it all is equal to Falstaff or Dumas's fat monk, Gorenflot, in "Chicot the Jester."

The next is all about family matters, but illustrated, and in his best style.

"In obedience to thy orders we proceed to indicate the positions of our satellites—12—exact time; *Thomas Theodosius Constantine* is at Bryckden—a place seven miles from everywhere. T. T. C. will make his triumphal entry into the Victoria Station at 4.20 p.m., followed by *all the game* he hath shot with his cross-bow, which we hope will not be more than the porters can conveniently carry.

"*Mary Ellen,* commonly called NELL, is at Gravesend, whacking into and keeping in order some juvenile cousins— the progeny of the Rev. Sleap, Bp. Designative of Alsatia, but at present holding forth at the parish Church of Ware.

"*Louisa Harriett,* commonly called Loo, is with her Aunt Loo, at Gosport, superintending the getting up of the festivities necessary on the Marriage of their Cousin the Daughter of Col. Wright (who has been where you tried to get and didn't as mentioned in your Boke), and who has a great desire to hold speech thereon.

" For ourselves we are supposed to be in charge of the house, and Thomas Thedosius Constantine will perhaps be at the Dovers, so that the probability is we shall be as thus—

" Thus much for this week; next week all the Chicks will be beneath our wing, and probably able and willing to de- molish, or assist in demolishing, any larder however large.

" SECULAR.

" We cancel that part relating to Loo and Aunt Loo—there is a screw loose. Nell was to have been one of Six! Brides- maids; but our hilarity on hearing of the absurdity hath given offence. Nell therefore retires, and *we* are under a Kibosh!

" As soon as we are able to breathe, we will communicate.
" Thine,
" DENTATUS."

Here's another

Cat — as — trophe

An old woman quite deaf — living in a sky parlour — hath a Cat — that touches her hand whenever the Bell rings — a mild Cat — not Bell — in ears — but considered the Bell of the house — hope she gets her Bell-in-full!

Thine

Nasco

MY LAST LETTER FROM DR. PURLAND.

[*To face p.* 81, VOL. II.

The next letter refers chiefly to an eccentric friend of his, Mr. Morgan Kavanagh, author of a work on "The Origin of Language and of Myths," and always referred to by Purland as "The Great O," on account of his fundamental idea that (O) was the sign of the sun, the only permanently circular object in nature, and that the word " O " was the original name of the sun (from making the figure with the lips), and was thus the origin of all language. The book, however, is full of the most ingenious and suggestive derivations from Sanscrit and the Eastern languages.

"Sept. 24, 1872.

"No! can't be a bigger man than you—19 stone. Will warm the only bed we have—as spare! But the nights are fine, and a walk home after the *Jaw* won't hurt you.

"You can grub if you like on what we have. As to the great O, he was here on Saturday—Och Murther—as usual, full of his diskivery—but it is all bosh.

"The true thing is this. Originally, man spoke by signs, and no wonder—Adam and Eve spoke by signs only, until one day Adam refused to go round the corner for some hard-bake, which put Eve into a passion, and in her rage she broke Adam's head with the bedpost, which made him cry 'O!' and Eve, alarmed at opening his head and mouth at one blow, cried 'O' too. That's the origin of Language!

"Some think Adam said 'O Crikey,' but as he was Crackey at the time it is uncertain.

"Thine,

"NASO."

The last I have was an anecdote of animal sagacity, a subject then being discussed in the papers, and of which he had given me some examples. I give a print of it, as it is a good example of his caricature drawing and of one of his fantastic signatures.

Our pleasant intimacy came to an end in a most absurd manner. Dr. Purland was, as I have said, a powerful and enthusiastic mesmerist, and had given his services for many surgical operations. Just as the opposition of the chiefs of the medical profession was dying away, and they were beginning to acknowledge the great value of the mesmeric sleep in alleviating pain and greatly facilitating serious operations, the discovery of anæsthetics offered a rival, which, though much more dangerous, was more certain and more easily applied in emergencies, and this led to the discontinuance of the use of mesmerism as a remedial agent. This naturally disgusted Dr. Purland, who, with the whole energy of his character, hated chloroform, ether, and nitrous-oxide gas, and would have nothing to do with them in his profession. Besides, he despised any one who could not bear the pain of tooth-drawing, and would turn away any patient who required the gas to be administered. A year or two after the date of his last letter my teeth were in a very bad state, and I had a number of broken stumps which required to be extracted preparatory to having a complete set of artificials. Entirely forgetting his objections, which, in fact, I had hardly believed to be real, after making an appointment I asked him to get a doctor to administer nitrous-oxide, as I could not stand the pain of three or four extractions of stumps of molars in succession. This thoroughly enraged him. He wrote me a most violent letter, saying he could not continue to be the friend of a man who could ask him to do such a thing, and gave me the name of an acquaintance of his who had no such scruples and whose work was thoroughly good. And that was the last communication I ever had from Dr. Purland.

The dentist to whom he recommended me was really a good workman, and made me a set of teeth which I wore almost constantly for thirty years, and which I have never had equalled since. While going about lecturing, and especially when going to America in 1886, I had new sets made, and I think I have had altogether four complete sets besides the first, but not one of them has been comfortable or even wearable without great pain ; with none could I eat satisfactorily or speak distinctly, and though I pointed out to each new dentist how well these old ones fitted me, and how comfortable they were, and begged each of them to make the new ones as nearly as possible the same shape, yet each one made them differently, and some were so totally unlike that, when placed side by side, no one would believe they could have been made for the same mouth. My experience of modern dentists is that they all want to improve upon nature, and care nothing for the comfort of those who are to use the teeth.

I will occupy the remainder of this chapter with a few particulars of my relations with persons of some eminence, but with whom I had very few opportunities of personal intercourse.

I made the acquaintance of Mr. Samuel Butler, the author of "Erewhon," through my friend Miss Buckley, at whose father's house on Paddington Green I met him two or three times. He was so good as to send me that wonderfully clever and original book, and also his less known satirical religious story, "The Fair Haven," which was reviewed with approval by some of the Church newspapers as a genuine piece of biography, which it purports to be. He also sent me "Life and Habit," and "Evolution Old and New," both of which I reviewed in *Nature* in the year 1879. The former is a wonderfully ingenious, brilliant, and witty application of the theory of Haeckel and others, that every animal cell, or even every organic molecule, is an independent conscious organism, with its likes and dislikes, its habits and instincts like the higher animals. He explains instincts as inherited memories, which, at the time he wrote, was a permissible hypothesis, but is now almost universally rejected as implying the inheritance of acquired characters, which all the available evidence is opposed to. The book, however, is well worth reading for its extreme ingenuity, logical arrangement, and all-pervading wit and humour.

The other work is a very full and careful exposition of the doctrines, as regards evolution, of Buffon, Lamarck, Dr. Erasmus Darwin, Mr. Patrick Matthew, and some more recent writers, with copious quotations from their works, and an attempt to show not only that their views were of the same general nature as those of Darwin, but were also of equal if not greater importance. After reading the volume I wrote the following letter to the author, which may be of interest to those naturalists who either have not seen the work or who have forgotten its essential features :—

"Waldron Edge, Duppas Hill, Croydon,
"May 9, 1879.

"My dear Sir,

"Please accept my thanks for the copy of 'Evolution Old and New,' and of 'Life and Habit,' which you were so good as to send me.

"I have just finished reading the former with mixed feelings of pleasure and regret. I am glad that a connected account of the views of Buffon, Dr. Darwin, and Lamarck, and especially of Mr. Patrick Matthew, should be given to the world ; but I am sorry that you should have, as I think, so completely failed in a just estimation of the value of their work as compared with that of Mr. Charles Darwin,—because it will necessarily prejudice naturalists against you, and will cause 'Life and Habit' to be neglected ; and this I should greatly regret.

"To my mind, your quotations from Mr. Patrick Matthew are the most remarkable things in your whole book, because he appears to have completely anticipated the main ideas both of the 'Origin of Species' and of 'Life and Habit.'

"I should have to write a long article to criticize your book (which perhaps I may do). In your admiration of Lamarck you do not seem to observe that his views are all pure conjecture, utterly unsupported by a single fact. Where has it been proved that, *in any one case, desires* have caused variation ? It is pure theory, with no fact to support it. And even if desires might, in a long course of generations, produce *some effect*, it can be demonstrated that in the same time 'natural selection' or 'survival of the fittest' would produce so much greater an effect as to overpower the other unless the two worked together.

"I am sorry to see also much that seems to me mere verbal quibbles. For instance, at p. 388 (last par.) you turn 'spontaneous variability' into 'unknown causes,' and then, of course, make nonsense of Mr. Darwin's words. In this way I will undertake to make nonsense of any argument. 'Spontaneous variability' is a FACT, as explained, for example, in my review of Mr. Murphy's book (along with yours) in *Nature*. It is an absolutely *universal fact* in the organic world (and for all I know in the inorganic too), and is probably a *fundamental fact*, due to the *impossibility* of any *two organisms* ever having been subjected to exactly *identical* conditions, and the extreme complexity both of organisms and their environment. This *normal variability* wants no other explanation. Its absence is *inconceivable*, because it would imply that *diversity* of conditions produced *identity* of result. The *wishes* or *actions* of individuals may be one of the causes of variability, but only one out of myriads. Now to say that such an universal *fact* as this cannot be taken as a basis of reasoning because the *exact* causes of it are unknown in each ease, is utterly illogical. The causes of *gravitation*, of *electricity*, of *heat*, of *all* the forces of nature are unknown. Can we not, then, reason on them, and explain other phenomena by them, without having the words 'unknown causes' substituted, and thus making nonsense?

"I am no blind admirer of Mr. Darwin, as my works show; but I must say your criticism of him in your present work completely fails to reach him.

"The mere fact that Lamarck's views, though well put before the world for many years by Sir Charles Lyell (and other writers) converted no one, while Darwin has converted almost all the best naturalists in Europe, is a pretty good proof that the one theory *is* more complete than the other.

"Yours very faithfully,

"ALFRED R. WALLACE."

In *Nature* (June 12) I reviewed this book more fully, showing by numerous quotations how completely Mr. Butler has failed to grasp the essential features of natural selection, while a large portion of his criticism of Mr. Darwin's work is purely verbal and altogether erroneous and misleading.   I received no reply either to my letter or to the review.

When I was at Montreal in 1887, Mr. Iles, the manager of the Windsor Hotel in that city, called my attention to a most humorous critical rhapsody which Mr. Butler had written after his recent visit to Canada and sent to the *Spectator*. As I do not think it has appeared elsewhere, and is a good example of his fantastic genius, I here give it from a copy furnished me by Mr. Iles.

A PSALM OF MONTREAL.

[The city of Montreal is one of the most rising and, in many respects, most agreeable on the American continent, but its inhabitants are as yet too busy with commerce to care greatly about the masterpieces of old Greek Art.   A cast, how-ever, of one of these masterpieces—the finest of the several statues of Discoboli, or Quoit-throwers—was found by the present writer in the Montreal Museum of Natural History ; it was, however, banished from public view, to a room where were all manner of skins, plants, snakes, insects, etc., and in the middle of these, an old man stuffing an owl.   The dialogue—perhaps true, perhaps imaginary, perhaps a little of one and a little of the other—between the writer and the old man gave rise to the lines that follow.]

Stowed away in a Montreal lumber-room,
The Discobolus standeth, and turneth his face to the wall ;
Dusty, cobweb-covered, maimed and set at naught,
Beauty crieth in an attic, and no man regardeth.
                    Oh God ! oh Montreal !

Beautiful by night and day, beautiful in summer and winter,
Whole or maimed, always and alike beautiful,
He preacheth gospel of grace to the skins of owls,
And to one who seasoneth the skins of Canadian owls.
                    Oh God ! oh Montreal !

When I saw him, I was wroth, and I said, " O Discobolus !
Beautiful Discobolus, a Prince both among gods and men,
What doest thou here, how camest thou here, Discobolus,
Preaching gospel in vain to the skins of owls ? "
                    Oh God ! oh Montreal !

And I turned to the man of skins, and said unto him, " Oh ! thou man of
    skins,
Wherefore hast thou done this, to shame the beauty of the Discobolus ? "
But the Lord had hardened the heart of the man of skins,
And he answered, " My brother-in-law is haberdasher to Mr. Spurgeon."
                    Oh God ! oh Montreal !

" The Discobolus is put here because he is vulgar,—
He hath neither vest nor pants with which to cover his limbs ;
I, sir, am a person of most respectable connections,—
My brother-in-law is haberdasher to Mr. Spurgeon."
                    Oh God ! oh Montreal !

Then I said, " O brother-in-law to Mr. Spurgeon's haberdasher !
Who seasonest also the skins of Canadian owls,
Thou callest ' trousers ' ' pants,' whereas I call them ' trousers,'
Therefore thou art in hell-fire, and may the Lord pity thee !
                    Oh God ! oh Montreal !

" Preferrest thou the gospel of Montreal to the gospel of Hellas,
The gospel of thy connection with Mr. Spurgeon's haberdasher to the
    gospel of the Discobolus ? "
Yet none the less blasphemed he beauty saying, " The Discobolus hath
    no gospel,—
But my brother-in-law is haberdasher to Mr. Spurgeon."
                    Oh God ! oh Montreal !

In June, 1863, an article appeared in the *Annals and
Magazine of Natural History* by the Rev. S. Haughton,
entitled " On the Bee's Cell and the Origin of Species."   At
that time I was eager to enter the lists with any one who
attacked natural selection or Darwin's exposition of it.   This
article was full of the usual errors and misconceptions, some
of the most absurd nature, but all set forth as if with the
weight of authority in a scientific periodical.   I accordingly
replied in the October number of the *Annals*, and criticized
the critic rather severely.   Mr. Haughton had written : " The
true cause of the shape of the cell is the crowding together of
the bees at work, as was first shown by Buffon "—a view
which Darwin had disproved both by observation of many
distinct species of bees, and by careful experiment with the
honey-bee, as I explained in the article.   He then argues that
" if economy of wax " was the essential cause of the bees
forming hexagon cells out of circular ones, by gnawing away
the solid angles, as Darwin observed them doing, we ought
to find a series of species, some making *triangular*, others
square cells, because these are the forms which geometrically
come next to the hexagon in economy of wax to a given area !
quite overlooking the fact that the primitive cells are proved

to be circular, and that circles in contact cannot be changed by any gradual process of modification involving saving of wax into triangles or squares.

He then charges Darwin with three unwarrantable assumptions, which he declares he "brings to the ground like a child's house of cards." These are (1) " The indefinite variation of species continuously in the one direction ; " (2) " That the causes of variation, viz. natural advantage in the struggle for existence (Darwin), are sufficient to account for the effects asserted to be produced ; " and (3) " That succession implies causation ; that the Palæozoic Cephalopoda produced the red-sandstone fishes ; that these in turn gave birth to the Liassic reptiles, etc." I easily showed that all these alleged "assumptions" of Darwin are absurd misrepresentations of his real statements ; and I concluded by applying his own words with regard to Darwinians as being really applicable to himself : " No progress in natural science is possible so long as men will take their rude guesses at truth for facts, and substitute the fancies of their imagination for the sober rules of reasoning."

This criticism gave great offence to Dr. John Edward Gray, of the British Museum, who, when I next met him, told me that I ought not to have written in such a tone of ridicule of a man who was much older and more learned than myself.

Mr. Haughton, however, seems to have taken it in good part and to have forgotten it, for eighteen years later, when he was F.R.S., Senior Lecturer of Trinity College, and Professor of Geology in Dublin University, he sent me a copy of his " Lectures on Physical Geography," inscribed " With the best respects of the author."

A little later I received from him the following letter :—

" Trinity College, Dublin, April, 25, 1882.

" MY DEAR MR. WALLACE,

" I have received your kind letter of 20th inst., for which I feel much obliged. If the statements about gulf-streams in my last paper support your own views rather than mine, no one will admit the result more readily than myself.

"I fear that I shall not have the pleasure of seeing your degree conferred on the 29th June, as I shall have to attend the General Medical Council in London on the 27th June.

"I was asked by the Provost and Senior Fellows to recommend two names for Honorary Degrees in Physical and Natural Science, and I chose Dr. Siemens and yourself as worthy representatives of the two 'poles of science.'

"I am, yours very truly,

"SAML. HAUGHTON."

Dr. Haughton did, however, return before I left Dublin, and I had the great pleasure one morning of breakfasting with him and the other members of the managing committee at the Zoological Gardens, and of enjoying his instructive and witty conversation. The brilliant midsummer morning, the cosy room looking over the beautiful gardens, and the highly agreeable and friendly party assembled rendered this one of the many pleasant recollections of my life.

CHAPTER XXIX

HAVING now lived in London eight years, and having finished, as I then thought, my chief literary work—my "Malay Archipelago"—I had a great longing for life in the country where I could devote much of my time to gardening and rural walks. My wife also was very fond of country life, so I began to look about for a place in which to settle. At this time it had been decided to build a museum in East London to illustrate both art and nature, and having the strong support and influence of Sir Charles Lyell, and through him of Lord Ripon, I felt much too confident of obtaining the directorship of it. I therefore determined to look out for a suitable place in Essex, where I should have easy access to the museum at Bethnal Green if I obtained the post, while, at all events, land would be cheaper there than in the more fashionable districts of the south and west.

As a kind of halfway house, I took an old cottage at Barking—Holly Lodge—to which we moved in March, 1870, and where I was still almost in London. Though Barking was a miserable kind of village, surrounded by marshes and ugly factories, there were yet some pleasant walks along the Thames and among the meadows, while within a quarter of a mile of us was a well-preserved tumulus close to an old farmhouse. Here, too, we had some very pleasant neighbours. Sir Antonis Brady at Stratford, whom I had often visited with my friend Silk, and who had a fine collection of fossils from the gravels of the district; Mr. C. M. Ingleby, the Shakespearean commentator, who was interested in spiritualism;

and more especially Colonel Hope, V.C., who was living at Parsloes, an old manor house within an easy walk, and with whose amiable and intellectual family we spent many pleasant Sunday afternoons. Colonel Hope had here laid out a large sewage farm, and had for years carried out experiments demonstrating the fact that many agricultural crops could be grown on absolutely sterilized sand by the application of sewage in proper quantities. He had urged that the whole of the London sewage, instead of being emptied into the Thames near Barking, should be carried on to the Maplin sands, where about ten thousand acres of land could be reclaimed and fertilized so as to grow a large portion of the vegetable food for London. This would have been the cheaper method in the end, saving the pollution of the whole tidal course of the Thames and the enormous annual cost of dredging required to partially remedy that pollution. Instead of this wasteful expenditure, the rental of the reclaimed land, with the fertilizing sewage, might have been so large as to fully repay the extra expenditure, and at the same time give us an unpolluted stream in our capital city. But the plan was too grand to be accepted, and we continue to pay the penalty.

In the following year I found near the village of Grays, on the Thames, twenty miles from London, a picturesque old chalk-pit which had been disused so long that a number of large elms and a few other trees had grown up in its less precipitous portions. The chalk here was capped by about twenty feet of Thanet sand and pleistocene gravel, and from the fields at the top there was a beautiful view over Erith to the Kent hills and down a reach of the Thames to Gravesend, forming a most attractive site for a house. After some difficulty I obtained a lease for ninety-nine years of four acres, comprising the pit itself, an acre of the field on the plateau above, and about an equal amount of undulating cultivable ground between the pit and the lane which gave access to it. I had to pay seven pounds an acre rent, as the owner could not sell it, and though I thought it very dear, as so much of it was unproductive, the site was so picturesque, and had such capabilities of improvement, that I thought it would be a fair investment. The owner lived at Winchester, and when I went down there to see him and arrange the terms, I recall one little incident illustrating *one* of the great social changes of the last thirty-five years. After our business was settled and we had had some lunch, he offered to show me the cathedral, and on our way there a gentleman passed us on one of the early bicycles, which were then a comparative novelty. As the cyclist passed, my companion remarked, " There goes a fool upon rollers "—expressing a very common opinion among the older portion of the community.

As there was a deep bed of rough gravel on my ground and there were large cement works at Grays, I thought it would be economical to build of concrete, and I found an architect of experience, Mr. Wonnacott, of Farnham, who made the plans and specifications, while I myself saw that the gravel was properly washed. In order to obtain water in ample quantity for building and also for garden and other purposes, I had a well sunk about a hundred feet into a water-bearing stratum of the chalk, and purchased a small iron windmill with a two-inch force pump to obtain the water. I made two small concrete ponds in the garden—one close to the windmill—and had a large tank at the top of a low tower to supply house water. My friend Geach, the mining engineer whom I had met in Timor and Singapore, was now at home, and took an immense interest in my work. He helped me to find the windmill—the only one that we could discover in any of the engineering shops in London—and the well being completed, he and I, with the assistance of my gardener, did all the work of fitting the pump at the bottom of the well with connecting-rods and guides up to the windmill, which also we erected and set to work ourselves. As the windmill had no regulating apparatus, and, when the wind became strong, revolved far too rapidly, and even bent the connecting-rod, I attached to the ends of the iron vanes pieces of plate iron about a foot square, fixed at right angles to the line of motion. These acted as brakes as soon as the revolution became moderately rapid, but had little effect when it was slow ; and the arrangement worked very well.

"THE DELL," GRAYS, ESSEX.

[To face p. 92, Vol. II.

With the help of another labourer I also myself laid down
1¼-inch galvanized water-pipes to the house, with branches and
taps where required in the garden. I also built concrete walls
round the acre of ground at top, the part facing south about
nine feet high for fruit trees, the rest about five feet ; and also
laid out the garden, planted mounds for shelter, made a
winding road from below, which, when the shrubs had grown
up, became exceedingly picturesque ; and helped to sift out
hundreds of cubic yards of gravel to improve the land for
the kitchen garden.   All this work was immensely interesting,
and I have seldom enjoyed myself more thoroughly, especially as my friend Geach was a continual visitor, was always
ready with his help and advice, and took as much interest in
the work as I did myself.   We got into the house in March,
1872, and I began to take that pleasure in gardening, and
especially in growing uncommon and interesting as well as
beautiful plants, which in various places, under many difficulties and with mingled failures and successes, has been a
delight and solace to me ever since.

During my four and a half years' residence at Grays I
received visits from several foreigners of eminence, among
whom I especially recollect three Russians—Hon. Alexander
Aksakoff, who may almost be called the Myers of Russian
and German spiritualism ; Professor Boutleroff, a biologist
and also a spiritualist ; and V. S. Solovyoff, also a spiritualist.
These were all delightful people, and they somewhat amused
my wife and myself by their enjoyment of the few delicacies
we were able to give them.   On one of the occasions we
had a fine crop of peaches on our concrete wall, small, but
very delicious, and we had feasted on them for some time.
So we put a handsome dish containing a dozen or more
on the tea-table, and as our Russian visitor seemed greatly
to appreciate them, we pressed him to eat as many as he
liked, and he took us at our word and finished the dish.
Another time we had some very good orange-marmalade
on the table, which we offered with bread and butter, but
our guest said, "No ; with my tea"—so he asked for half
a cup of tea, of course without milk or sugar, in the Russian

fashion, and then put spoonful after spoonful of marmalade in, till the cup was full. " That is very nice," he said ; and he had another cup of the same mixture. I love delicacies myself, and these little eccentricities interested me ; but I draw the line at marmalade and tea.

At this time I was somewhat doubtful in what particular direction to work, as I found that I could not now feel sufficient interest in any branch of systematic zoology to devote myself to the minute study required for the classification and description of any important portion of my collections. There were many other men who could do that better than I could, while my special tastes led me to some work which involved a good deal of reasoning and generalization. It was, I think, my two friends, Professor A. Newton and Dr. Sclater, who urged me to undertake a general review of the geographical distribution of animals, and after a little discussion of the subject I came to the conclusion that I might perhaps be able to do it ; although, if I had been aware of the difficulties of the task, I should probably not have undertaken it.

As this was the largest and perhaps the most important scientific work I have done, I may perhaps be allowed here to say a few words as to its design and execution. I had already, in several of my papers and articles, explained my general views of the purport and scope of geographical distribution as a distinct branch of biological science. I had accepted and supported Dr. P. L. Sclater's division of the earth's surface into six great zoological regions, founded upon a detailed examination of the distribution of birds, but equally applicable to mammalia, reptiles, and several other great divisions, and best serving to illustrate and explain the diversities and apparent contradictions in the distribution of all land animals ; and I may now add that the additional facts accumulated, and the various divisions suggested during the thirty years that have since elapsed, have not in the least altered my opinions on this matter.

In whatever work I have done I have always aimed at

systematic arrangement and uniformity of treatment throughout. But here the immense extent of the subject, the overwhelming mass of detail, and above all the excessive diversities
in the amount of knowledge of the different classes of animals,
rendered it quite impossible to treat all alike. My preliminary studies had already satisfied me that it was quite
useless to attempt to found any conclusions on those groups
which were comparatively little known, either as regards the
proportion of species collected and described, or as regards
their systematic classification. It was also clear that as the
present distribution of animals is necessarily due to their past
distribution, the greatest importance must be given to those
groups whose fossil remains in the more recent strata are the
most abundant and the best known. These considerations
led me to limit my work in its detailed systematic groundwork, and study of the principles and laws of distribution, to
the mammalia and birds, and to apply the principles thus
arrived at to an explanation of the distribution of other
groups, such as reptiles, fresh-water fishes, land and fresh-
water shells, and the best-known insect-orders.

There remained another fundamental point to consider.
Geographical distribution in its practical applications and
interest, both to students and the general reader, consists of
two distinct divisions, or rather, perhaps, may be looked at
from two points of view. In the first of these we divide the
earth into regions and subregions, study the causes which
have led to the differences in their animal productions, give a
general account of these, with the amount of resemblance to
and difference from other regions; and we may also give lists
of the families and genera inhabiting each, with indications
as to which are peculiar and which are also found in adjacent
regions. This aspect of the study I term zoological geography,
and it is that which would be of most interest to the resident
or travelling naturalist, as it would give him, in the most
direct and compact form, an indication of the numbers and
kinds of animals he might expect to meet with.

But a large number of students now limit themselves to
a study of one of the classes, or even orders, of the higher

animals from all parts of the world, and it is of special
interest to him to be able to see at a glance how each family
and genus is distributed, with the number of known species.
He thus see what are the deficiencies in his collection, and
from what countries he most needs additional species ; and all
this information I wished to give him, as I had often felt
the want of it myself. This part of the work I termed
" geographical zoology," and to this I gave special attention,
and have given for every family of mammals, birds, and reptiles
a diagram, which in a single line exhibits its distribution in
each of the four subregions of the six regions. To give the
reader some idea of this compact method of summarizing
information, I will give here its application to one family of
mammalia :—

FAMILY 50—CERVIDÆ (8 genera 52 species).

Neotropical Subregions.	Nearctic Subregions.	Palæarctic Subregions.	Ethiopian Subregions.	Oriental Subregions.	Australian Subregions.
1, 2, 3, –	1, 2, 3, 4,	1, 2, 3, 4,	– – – –	1, 2, 3, 4,	1, – – –

Here the distribution of the true deer over the earth is
shown at a glance when once the limits of the regions and
subregions are learnt, as marked on the general and special
maps by which the book is illustrated. The work was
published in 1876, in two thick volumes, and it had occupied
a good deal of my time during the four years I lived at Grays.
As this book, being very costly and technical, is less known
to English readers than any of my other works, I will here
give the titles of the chapters, which will sufficiently indicate
the range of subjects treated in its eleven hundred pages :—

PART I.—THE PRINCIPLES AND GENERAL PHENOMENA OF
DISTRIBUTION.

Chap. I. Introductory.
„ II. The means of Dispersal and the Migrations of Animals.
„ III. Distribution as affected by the Conditions and Changes of the
Earth's Surface.
„ IV. The Zoological Regions.
„ V. Classification as affecting the Study of Geographical Dis-
tribution.

I devoted a large amount of labour to making a fairly
complete index, which comprises more than six thousand
entries.

No one is more aware than myself of the defects of the
work, a considerable portion of which are due to the fact that
it was written a quarter of a century too soon—at a time when

ALFRED R. WALLACE. 1878

both zoological and palæontological discovery were advancing with great rapidity, while new and improved classifications of some of the great classes and orders were in constant progress. But though many of the details given in these volumes would now require alteration, there is no reason to believe that the great features of the work and general principles established by it will require any important modification. Its most severe critics are our American cousins, who, possessing a "region" of their own, have been able to explore it very rapidly ; while from several references made to it, I think it is appreciated on the European continent more than it is in our own country.

While this work was in progress I wrote a considerable number of reviews and articles, published my book on "Miracles and Modern Spiritualism," and wrote the article "Acclimatization" for the "Encyclopædia Britannica."

In 1876 I sold the house at Grays and removed to Dorking, where we lived two years. But finding the climate relaxing, we moved next to Croydon, chiefly in order to send our children first to a kindergarten, and then to a high school, and remained there till May, 1881.

During this period, besides my usual reviews and articles, I prepared my address as president of the Biological Section of the British Association at Glasgow, wrote the article on "Distribution—Zoology" for the "Encyclopædia Britannica," and prepared a volume on "Tropical Nature," which was published in 1878. In this work I gave a general sketch of the climate, vegetation, and animal life of the equatorial zone of the tropics from my own observations in both hemispheres. The chief novelty was, I think, in the chapter on "climate," in which I endeavoured to show the exact causes which produced the great difference between the uniform climate of the equatorial zone, and, say, June and July in England, although at that time *we* receive actually more of the light and heat of the sun than does Java in June or Trinidad in December. Yet these places have then a mean temperature very much higher than ours. It contained also a chapter on humming birds, as illustrating the luxuriance of tropical nature ; and others on the colours of animals and of plants, and on various biological problems.

As soon as we were settled at Croydon, I began to work at a volume which had been suggested to me by the necessary limitations of my "Geographical Distribution of Animals." In that work I had, in the first place, dealt with the larger groups, coming down to families and genera, but taking no account of the various problems raised by the distribution of particular *species*. In the next place, I had taken little account of the various islands of the globe, except as forming subregions or parts of subregions. But I had long seen the great interest and importance of these, and especially of Darwin's great discovery of the two classes into which they are naturally divided—oceanic and continental islands. I had already given lectures on this subject, and had become aware of the great interest attaching to them, and the great light they threw upon the means of dispersal of animals and plants, as well as upon the past changes, both physical and biological, of the earth's surface. In the third place, the means of dispersal and colonization of animals is so connected with, and often dependent on, that of plants, that a consideration of the latter is essential to any broad views as to the distribution of life upon the earth, while they throw unexpected light upon those exceptional means of dispersal which, because they are exceptional, are often of paramount importance in leading to the production of new species and in thus determining the nature of insular floras and faunas.

Having no knowledge of scientific botany, it needed some courage, or, as some may think, presumption, to deal with this aspect of the problem ; but, on the other hand, I had long been excessively fond of plants, and was always interested in their distribution. The subject, too, was easier to deal with, on account of the much more complete knowledge of the detailed distribution of plants than of animals, and also because their classification was in a more advanced and stable condition. Again, some of the most interesting

of the islands of the globe had been carefully studied botani-
cally by such eminent botanists as Sir Joseph Hooker, for
the Galapagos, New Zealand, Tasmania, and the Antarctic
islands; Mr. H. C. Watson for the Azores; Mr. J. G. Baker
for Mauritius and other Mascarene islands; while there were
floras by competent botanists of the Sandwich Islands, Ber-
muda, and St. Helena. With such excellent materials, and
with the further assistance of Sir Joseph Hooker's invaluable
essays on the relations of the southern and northern floras, I
felt that my work would be mainly of a statistical nature, as
interpreted by those general principles of organic evolution
which were my especial study.

But I also found it necessary to deal with a totally distinct branch of science—recent changes of climate as dependent on changes of the earth's surface, including the causes
and effects of the glacial epoch, since these were among the
most powerful agents in causing the dispersal of all kinds of
organisms, and thus bringing about the actual distribution
that now prevails. This led me to a careful study of Mr.
James Croll's remarkable works on the subject of the astronomical causes of glacial and interglacial periods, and I had
much correspondence with him on difficult points of his
theory. While differing on certain details, I adopted the
main features of his theory, combining with it the effects of
changes in height and extent of land which form an important
adjunct to the meteorological agents. To this subject I
devoted two of my longest and most argumentative chapters,
introducing many considerations not before taken account of,
and leading, I still think, to a more satisfactory explanation
of the causes that actually brought about the glacial epoch
than any which have since been put forth.

Besides this partially new theory of the causes of glacial
epochs, the work contained a fuller statement of the various
kinds of evidence proving that the great oceanic basins are
permanent features of the earth's surface, than had before
been given; also a discussion of the mode of estimating the
duration of geological periods, and some considerations leading to the conclusion that organic change is now less rapid

than the average, and therefore that less time is required for this change than has hitherto been thought necessary. I was also, I believe, the first to point out the great differences between the more ancient continental islands and those of more recent origin, with the interesting conclusions as to geographical changes afforded by both; while the most important novelty is the theory by which I explained the occurrence of northern groups of plants in all parts of the southern hemisphere—a phenomenon which Sir Joseph Hooker had pointed out, but had then no means of explaining.

This volume, on "Island Life," involved much detailed work as regards the species of plants and animals, information on which points I had to obtain from numerous specialists, involving a great amount of correspondence; while it was illustrated by a large number of maps and diagrams, most of which were drawn by myself. The preparation and writing this book occupied me for about three years, and it was published in 1880. It has gone through three editions, which have involved a large amount of corrections and additions; and it is a work which seems to have opened up a new world of interesting fact and theory to a large number of readers, from several of whom I have received letters expressing the delight and instruction it has given them.

In 1878 I wrote a volume on Australasia for Stanford's "Compendium of Geography and Travel," in which I gave a fuller account than usual of the physical geography, the natural history, and the geology of Australia. In a later edition of this work, in 1893, I gave a much fuller account of the natives of Australia, and adduced evidence for the theory that they are really a primitive type of the great Caucasian family of mankind, and are by no means so low in intellect as has been usually believed. This view seems now to be generally accepted.

In 1878 Epping Forest had been acquired for the public, and its care and management were given to a committee formed mainly of members of the Corporation of the City of London. I was a candidate for the post of superintendent, and obtained testimonials from the presidents of all the

natural-history societies of London, and from many eminent
men, but was not chosen. At the time this was a great disappointment, but I have reason to believe now that it was
" all for the best."

In 1881 a society was formed for advocating the nationalization of the land, of which I was elected president, and in
1882 I published a volume, entitled " Land Nationalization :
its Necessity and its Aims." Some account of this movement will be given in a future chapter. Its publication
brought me letters of sympathy and general agreement from
Sir David Wedderburn, M.P., Lord Mount-Temple, and
many other friends and correspondents. In this year, on
June 29, the Dublin University gave me the honorary degree
of LL.D., as already mentioned in the last chapter. I will
here give the very short but flattering Latin speech of the
public orator in introducing me, with a translation by my
friend Mr. Comerford Casey—

"Introduco quoque ALFREDUM RUSSEL WALLACE, Dar-
winii aemulum, immo Darwinium alterum. Neque hunc
neque illum variae eluserunt species atque ora ferarum. Dar-
winius nempe lauri foetus auricomos decerpsit primus. Sed
quid querimur ?

" ' Primo avulso non deficit alter
   Aureus, et simili frondescit virga metallo.' "

"I introduce also Alfred Russel Wallace, the friendly
rival of Darwin. Equally familiar to both are the different
species and varieties of animals. Darwin, indeed, was the
first to pluck the golden laurel-branch. Yet through this did
Wallace suffer no eclipse ; for as Virgil sang—

" 'One branch removed, another was to hand :
   Another, bright and golden as the first.' "

In this year, too, the world was made poorer by the
death of my kind friend and teacher, Charles Darwin, and I
was honoured by an invitation to his funeral (on April 26)
in Westminster Abbey, as one of the pall-bearers, along
with nine of his most distinguished friends or admirers,

"NUTWOOD COTTAGE," GODALMING.

[*To face p.* 103, VOL. II.

among whom was J. Russell Lowell, as the representative of American science and literature. Among the many obituary notices of Darwin, that by Huxley (in *Nature*, of April 27) is one of the shortest, most discriminating, and most beautiful. It is published also in the second volume of his "Collected Essays." For those who have not read this true and charming estimate of his friend, I may quote one passage : "One could not converse with Darwin without being reminded of Socrates. There was the same desire to find some one wiser than himself; the same belief in the sovereignty of reason; the same ready humour; the same sympathetic interest in all the ways and works of men. But instead of turning away from the problems of nature as wholly insoluble, our modern philosopher devoted his whole life to attacking them in the spirit of Heraclitus and Democritus, with results which are as the substance of which their specu-lations were anticipatory shadows."

In the year 1881 I removed to Godalming, where I had built a small cottage near the water-tower and at about the same level as the Charterhouse School. We had been partly induced to come here to be near my very old friend Mr. Charles Hayward, whom I had first known during my residence at Neath about forty years before. He was living with his nephew, the late C. F. Hayward, a well-known archi-tect, whose children were about the same age as my own. We found here some very pleasant friends among the masters at Charterhouse School, as well as among residents who had come to the place for its general educational advantages or for the charm of its rural scenery. We had here about half an acre of ground with oak trees and hazel bushes (from which I named our place "Nutwood Cottage"), and during the eight years we lived there I thoroughly enjoyed making a new garden, in which, and a small greenhouse, I cultivated at one time or another more than a thousand species of plants. The soil was a deep bed of the Lower Greensand formation, with a thin surface layer of leaf-mould, and it was very favourable to many kinds of bulbous plants as well as half-hardy shrubs, several of which grew there more freely and flowered better than in any of my other gardens.

In 1884 Messrs. Pears offered a prize of £100 for the best essay on "The Depression of Trade," and Professor Leone Levi had agreed to be one of the judges. As I had been for some time disgusted with the utter nonsense of many of the articles on the subject in the press, while what seemed to me the essential and fundamental causes were never so much as referred to, I determined to compete, though without any expectation of success. The essay was sent in some time during the summer of 1885, and in July I received a letter from Professor Leone Levi, in which he writes : "My colleague and myself were greatly pleased with the essay bearing a motto from Goldsmith. We, however, did not see our way to recommend it for the prize, especially on account of disagreement as to the remedies suggested. But, the essay having great merit, we thought it proper to open the envelope in order to correspond with the author."

He then asked me if I would allow the first part of my essay, upon "Conditions and Causes," to be printed with the other essays.

As my proposed remedies were the logical conclusion from the "Conditions and Causes," which I had detailed, and of which the validity seemed to be admitted, I of course declined this offer, and Messrs. Macmillan agreed to publish it under the title, "Bad Times: An Essay on the Present Depression of Trade, tracing it to its Sources in enormous Foreign Loans, excessive War Expenditure, the Increase of Speculation and of Millionaires, and the Depopulation of the Rural Districts ; with Suggested Remedies."

This little book was widely noticed, but most of the reviewers adverted to the fact that I was an advocate of land-nationalization, and therefore that my proposed remedies were unsound. But a few were more open-minded. The *Newcastle Chronicle* declared it to be "the weightiest contribution to the subject made in recent times." *The Freeman's Journal* thus concluded its short notice: "Every point is driven home with vigour and directness, and the little book is well calculated to assist in the formation of sound views upon the urgent question of which it treats." *The Beacon* (Boston, U.S.A.) termed it "a very important little book," and gave it a wholly favourable review; but the notice that pleased me most was that in *Knowledge*, then edited by Richard Proctor, a man of originality and genius. He declared that my book was remarkable as being the application of scientific method to a complex problem of political economy, which, of course, rendered it impossible for the official representatives of that science to accept its conclusions. The book, however, had very little sale, and after a few years the publishers sent me about a hundred copies, which remained an incumbrance to their shelves, and which I gave away. It is, therefore, at present, one of the rarest of my books. In the same year I wrote my first small contribution to the literature of anti-vaccination, entitled "Forty-five Years of Registration Statistics, proving Vaccination to be both Useless and Dangerous;" but this subject will be referred to in a future chapter.

Towards the close of the year I received an invitation from the Lowell Institute of Boston, U.S.A., to deliver a course of lectures in the autumn and winter of 1886. After some consideration I accepted this, and began their preparation, taking for my subject those portions of the theory of evolution with which I was most familiar. At this time I had made the acquaintance of the Rev. J. G. Wood, the well-known writer of many popular works on natural history. He had been twice on lecturing tours to America, and gave me some useful information, besides recommending an agent he had employed, and who had arranged lectures for him at various schools and colleges. I had already lectured in many English towns on the permanence of the great oceans, on oceanic and continental islands, and on various problems of geographical distribution. To these subjects I now added one on "The Darwinian Theory," illustrated by a set of original diagrams of variation. I also wrote three lectures on the "Colours of Animals (and Plants)," dwelling especially on protective colours, warning colours, and mimicry, and for these I had to obtain a series of lantern slides coloured from nature, so as to exhibit the most striking examples of these curious and beautiful phenomena. All this took a great deal of time, and the maps and diagrams forming a large package, about six feet long in a waterproof canvas case, caused me much trouble, as some of the railways refused to take it by passenger trains, and I had to send it as goods ; and in one case it got delayed nearly a week, and I had to give my lectures with hastily made rough copies from recollection.

The lectures I finally arranged for the Lowell course were eight in number, to be given twice a week in November and December. As these lectures formed the groundwork for my book on Darwinism, I will here give their titles—

1. The Darwinian Theory : what it is, and how it has been demonstrated.

2. The Origin and Uses of the Colours of Animals.

3. Mimicry, and other exceptional modes of Animal Coloration.

4. The Origin and Uses of the Colours of Plants.

5. The Permanence of Oceans, and the relations of Islands and Continents.

6. Oceanic Islands and their Biological History.

7. Continental Islands : their Past History and Biological Relations.

8. The Physical and Biological Relations of New Zealand and Australia.

Shortly before I left England I gave the lecture on "Darwinism" to the Essex Field Club in order to see how my diagrams of variation struck an intelligent audience, and was fairly satisfied with the result.

I left London on October 9 in a rather slow steamer in order to have a cabin to myself at a moderate price, and landed at New York on the 23rd, after a cold and disagreeable passage. A sketch of my American tour will be given in the following chapters.

CHAPTER XXX

AN AMERICAN LECTURE TOUR—BOSTON TO WASHINGTON

WHEN I left home I had some idea of extending my journey across the Pacific, lecturing in New Zealand and Australia, perhaps also in South Africa, on my way home. But my voyage out was so disagreeable, making me sick and unwell almost the whole time, that I concluded it would not be wise to extend my sea voyages except under very favourable conditions, which did not occur. One of these was the success of my American tour, but owing to my agent not being a good one, or, perhaps, to my not being sufficiently known in America, I was kept throughout the winter in Washington waiting for lecture engagements, which did not come till March and April.

On reaching New York (October 23), I had my first experience of American prices by having to pay two dollars for a cab to the American Hotel, not a mile off, where I was obliged to go for the night. The next morning (Sunday) I went to stay for a few days with Mr. A. G. Browne, a gentleman on one of the New York daily papers who had called on me at Godalming in the summer. On the way to his house we drove to the picturesque Central Park, in the company of Henry George, the well-known author of "Progress and Poverty," who was then a candidate for the important post of Mayor of New York, and who had been invited by Mr. Browne to meet me. The next evening I attended one of his meetings, and was called upon to say a few words to an American audience. I tried my best to be forcible, praised George, and said a few words about what we were doing in England, but I could see that I did not impress them much.

As Mr. Browne's occupation was to summarize all the evening papers for the morning's issue, his work was from midnight till four in the morning. Then all the forenoon he had to do the same thing with the morning papers for the evening issue, getting his sleep in the early morning and afternoon. One day he got free in order to take me up the Hudson river as far as West Point, passing the celebrated "Palisades"—a continuous row of cliffs about two hundred feet high, and extending for nearly twenty miles on the south bank of the river. They look exactly like a huge fence of enormous split trees, placed vertically, side by side, but are really basaltic columns like those at the Giant's Causeway, crowning a slope of fallen rock. In places the well-wooded country was very beautiful, with the autumnal tints of bright red, purple, and yellow, though we were a little late to see them in perfection. Where we landed, I was delighted to see wild vines clambering over the trees, as well as the Virginia creeper, and there were also sumachs and other characteristic American plants. The situation of the great American Military College is splendid, on an elevated promontory in a bend of the Hudson, surrounded by rugged wooded hills, and with magnificent views up and down the river.

On the 28th I went to Boston to be ready for my first lecture on November 1. I had been recommended by Mr. J. G. Wood to go to the Quincy House, as being moderate in charges, and celebrated for its excellent table. I stayed there nearly two months, and was, on the whole, very comfortable; but it was essentially a business man's hotel, and I made no interesting acquaintances there. My scientific friends told me I ought to have gone to a better hotel, but as these were all four or five dollars a day, with no better accommodation than I had at three dollars, I did not care to change. As I never had better meals at any hotel I stayed at in America (except, perhaps, in San Francisco), I may quote my description of them in a letter to my daughter while they were new to me. "You ought to see the meals at this hotel! The bill of fare at dinner (1 to 3 o'clock) has generally two kinds of soup, two of fish, about twenty to thirty different dishes of meat, poultry, and game, a dozen sorts of pastry, a dozen of vegetables, besides ices, and whatever fruits are in season. You can order anything you like in any combination, and they are brought in little dishes, which are arranged around your plate. Everything is good and admirably cooked. The pies and puddings are equally good. At breakfast and supper there is about half the number of dishes."

During the whole time I was in America I had a wonderful appetite, and ate much more than I did at home, and enjoyed excellent health. I imputed this at the time to the more bracing air, the novelty, and the excitement. But from subsequent events I am inclined to think that I really did not eat enough *nourishing* food at home, although I had what I liked best, and seemed to eat plenty of it.

At my first lecture on " The Darwinian Theory," I had a crowded and very attentive audience, and the newspaper notices the next morning showed that it was a success. One of the shortest and best of these was in *The Transcript*, and was as follows :—

" The first Darwinian, Wallace, did not leave a leg for anti-Darwinism to stand on when he had got through his first Lowell lecture last evening. It was a masterpiece of condensed statement—as clear and simple as compact—a most beautiful specimen of scientific work. Mr. Wallace, though not an orator, is likely to become a favourite as a lecturer, his manner is so genuinely modest and straightforward."

During the time my lectures were going on I occupied myself at the museums, libraries, and institutions of Boston, and paid a few visits in the country. I soon made the acquaintance of Dr. Asa Gray, the first American botanist, General Walker, the political economist, Messrs. Hyatt, Scudder, Morse, and other biologists ; while Mr. Houghton, the publisher, who was very polite, asked me to call at his office to read whenever I liked, and invited me to dinner to meet Oliver Wendell Holmes. I met the Autocrat of the

Breakfast Table several times afterwards, and once called at his house and had a two hours' private conversation. He was very interesting from his constant flow of easy conversation ; but when we were alone he turned our talk on Spiritualism, in which he was much interested and which he was evidently inclined to accept, though he had little personal knowledge of the phenomena.

The National Academy of Science was now sitting at Boston, and I attended several of its meetings, at one of which I heard Professor Langley explain his wonderful discovery of the extension of the heat-spectrome by means of his new instrument, the bolometer. At another meeting Professor Cope read a paper, while Professor Marsh was in the chair, evidently to his great annoyance, as the relations of these great palæontologists were much as were those of Owen and Huxley after 1860. At another meeting the question of geographical distribution came up, and Professor Asa Gray called on me to say something. I was rather taken aback, and could think of nothing else but the phenomena of seed dispersal by the wind, as shown by the varying proportion of endemic species in oceanic islands, and by the total absence in the Azores of all those genera whose seeds could not be air-borne (either by winds or birds), thus throwing light upon some of the most curious facts in plant-distribution. I think the subject, as I put it, was new to most of the naturalists present.

I went several times to Cambridge in order to examine carefully the two important museums there—the Agassiz Museum of Zoology and the Peabody Museum of Archæology. Both are admirable, and Mr. Alexander Agassiz kindly showed me over every part of the former museum, an account of which I have given in the second volume of my " Studies, Scientific and Social."

One day I spent at Salem on a visit to Professor Edward Morse and his pleasant family. He had lived several years in Japan, and had made a very extensive collection of Japanese *pottery*, ancient and modern. He has about four thousand specimens, all distinct, many of great rarity and value. I also dined with Professor Asa Gray to meet most of the biological professors of Harvard University. After dinner he asked me to give them some account of how I was led to the theory of natural selection, and this was followed by some interesting conversation.

One evening I was invited to a meeting of the New England Women's Club, where the Rev. J. G. Brooks gave an address on "What Socialists want." I could hardly make out whether he was a socialist or not, but I thought his views to be very vague and unpractical. I was not at that time a thorough socialist, but considered that a true "social economy" founded on land nationalization and equality of opportunity was what was immediately required. When called upon, I spoke in this sense for about half an hour. I afterwards wrote it out, treating it more systematically, and read it to a private meeting of my friends at Washington. Its substance is embodied in the chapter on "Economic and Social Justice" in my "Studies."

After my earlier Lowell lectures were over, I was free to give them elsewhere, and had a few very interesting engagements. On November 19 I went to Williamstown, in the extreme north-west corner of Massachusetts, to lecture at Williams College on "Colours of Animals." The lecture was appreciated, but, unfortunately, the lantern was so poor as not to show the coloured slides to advantage. Williamstown is in a fine mountainous country, and the next day one of the professors drove me in a buggy over very rough roads, and sometimes over snow, to a pretty waterfall, where I collected a few of the characteristic American ferns, which I sent home, and which lived for many years in my garden. I here first noticed the very striking effect of the white-barked birches and yellow-barked willows in the winter landscape. The fine *Cypripedium spectabile*, I was told, grew abundantly in the bogs of this district. I was hospitably entertained by President Carter, who invited me to visit him in the summer, when there are abundance of pretty flowers—an invitation, I much regret, I was unable to profit by.

Between my two last lectures in Boston I had to give one

at Meriden, a small manufacturing town in Connecticut, involving a railway journey of nearly two hundred miles each
way. I stayed with Mr. Robert Bowman, an Englishman
and a manufacturer of plated goods, who had been thirty years
in America. Much of the country I passed through, as well
as that round Meriden itself, was picturesque with rock and
mountain and rapid streams, yet the whole effect was, as I
noted in my journal, "scraggy as usual," while an American
writer declares that the whole country "has been reduced to
a state of unkempt and sordid ugliness." But I am pretty
sure that the more naturally picturesque parts of this New
England country must be very beautiful in spring and early
summer, when the abundant vegetation would conceal and
beautify that which is bare and ugly in winter. The climate,
too, is unfavourable to that amount of verdure which we can
show throughout the year; while the universality of old
irregular hedgerows in our lowland districts gives a finish
and a charm to our scenery which is wholly wanting
where straight lines of split-wood fences are almost equally
universal.

My next lecture was at Vassar College, Poughkeepsie, on
the way to which I had agreed to pay a visit to Professor
Marsh, at Newhaven, where I arrived on the evening of
November 26. My host, who was a bachelor and very
wealthy, had built himself an eccentric kind of house, the
main feature of which was a large octagonal hall, full of
trophies collected during his numerous explorations in the
far West, and used as a reception and dining-room, with
pretty suites of visitors' rooms opening out of it—a roomy
kind of solidly built bungalow. It is situated near the
Peabody Museum of Yale College, where there was at that
time the largest collection of fossil skeletons, chiefly of
mammals and reptiles of America, to be seen anywhere. The
next morning was devoted to seeing these wonderful remains
of an extinct world, among which were the huge bones of the
atlantosaurus, a reptile near a hundred feet long and thirty
feet high, supposed to be the largest land animal that has
ever existed. The remarkable horned dinosauria, the flying

pterodactyles of strange forms, as well as the almost complete series of links connecting the modern horse with the
very ancient eohippus and hyracotherium, were very interesting. These latter were very small animals with four
toes, which were succeeded by larger and larger forms with
fewer toes, till they culminated in the modern horses, asses,
and zebras, with a single toe, or hoof, on each foot.

In the evening I had the pleasure of meeting Professor
Dana, the first of American geologists, and one or two other
professors of Yale. The next morning was spent in a stroll
over the parks and gardens, and in admiring the grand elm
trees which line many of the streets of this picturesque city
and render it one of the most pleasing I visited in America.
In the afternoon I went by train to New York, and then on
to Poughkeepsie and to Vassar College, one of the most
extensive and complete ladies' colleges, where half the professors are ladies, while the president was Dr. J. M. Taylor.

I breakfasted in the hall with the lady principal, doctor,
professors, and students, of whom there are about three
hundred. Each student has a separate bedroom, and to
each three bedrooms there is a sitting-room, and so far as
possible they are allowed to group themselves. Students
enter at sixteen by a rather stiff examination in mathematics, Latin, either Greek, German, or French, history, etc.
The regular course of study includes natural history, physiology, chemistry, physics, and astronomy, all taught experimentally in laboratories, and an observatory which has
a meridian circle and a twelve-inch equatorial. There is also
a good natural history museum and art gallery. Anglo-
Saxon and moral philosophy are taught in the last term.
The grounds are over two hundred acres of rather rough
park-like country, containing a lake with boats and a gymnasium. In the evening I lectured on "Oceanic Islands" to
a good and very attentive audience.

The next morning I had to be up at 5 a.m. in order to
catch the train to New York and on to Baltimore, where I
lectured in the evening on "Darwinism." I gave here four
lectures to the Peabody Institute, and one, on "Island Life,"

at the Johns Hopkins University. The next morning I called on President Gilman, who showed me round the buildings, library, reading-room, etc., and introduced me to the professors, among whom was Dr. W. K. Brooks, the zoologist, who asked me to lunch with him, and afterwards took me to walk over the Druid Hill Park, a finely wooded hilly tract of 680 acres, close to the town, and forming one of the most picturesque recreation grounds I have seen. I also spent an evening with Professor Brooks, when we talked on Darwinian topics mainly. One day I dined with President Gilman, and met afterwards a host of professors, students, and ladies, and had a very pleasant evening. Another day I called on Professor Ely and had a long talk on the political and social outlook. In the evening he took me to a meeting of psychologists—professors and students—whose talk was so technical as to be almost unintelligible to me, and when they asked my opinion on some of their unsettled problems, I was obliged to say that I had paid no attention to them, and that I was only interested in the question of how far the intellectual and moral nature of man could have been developed from those of the lower animals through the agency of natural selection, or whether they indicated some distinct origin and some higher law ; and I gave them a sketch of my views as afterwards developed in the last eighteen pages of my " Darwinism."

After my last lecture (on December 9) I went to President Gilman's, where I met, among others, Professor Langley, the physicist. The talk was chiefly about Professor Sylvester, who had excited immense interest, not only by his wonderfully original mathematical genius, but also by his eccentricities and self-absorption. Many anecdotes were told of him. He had started to dine with a professor who lived not five minutes' walk from his own house, and whom he had repeatedly visited ; yet he wandered about the streets searching for it in vain, and came in a full hour late. After having lived several years in Baltimore, he was one day asked in the street to direct a person to one of the best-known public buildings, and hastily replied, " Pray excuse me; I am quite a stranger here." His genius for solving puzzles in mathematics gave him an interest in making rhymes. There was a remarkably pretty young lady who came to one of the University festivals whose name was suitable for rhyming purposes, and Sylvester started some complimentary verses to see how many successive rhymes he could make. His intimates declare that for weeks afterwards he would say on meeting them in the morning, " I have got another rhyme for Miss ——," and after all his friends had declared that no more were possible, he still kept on discovering new ones till they amounted to some incredible number.

On December 11 I returned to Boston, the whole country being snow-clad and the rivers all ice-bound. On calling upon my agent I found he had got no more engagements for me, so I determined to go to Washington at the end of the month. Considering that my lectures were so well received wherever I went and so well spoken of in the papers, I was puzzled to know why there was not more demand for them. But later on some of my friends told me that it was because I had been preceded for two years by Rev. J. G. Wood, who, though a very clever artist in colour on the blackboard and an excellent field naturalist, put very little into his lectures. Yet he had been well puffed by the same agent as a " great English naturalist," and had given lectures in most of the colleges in the United States. Hence, when the same agent announced another "great English naturalist," there were few bidders, as I was not at that time sufficiently well known in America. With one exception, I had no lectures whatever for three months !

I spent the three weeks in Boston studying the museums, reading at the public library, paying visits, etc. One evening I dined with the Naturalists' Club at the Revere House Hotel, with such well-known men as Hyatt, Hagen, Minot, Scudder, James, Gould, etc. ; and just before I left I was invited by a wealthy merchant and yachtsman, Mr. John M. Forbes, to a farewell dinner at Parker's Hotel to meet some of Boston's most eminent men. These were Oliver Wendell Holmes ; James Russell Lowell ; Edward Waldo Emerson, son of the philosopher; Dr. Asa Gray; Rev. James Freeman Clarke; Dr. William James; General Francis Walker, President of the Technological Institute; Sir William Dawson, the Canadian geologist, who was lecturing at the Lowell Institute; and two others less known. The dinner was luxurious in the extreme, the table covered over with delicate ferns, and roses with bouquets of violets and daffodils before each guest. I sat next to Lowell, and was rather awed, as I did not know much of his writings, and I think he had never heard of me. The condition of things was not improved by his quoting some Latin author to illustrate some remark addressed to me, evidently to see if I was a scholar. I was so taken aback that instead of saying I had forgotten the little Latin I ever knew, and that my special interests were in nature, I merely replied vaguely to his observation. However, the conversation soon became more general, and such subjects as politics, travel, Sir James Brooke, and even spiritualism, afforded some pleasant interchange of ideas. Fortunately there were no speeches, but I was not so much impressed by the Boston celebrities as I ought to have been.

A good deal of time during my last three weeks in Boston was spent in the society either of the professed men of science or the spiritualists, with both of whom I felt myself at ease; while for general intelligence the latter were quite equal to the former. I also attended some very remarkable séances, an account of which will be given in a future chapter. I had one good example of the sudden changes of temperature to which Boston is liable. On December 24 it was a very mild day, so much so that walking was quite oppressive, and in the evening I sat in my room with the window open to keep cool. At night it rained tremendously till 2 or 3 a.m., but Christmas Day was a hard frost, and the next day the greatest cold I felt in America. I was told that during the winter and spring the thermometer often falls 60° in two hours, and a Bostonian never goes out for a few hours, however mild it may be, without being provided with warm clothing against such sudden changes, which often produce serious effects.

I reached Washington on December 31, and after spending four days with Professor Riley, the State entomologist, I took a room at the Hamilton Hotel, where (with the exception of ten days in Canada) I lived till April 7. I found Washington a very pleasant residence on account of the large number of scientific men in the various Government departments and in the Smithsonian Institution, and also the presence of many literary men, as representatives of the great northern papers or as permanent or temporary residents. Among my earliest acquaintances was Dr. Elliott Coues, a man of brilliant talents, wide culture, and delightful personality, with whose ideas I had much in common, and with whom I soon became intimate. He was not only a practical but highly philosophical biologist, and was equally interested with myself in psychical research. I met many pleasant people at his house, where I often spent my Sunday evenings. I found another equally close friend in Professor F. Lester Ward, who divided his enthusiasms and his work between botany and sociology, both subjects which (as an amateur) interested myself. His writings on the latter subject are very numerous—his "Dynamic Sociology," in two large volumes, being a masterpiece of elaborate systematic study of almost every phase of social science. A more readable and more suggestive work is his "Psychic Factors of Civilization," published in 1893, and he has since contributed numerous papers and addresses of great value to periodicals or to the publications of scientific societies.

As soon as the earliest flowers appeared, he took me long Sunday walks in the wild country round Washington, our first being, on February 13, through the stretches of virgin forest called Woodley Park, now, I believe, a botanical and zoological reserve, where many interesting plants were gathered to send home—Goodyera, *Epigœa repens*, *Carex platyphylla*, and the curious leafless parasite called beech-drops, allied to our orobanche. One curious bog-plant, *Symplocarpus fœtidus*, was in flower, as was the pretty blue hepatica, also found in Europe. February and March were, however, very cold, and Washington was snow-covered and wintry, and so our first really good spring botanizing was on March 27, when we went a rather long walk of about nine miles to High Island, a locality for many rarities. Here we found several pretty or curious spring flowers, the most interesting to me being the strange little white-flowered umbelliferous plant, *Erigena bulbosa;* but other peculiar American plants—Claytonia, Podophyllum, Jeffersonia, etc.— I now saw in flower for the first time. During these excursions we had many long talks and discussions while taking our lunch. At that time I was not a convinced socialist, and in that respect Lester Ward was in advance of me, though he could not quite convince me. He was also an absolute agnostic or monist, and around this question our discussions most frequently turned. But as I had a basis of spiritualistic experiences of which he was totally ignorant, we looked at the subject from different points of view ; and I was limited to urging the inherent and absolute differences of nature between matter and mind, and that though, as a verbal proposition, it may be as easy to assume the eternal and necessary existence of matter and its forces as it is to assume mind as the fundamental cause of matter, yet it is not really so complete an explanation or so truly monistic, since we cannot actually conceive matter as producing mind, whereas we certainly can conceive mind as producing matter.

I also soon became very intimate with Major Powell, the head of the Geological Survey, and also with Captain Dutton, Mr. McGee, and other members of the survey. I spent a good deal of time in their library, reading up the history of the glacial phenomena and antiquity of man in America. At twelve o'clock we all lunched together, in a very informal way, on bread and cheese, fruit, cakes, and tea ; and at this time we had many interesting conversations, as Major Powell was a great anthropologist and psychologist, as well as a geologist, and we thus got upon all kinds of subjects.

I also spent a good deal of my time in the great collection of prehistoric remains, stone implements, weapons, etc., of early man in the National Museum, perhaps the most

XXX] BOSTON TO WASHINGTON 119

wonderful and interesting collection of such objects in the world. One of the gentlemen interested in such things, Dr. Hoffman, took me to a field in the suburbs which had been the site of an old Indian village and where arrow-heads were still often found, and I was able to pick up a few specimens myself.

I was also made free of Cosmos Club, where I went to read papers and magazines. Soon after my arrival Mr. Riley took me to one of the evening receptions, where I met most of the scientific men and women of Washington, and was intro-duced to many of them. Most of them told me they had read my books, and several said that my "Malay Archipelago" had first led them to take an interest in natural history and its more general problems. Here, at one time or another, I met almost all the scientific men of Washington and many of those from other States. One evening I was taken by Major and Mrs. Powell to a meeting of the Literary Society at the house of Mr. Nordhoff, author of an important work on the communistic societies of the United States, and a very advanced thinker. Here I met hosts of people who were really too polite and enthusiastic—"proud to meet me;" "honour and pleasure never expected;" "read my books all their life!" etc.—leaving me speechless with amazement!

The event of the evening was a paper by Mr. Kennan, describing his recent visit, on his return from Siberia, to Count Tolstoi, the great Russian novelist, philanthropist, and non-resisting nihilist. It was a very clever, sympathetic, and suggestive picture of a man described as "a true social hero— one of the Christ type." I often dined at Mr. Nordhoff's, and met many interesting people there, and spent several pleasant evenings with his highly intellectual family. Among the celebrities I met there were Mrs. Hodgson Burnett, none of whose works I had then read ; Captain Greely, the Arctic explorer ; and Senator and Mrs. Stanford, whom I afterwards visited in California.

When settled at the hotel I was allotted a place to take my meals, at a table where there were five other persons. Not

knowing the etiquette of such a position, I did not begin conversation till, I think, the second day, a gentleman and lady of
middle age introduced themselves as Mr. and Mrs. Armstrong,
and we soon became quite friendly. They had a private sitting-
room in the hotel, and I often had afternoon tea with them
or spent the evening ; and as they were educated people
interested in science and literature, while Mr. Armstrong
was a spiritualist, they were very agreeable acquaintances.
Through them I was introduced to the other occupants of
the table—Judge Holman, with his wife and daughter. The
judge was a member of Congress, as representative of Indiana,
and we had sometimes long conversations at breakfast or
dinner on political questions. One of the most interesting
was about the Irish in America. He said, " Why does your
Government drive the Irish out of their country by not letting
them govern themselves ? We find them among our best
citizens when they have a chance. I have known and
observed them for fifty years. Near me, in Indiana, is a
township which was settled about forty years ago by Irish
and Germans, all Catholics. The Germans have increased in
numbers, the Irish have diminished by emigrating further
west and other causes. Many of the Irish have became public
men of eminence, and many others rose to good positions.
Those that remain farmers cultivate their land as well as the
Germans, and show equal industry. Considering the low
class of Irish that usually come over, and their extreme
poverty as compared with the Germans and other immigrants,
it cannot be said that they are at all inferior in industry and
in success in life. That is the general experience all over our
country. They form a valuable portion of our citizens, yet
you English will have it they can't govern themselves, and
make that an excuse for keeping them down and driving
them to emigrate." That is the substance of his remarks,
which I noted down immediately ; and as he was a highly
intelligent man, and a good example of the moderate
American legislator, his opinion seemed to me especially
valuable, and should make our " Unionists " (as they call
themselves, but they are really " gaolers ") pause in their

endeavours to perpetuate the subjection of people who are in every respect as good as themselves.

But to my mind, the question of good or bad, fit or not fit for self-government, is not to the point. It is a question of fundamental justice, and the just is always the expedient, as well as the right. It is a crime against humanity for one nation to govern another *against its will*. The master always says his slaves are not *fit* for freedom; the tyrant, that subjects are not *fit* to govern themselves. The fitness for self-government is inherent in human nature. Many savage tribes, many barbarian peoples, are really better governed to-day than the majority of the self-styled civilized nations. America deserves the gratitude of all upholders of liberty by founding her own freedom on the principle of immutable *right* to self-government—that Governments derive their just powers only from the consent of the governed. To-day, however, America has taken leave of this high ideal, and has become, like ourselves, a tyrant, ruling the Philippinos against their will as we have so long ruled the Irish.

Among the visitors to Washington was the Rev. J. A. Allen, of Kingston, Canada (the father of our Grant Allen), who, with his wife and two daughters, was living in apartments nearly opposite my hotel. I soon became intimate with this amiable and very intellectual family, and spent many pleasant evenings with them; while Mr. Allen sometimes went for walks with me and took me over the Patent Museum, where there is a most wonderful exhibition of models of all the successful and unsuccessful inventions that have been patented in the States. From him I first learnt that his son was a poet, and he gave me a copy of his marvellous poem entitled " In Magdalen Tower," written when he was an undergraduate, describing with wonderful ingenuity and picturesqueness the appearance of the city on a moonlight October night, but going on to discuss the deepest problems of philosophy and their attempted solutions. Take as a sample these two verses on law in the universe—

> " We yearn for brotherhood with lake and mountain ;
>   Our conscious soul seeks conscious sympathy,
>   Nymphs in the coppice, Naiads in the fountain,
>   Gods in the craggy heights and roaring sea.
>   We find but soulless sequences of matter,
>   Fact linked to fact by adamantine rods,
>   Eternal bonds of former sense and latter,
>   Blind laws for living gods.

> " They care not any whit for pain or pleasure
>   That seem to men the sum and end of all ;
>   Dumb force and barren number are their measure ;
>   What can be, shall be, though the great world fall.
>   They take no heed of man or man's deserving,
>   Reck not what happy lives they make or mar,
>   Work out their fatal will unswerv'd unswerving,
>   And know not that they are ! "

The poem consists of twenty-one verses, every one of them perfect in rhyme and rhythm, and each carrying on the argument and illustration to the conclusion. This gifted writer would have been a great naturalist, and perhaps also a great poet, had he not been obliged to write novels and magazine articles for a livelihood.

Another interesting character was Mrs. Beecher Hooker, sister to Henry Ward Beecher and Harriet Beecher Stowe. She was a fine lecturer on social, ethical, and spiritual subjects, and was also a spiritualist and trance speaker, well known throughout America. One evening she gave a reception, to which she invited her friends to meet me. Many of the clergy and a large number of the senators and congressmen, with their wives and daughters, were present, and she would insist on introducing me to a number of them, so that I had to shake hands with fifty or sixty people. They seemed quite puzzled. I heard one say to another, " I guess he's some Western man, but I never heard of him." " No," said his friend ; " he's an Englishman, lecturing on biology and Darwin, and such things." " Wal," said the first, " he hasn't much of the English accent." Mrs. Hooker was very anxious that we should come to live in America (she had visited us in England) and form a kind of home colony, being sure that she could get many advanced thinkers to join ; and some years after she wrote to me about it. But my work was at home.

Many of my most interesting and most intellectual friends were spiritualists. Besides Professor Coues, a man of the mental calibre of Huxley with the charming personality of Mivart, I saw most of General Francis Lippitt, a man who was a lawyer as well as a soldier, and had held many high offices under the Government. He was highly educated and had seen much of the world, and we spent many pleasant hours together. He introduced me to Mr. Daniel Lyman, solicitor to the Treasury, a man of powerful physique and strong character, who had for many years made a study of spiritualistic phenomena, and, like Sir W. Crookes, had had mediums to live with him and be wholly subject to his own conditions. Under such circumstances he had obtained phenomena of a more astounding, yet more convincing nature than any person I have met. He took us over the Treasury, showed us the beautiful machinery for engraving bank-notes, so that every fresh issue—and they are continually being made—may have a new and highly complex pattern. We were also taken to the Treasury vaults—some filled to the roof with bags of dollars, others with gold in interminable ranges. One huge vault, about sixty feet by thirty feet, with iron partitions, was filled from floor to ceiling with bags of dollars, one thousand in each bag. The total amount was fifty-seven millions, and in another vault there was twenty-five millions in gold. The large double doors closing these vaults are of steel, strengthened by massive cross-bars and with huge cylindrical bolts at top, bottom, and sides, all connected by a clockwork arrangement, which prevents the bolts from being moved till the hour at which the clock was previously set. The doors and locks are highly finished pieces of engineer-ing, and must have cost a very large sum each. These enor-mous stores of coin, and the complex and costly arrangements for keeping them safely, afford a striking object-lesson to the socialist of the waste and absurdity of our existing systems of currency, which would be completely unnecessary under a more rational social organization.

One day Mr. Allen went with me to the House of Representatives, where we heard part of a debate on the Pleuro-pneumonia Bill, State rights, etc. The arrangements differ widely from ours. The whole building seems to be open to the public. There is a very broad gallery all round the chamber with comfortable seats, accommodating perhaps several thousand people. Every member has a separate desk and chair, and most of them write or read at their ease while the speeches are going on. Dozens of messenger-boys are always running about, taking letters, telegrams, or messages to friends. To call a boy the member claps his hands. There is much more energy and gesticulation in the speeches than with us. The Capitol is a very fine building, standing on a small hill in a fine park. It is in the classical style, with very broad flights of steps, great numbers of columns, and a beautiful central dome, as graceful in form as that of St. Paul's, and over three hundred feet high. The whole building is pure white, part painted stone, the rest white marble. The general effect is really magnificent. The inside is equally fine, the central hall under the dome forming a kind of public lounge. Owing, however, to its being situated in a city which is not a great business centre, it is rarely crowded.

The Corcoran Art Gallery occupied an afternoon. The most remarkable pictures were Church's "Niagara," Bierstadt's grand view in the Sierra Nevada, and Muller's "Charlotte Corday." One morning I went by invitation to the Naval Observatory to see the instruments and the wonderfully in-genious electrical arrangement by which clocks all over the country are automatically set right at noon, both second and minute hands being moved back or forward as required. I also saw the great equatorial, with twenty-six-inch object glass and of thirty-feet focal length; at that time the finest telescope in the world. A week later, on a frosty night, I went again, and was shown Saturn, with powers of four hundred and six hundred. The division of the ring was very sharp, but the dark ring was barely visible as a shadow on the two ends. The white equatorial belt was, however, very distinct. I was then shown the great nebula in Orion, the double star Castor, and a fine cluster in Perseus, the most beautiful object I saw. The telescope is not quite achromatic, but it is wonderfully steady, and the clockwork motion very perfect. The night, though very clear, was not one for what is termed "good seeing;" hence high powers could not be used, and the result was somewhat disappointing. A really good telescope of moderate size, say four-inch or six-inch object glass, properly mounted, and which can be used whenever the conditions are good, will afford more pleasure and instruction than chance visits to the largest instruments.

Early in January I had an engagement to lecture before the American Geographical Society at New York, the subject being, "Oceanic Islands and the Permanence of Continental and Oceanic Areas." I stayed with my kind friend Mr. A. G. Browne, who took me after the lecture to the Century Club, where I met Clarence King, the geologist, and some other scientific men. Next morning I visited the American Museum of Natural History, where I met Dr. J. B. Holder, Mr. J. A. Allen, the well-known writer on birds and mammals, and some other naturalists; and returned to Washington in the afternoon.

On Sunday evening, March 6, I started on a ten days' visit to Canada to fulfil some lecture engagements. I went by a circuitous route by Williamsport, where I breakfasted; then on by Seneca lake and Rochester to Niagara. All this country was very picturesque—much like Wales, but no walls or hedges, and wooden houses. Willows with bright yellow bark were conspicuous, and very handsome. Near the lake were abundant vineyards, deep gullies in horizontal shaly rock, with numerous waterfalls. I reached the Niagara old suspension bridge at 5 p.m., and had just time to see the rapids by going down the cliffs in an elevator about two hundred feet. The leaping, irregular waves were fine, but hardly up to my expectation. I had an excellent supper at a small hotel, and then went on to Toronto, which I reached at 12.20, going on next morning to Kingston, which I reached at 2.30 p.m., where Principal Grant met me and took me

in a sleigh to the college. In the evening I lectured on
"Darwinism" to a good and attentive audience.

After the lecture some friends of Principal Grant came in,
and we had much conversation. A lady who was interested
in spiritualism spoke to me, and asked me if I knew that
Romanes was a spiritualist, and had tried to convert Darwin.
I told her that I knew he was interested in the phenomena
of spiritualism, but that I thought it most improbable that he
had said anything to Darwin. "But," said she, "Professor
Romanes's brother is a great friend of mine, and he gave me
the drafts of the letters they jointly wrote to Darwin. Would
you like to see them?" I said I certainly should, and she
promised to bring them the next morning. She did so, and
I read them with great interest and surprise, as he had never
mentioned them to me when he had come to see me expressly
to discuss spiritualism. On asking, she said I might take notes
of the contents, as they were given to her without any restriction, and the Canadian Romanes was a thorough spiritualist.
This curious episode, and what it led to, will be explained in
a future chapter.

In the afternoon I left for Toronto, where I arrived about
11 p.m., and drove to Professor Wright's house. We lunched
next day with Dr. Wilson, and met Mr. Hale, the well-known
anthropologist. In the afternoon there was a reception at
Professor Wright's, and in the evening I gave my lecture on
the Darwinian theory, which gave the argument as afterwards
developed in the first five and the last chapters of my book
on "Darwinism." When I had finished, the Bishop of Toronto
made a few remarks, and expressed his relief when he heard
my concluding observations. The next day I gave a combined
lecture on "Animal Colours and Mimicry," which occupied an
hour and three-quarters; but the crowded audience seemed
much interested, and the lantern was an excellent one, and
showed the coloured slides to perfection. A Mr. Smith, the
head of a veterinary college, who had heard my first lecture,
wished me to repeat it to his pupils, which I did the next
day to a very attentive audience of three hundred young
men.

In the evening I dined with Professor Goldwin Smith and a party of scientific men in his fine old house, with black walnut staircase and furniture. Afterwards we adjourned to his spacious library, where we discussed politics and literature. The next evening was spent at Mr. Allen's, where I saw a fine collection of Canadian birds, and was struck by the large number of handsome woodpeckers and other bright-coloured birds as compared with Europe. On my way back to Washington I spent four days at Niagara, living at the old hotel on the Canadian side, in a room that looked out on the great fall, and where its continuous musical roar soothed me to sleep. It was a hard frost, and the American falls had great ice-mounds below them, and ranges of gigantic icicles near the margins. At night the sound was like that of a strong, steady wind at sea, but even more like the roar of the London streets heard from the middle of Hyde Park. When in bed a constant vibration was felt. I spent my whole time wandering about the falls, above and below, on the Canadian and the American sides, roaming over Goat Island and the Three Sisters Islands far in the rapids above the Horse-shoe Fall, which are almost as impressive as the fall itself. The small Luna Island dividing the American falls was a lovely sight ; the arbor-vitæ trees (*Thuya Americana*), with which it is covered, young and old, some torn and jagged, but all to the smallest twigs coated with glistening ice from the frozen spray, looked like groves of gigantic tree corals—the most magnificent and fairy-like scene I have ever beheld. All the islands are rocky and picturesque, the trees draped with wild vines and Virginia creepers, and afford a sample of the original American forest vegetation of very great interest. During these four days I was almost entirely alone, and was glad to be so. I was never tired of the ever-changing aspects of this grand illustration of natural forces engaged in modelling the earth's surface. Usually the centre of the great falls, where the depth and force of the water are greatest, is hidden by the great column of spray which rises to the height of four hundred or five hundred feet ; but occasionally the wind drifts it aside, and allows the great central gulf of falling water to be seen nearly from top to bottom—a most impressive sight.

When I got back to Washington it was snowing hard, and the whole country was more wintry-looking than at Niagara, four degrees further north. I at once went to the Geological Survey Library to look up recent works on Niagara, and had an interesting talk with Mr. McGee about it. He told me that the centre of the Horse-shoe Fall has receded about two hundred feet in forty years. The Potomac falls, which are in gneiss rock, have receded quite as fast. The conditions that combine to produce the recession of waterfalls are numerous, and so liable to change, that it is impossible to trust to conclusions drawn from observations during limited periods. It is evident, for example, that while the Canadian falls have receded nearly one-third of a mile, the American falls have not receded more than ten or twenty feet.

Although I did not have a single lecture engagement at Washington, I read two short scientific papers there. There was a Woman's Anthropological Society, which invited me to address them, and being rather puzzled what to talk about, I made a few remarks on " The Great Problems of Anthro-pology." These I defined as the problem of race and the problem of language. On the first point I stated that there are three great races or divisions of mankind clearly definable —the black, the brown, and the white, or the Negro, Mon-golian, and Caucasian. If we once begin to subdivide beyond these primary divisions, there is no possibility of agreement, and we pass insensibly from the five races of Pritchard to the fifty or sixty of some modern ethnologists. The other great problem, that of language and its origin, was important, because it was, above all others, the human characteristic, and was the greatest factor in man's intellectual develop-ment. I then laid down the outlines of the theory of mouth-gestures, which I afterwards developed in my article on " The Expressiveness of Speech," showing how greatly it extends the range of mere *initiative* sounds (which had been ridiculed by some great philologists) and affords a broad and secure foundation for the development of every form of human speech.

The other paper was on "Social Economy *versus* Political Economy," and was given at the request of Major Powell and a few other scientific friends to a large audience of gentlemen and ladies. It was an attempt to show how and why the old "political economy" was effete and useless, in view of modern civilization and modern accumulations of individual wealth. Its one end, aim, and the measure of its success, was the accumulation of wealth, without considering who got the wealth, or how many of the producers of the wealth starved. What we required now was a science of "social economy," whose success should be measured by the good of all. Under this system, not only should no worker ever be in want, but labour must be so organized that every worker, without exception, must receive as the product of his labour all the essentials of a healthy and happy life; must have ample relaxation, adequate change of occupation, the means of enjoying the beauty and the solace of nature on the one hand, and of literature and art on the other. This must be a first charge on the labour of the community; till this is produced there must be no labour expended on luxury, no private accumulations of wealth in order that unborn generations may live lives of idleness and pleasure.

This paper was altogether too revolutionary for many of my hearers, and the general feeling was perhaps expressed in the following passage from the *Washington Post:* "It is astounding that a man who really possesses the power of induction and ratiocination, and who, in physical synthesis has been a leader of his generation, should express notions of political economy, which belong only or mainly to savage tribes." At that time, however, there was hardly a professed socialist in America. In the eighteen years that have elapsed since this paper was read an enormous advance in opinion has occurred, and to-day, not only to a large proportion of the workers, but to thousands of the professional classes, the views therein expressed would be accepted as in accordance with justice and sound policy.

Another evening I was asked by Dr. T. A. Bland, editor of *The Council Fire*, and friend of the Indians, who had seen

the evils of land-speculation in leading to the robbery of
land granted as Indian reserves, to give some of his friends
a short address, explaining my views on land reform.   I note
in my journal, "preached on 'Land Nationalization,' talk
afterwards."   At this time, however, the one subject of
private interest everywhere in America was land-speculation,
and nobody could see anything bad in it.   My ideas, therefore, seemed very wild, and I don't think I made a convert.

One of the most interesting visits I made in Washington
was to the National Deaf-Mute College, founded in 1857,
and one of the best institutions of the kind in the world.   The
president, Dr. Galaudet, learnt to speak by signs before he
spoke audibly, his mother being deaf and dumb, while his
father was the first teacher of deaf-mutes in the United States.
There are about one hundred and twenty students from
all parts of the Union, and the buildings stand in one hundred
acres of beautifully wooded grounds, within ten miles of the
Capitol.   The more advanced students learn every subject
taught in the best colleges, such as mathematics, the ancient
and modern languages, the various sciences, moral philosophy, etc., and all these subjects are taught as thoroughly
and as easily as to those who possess the power of speech.

But besides being taught to use the gesture language as
easily and as quickly as we use ordinary speech, and to read
and write as well as we do, they are also now taught to
speak—a much more difficult thing, and long thought impossible, because, not being able to hear either the teacher's voice
or their own, they have to be taught by watching their
tutor's mouth while speaking, and then trying to imitate the
movements of the lips and tongue, aided by feeling the
throat with their fingers.   It is a very slow process, and
success depends much on the special imitative faculties and
vocal organization of the learner.   Even in the best cases
there is a hardness and want of modulation in the voice, but
they learn to say everything, even to make a speech in
public, and at the same time they learn what is termed lip-
reading—that is, to know what a person is saying by watching
the motions of the lips and throat.   But in this there is,

of course, a good deal of guesswork, and unless they know the subject of conversation, they are likely to make great mistakes.

Many persons cannot understand how it is possible to convey all kinds of abstract ideas by means of gestures or signals as quickly and as certainly as by vocal sounds. But in reality the former has some advantages over the latter, and is equally capable of unlimited extension and the expression of new ideas, by a modification of familiar symbols. If we consider how easily we convey the idea, "Don't speak," by putting the fingers to the lips; "yes" or "no" by the slightest motions of the head; "come" or "go" by motions of the hand; "joy" or "sorrow" by the expression of the face; "a child" or "a man" by holding the hand at the corresponding height; weakness of mind by tapping the forehead with the finger, we can see how a system of signs and gestures may be gradually built up as surely as have been the vocal sounds of all the various languages of the world. And such a system *has* been built up, and is so complete that a spoken lecture upon any subject whatever can be translated into gestures so as to be perfectly understood and enjoyed by an audience of deaf-mutes. Of course, proper names and the less common technical terms are given by rapidly spelling out the word by letter-signs. No doubt the power of speaking and lip-reading is by far the more valuable for the deaf-mute, since it enables him to communicate with the outside world; but as a means of familiar intercourse with each other, the gesture language is the most certain and the most enjoyable. Both require light, but the latter, involving motions of the limbs and body, can be understood at a greater distance and with less strained attention.

The students trained in this college have no difficulty in finding employment. Some become teachers to the deaf, but others are editors or journalists, clerks, surveyors, draughtsmen, mechanics, etc. I saw one of the younger pupils being taught to speak, which requires immense patience and perseverance in the teacher, and in some cases is almost impossible, except to a very limited extent. Others, on the contrary, learn with comparative rapidity, just as some who can hear acquire foreign languages with a rapidity which seems almost incredible to those without the special faculty. Those who are familiar with the gesture language, and can read and write with facility, seem to enjoy their lives as well as we hearers and speakers enjoy ourselves. They are seen walking together, laughing and gesture-talking with each other, or engaged in the various sports and occupations of their age without any indication of the loss of the means of communication which seems to us so essential. As now taught, the deaf-mute is in a far less painful position than the blind ; indeed, Professor Newcomb told them, in an address he gave at the college in 1885, that they were peculiarly fortunate in being so situated as to escape much of the idle, useless talk that is going on in the world. His own time, he said, was largely taken up by people who had nothing to say. Almost everything worth knowing that has been said is now to be found in print.

While at Washington I was asked by two American papers—*The Nation* and *The Independent*—to review a book just published by Professor Cope, with the rather catching title, " The Origin of the Fittest," made up by combining Darwin's title, " The Origin of Species," and Herbert Spencer's, " The Survival of the Fittest." With such a title from a man who, owing to his extensive knowledge of anatomy and palæontology, was looked up to as a kind of American Haeckel, a really important work might naturally be expected. But this volume consisted almost entirely of a collection of lectures, addresses, and magazine articles, printed just as they were written or delivered, some in the first, some in the third person, with whole pages of the same matter repeated in different chapters, some of the illustrations having no reference in the text. In fact, a more egregious case of bookmaking with a misleading title was never perpetrated. Of course, there was good and original matter in it ; but all

those parts which attempted to justify the title by propound-
ing a new theory of evolution were either quite unsound in
reasoning or wholly unintelligble. When the second application came, I told the editor that I had already agreed to
write one, but could easily write another from a different
point of view. This was accepted, and as the reviews were
unsigned, it was not difficult to make them appear to be by
distinct writers. In the first (which appeared in *The Nation*,
February 10, 1887) I gave a careful summary of the most
important contents of the volume, pointing out the novel
views, and stating how he differed from the Darwinians and
from the chief other schools of biologists. Only in one paragraph at the end I pointed out the great imperfections of the
work, due to the absence of any attempt to weld the mass of
heterogeneous matter into a consistent whole.

In the other article (*The Independent*, March 17, 1887) I
presented my readers with a severe but, I think, perfectly
fair criticism, pointing out the extraordinary incongruity of
the materials, the numerous repetitions, the illustrations without explanation on the plates or any reference to them in
the text, and many other deficiencies. I showed how con-
temptuously he spoke of Darwin as a mere compiler of facts
which every one knew, and the inventor of a theory that
proved nothing, until he himself had now supplied the missing
link—the new conception which cleared up everything! Then
I dealt with his own supposed discoveries, his growth-force,
his law of acceleration and retardation, and other such matters,
showing that, so far as they were intelligible, they were all
included in Darwin's writings ; and I concluded by expressing regret that the talented author should have issued so
incomplete a work. I do not think that any of my friends in
Washington suspected that I was the author of either of the
articles, which I heard spoken of as fair criticisms.

Before I left Washington, Judge Holman took me one
morning to call upon the President, Mr. Cleveland. The
judge told him I was going to visit California, and that
turned the conversation on wine, raisins, etc., which did not

at all interest me. There was no ceremony whatever, but, of course, I had nothing special to say to him, and he had nothing special to say to me, the result being that we were both rather bored, and glad to get it over as soon as we could. I then went to see the White House, some of the reception rooms being very fine; but there was a great absence of works of art, the only painting I saw being portraits of Washington and his wife.

Washington itself is a very fine and even picturesque city, owing to its designer having departed from the rigid rectangularity of most American cities by the addition of a number of broad diagonal avenues crossing the rectangles at different angles, and varying from one to four miles long. The broadest of these are one hundred and sixty feet wide, planted with two double avenues of trees, and with wide grassy spaces between the houses and the pavements. Wherever these diagonal avenues intersect the principal streets, there are quadrangular open spaces forming gardens or small parks, planted with shrubs and trees, and with numerous seats. Conspicuous in these parks are the many specimens of the fine *Paulowina imperialis*, one of the handsomest flowering trees of the temperate zone, but which rarely flowers with us for want of sun-heat. It has very large cordate leaves and erect panicles of purple flowers, in shape like those of a foxglove. It was a great regret to me that I had to leave before the flowering season of these splendid trees.

It is, however, a great pity that when the city was founded it was not perceived that the whole of the land should be kept by the Government, not only to obtain the very large revenue that would be sure to accrue from it, but, what is much more important, to prevent the growth of slums and of crowded insanitary dwellings as the result of land and building speculation. As it is now, some of the suburbs are miserable in the extreme. Any kind of huts and hovels are put up on undrained and almost poisonous ground, while in some of these remoter streets I saw rows of little villas closely packed together, but each house only fifteen feet wide.

My three months' sojourn in Washington, though a considerable loss to me financially, was in all other respects most
enjoyable. I met more interesting people there than in any
other part of America, and became on terms of intimacy, and
even of friendship, with many of them. There was a very
good circulating library of general literature to which I sub-
scribed for a quarter, and was thus enabled to read many of
the gems of American literature which I had not before met
with. Among these I read a good many of the works of
Frank Stockton, perhaps the most thoroughly original of
modern story-writers. " Rudder Grange " and " The Adventures of Mrs. Leck and Mrs. Aleshine " are among the
best known; but I found here quite a small book, called
" Every Man his own Letter-Writer," which professes to
supply a long-felt want in giving forms of letters adapted
to all the varied conditions of our modern civilization. The
result is that these conditions are found to be so complex
that to merely state them from " so-and-so " to " so-and-so "
takes up much more space than the letter itself, and is made
so humorously involved that I was, and am still, quite unable
to read them for laughter. One day a small, active-looking
man was pointed out to me as this very clever writer, and
though I did not speak to him, it is a pleasure to recall his
appearance when I read any of his delightfully fantastic
works. For many reasons I left Washington with very great
regret.

CHAPTER XXXI

LECTURING TOUR IN AMERICA—WASHINGTON TO SAN FRANCISCO

I HAD two lecture engagements at Cincinnati, and had also an invitation to visit Mr. W. H. Edwards, the lepidopterist, whose book induced Bates and myself to go to Para, and who resided at Coalburg, in West Virginia. I was also very anxious to see a new cavern which had been discovered about ten years before, and which was said to be far superior to the Mammoth Cave in the variety and beauty of its stalagmitic formations, though not so extensive. I therefore took a rather circuitous route in order to carry out this programme.

Leaving Washington April 6 at 3 p.m., I reached Harper's Ferry about 5.30, through a fairly cultivated country, a few fields green with young wheat and a few damp meadows with grass, but otherwise very wintry looking. Changing to a branch line up the Shenandoah Valley, I passed through a picturesque country like the less mountainous parts of Wales, but mostly uncultivated, and reached Luray station about 9 p.m. There was a rather rough hotel here, where I had supper and bed, and the next morning after breakfast a waggon took myself and a few other visitors to the cavern about a mile away, for seeing which we paid a dollar each, and it was very well worth it. We walked through the best parts (which are lit up with electric lights) for about two hours, through a variety of passages, galleries, and halls, some reaching a hundred feet high, some having streams or pools of water, and some chasms of unknown depth, like most caves in the limestone. But everywhere

THE SARACEN'S TENT, LURAY CAVERN.

there are stalactites of the most varied forms, and often of the most wonderful beauty. Usually they form pillars like some strange architecture, sometimes they hang down like gigantic icicles, and one of these is over sixty feet long, the dripping apex being only a few inches from the floor. In some places the stalactites resemble cascades, in others organs, and several are like statues, and have received appropriate names. Many of them are most curiously ribbed ; others, again, have branches growing out of them at right angles a few inches long—a most puzzling phenomenon. There is a Moorish tent, in which fine white drapery hangs in front of a cave, a ballroom beautifully ornamented with snow-white stalactitic curtains, etc. Some of these, when struck, give out musical notes, and a tune can be played on them. A photograph of the Moorish tent and the curious pillars near it is here reproduced. The curtain is like alabaster, and when a lamp is held behind it, the effect is most beautiful. In many places there are stalagmitic floors, beneath which is clay filled with bones of bats, etc., and at one spot human bones are embedded in the floor under a chasm opening above. The print of an Indian mocassin is also shown petrified by the stalagmite. Rats and mice are found with very large eyes ; and there are some blind insects and centipedes, as in the Mammoth Cave. Several miles of caverns and passages have already been explored, but other wonders may still be hidden in its deeper recesses. The only caves in the world which appear, from the descriptions, to surpass those of Luray are the Jenolan caves in New South Wales. The latter have all the curious and elegant forms of stalactites found at Luray, and in addition others of beautiful colours, such as salmon, pink, blue, yellow, and various tints of green, a peculiarity, so far as I am aware, found nowhere else.

Returning to the station, I went on to Waynesboro' Junction, where I dined, and had to wait two or three hours for the train on at 5 p.m. I took a walk on a wooded hill close by, but the only flower I could find was the little *Epigœa repens*, the only indication of spring. The appearance of the woods was no more advanced than with us in February ; yet it was in the latitude of Lisbon ! I reached Clifton Forge, where I had to stay the night, at 8 p.m., and found the hotel full, and was sent to another—small, dirty, and ruinous. Next morning I was so unlucky as to lose my train by getting into the wrong one, which was standing ready on the line with steam up. The conductor, after seeing my ticket, stopped the train and set me down, telling me that if I walked back quickly I might be in time. But I had a heavy bag to carry, and a mile to walk, and arrived dripping with perspiration to find that the train had gone, and there was no other till the same hour next day. I therefore wired to Mr. Edwards, and spent the day exploring the country for several miles around. Two or three miles up the valley I came to a fine gorge, where there was a good specimen of arched stratification. I came across a thicket of rhododendrons, the first I had seen wild. There were also some tulip trees with dry capsules, and the brilliant red maple in flower, as well as the yellow-flowered spice bush, *Benzoin odoriferum*. There was an undergrowth of kalmia, and some of the deciduous trees were in leaf, but there were no herbaceous spring flowers and very few showing leaf in the woods.

The next day I left at 7 a.m., passing through a very interesting country, first among iron works in a rather flat, open valley, then along narrow winding valleys, then into a dry valley always rising towards the ridge of the Alleghanies, then through a tunnel into another valley, still going up among woods of firs and oaks, rather small and scraggy, till at 8.30 a.m. we passed the summit level by a tunnel, and soon got into a rather wide, deep valley with a stream flowing west, and at 8.40 reached White Sulphur Springs, in a pleasant basin surrounded by mountains, with a pretty church, neat houses, good roads, and gardens with painted wood fences ! the first bit of an attempt at neatness I had seen since leaving Washington. Here were some fine pine trees, and the grand ridges and mountains, wooded to the summits, reminded me of Switzerland, without its great charms—the lowland and upland pastures and snow-capped peaks. Soon the valley widens, the rock becomes a highly inclined schist or slate, cultivated fields are more numerous, but often still full of tree-stumps. Men are seen ploughing with very small light ploughs, which can turn easily among the stumps ; ugly snake fences are present everywhere ; queer little wooden huts are dotted about ; and ragged, dirty children abound—a regular bit of backwoods life.

Passing through a long tunnel, we come out upon the Greenbriar river, a quiet stream whose greenish waters are full of logs cut in the surrounding mountains and being floated down to the Ohio. At Hinton the New River joins our stream, the valley gradually narrows till we are walled in by grand crags and precipices, there are enormous fallen boulders, and the river foams over ledges and down whirling rapids. We passed a fine lofty point called Hawk's Nest, and soon after reached the Kanahwha river, which is navigable down to the Ohio. Here we saw one of the old-fashioned stern-paddle steamboats ; the climate became warmer, a peach tree was in full blossom, and I even saw that rarity in America, a greenhouse attached to a small country house. All down the valley in alluvial flats the Western plane tree (*Platanus occidentalis*) had a remarkable appearance, its upper half being pure white, exactly as if whitewashed. This is the colour of the young bark before it flakes off, as it does on the trunk and larger limbs. The peculiar appearance is not noticed by Loudon, so perhaps it is not produced in our less sunny climate.

I reached Coalburg at 3 p.m., where Mr. Edwards met me and took me to his pleasant house with a broad verandah in a pretty orchard at the foot of the mountain, which rises in a steep forest-clad slope close behind. The grass of the orchard was full of the beautiful white flowers of the blood-root (*Sanguinaria canadensis*), together with yellow and blue violets, and there were fine views of the river and high sloping hills, which, together with the tramways and coal trucks on the railway, and here and there the chimneys of a colliery engine, reminded me of some of the South Wales valleys. I spent four days here roaming about the country, seeing my host's fine collection of North American butterflies and his

elaborate drawings of the larvæ at every moult, from their
first emergence from the egg up to the pupa stage, which
often served to determine otherwise too closely allied species.
We had only met once forty years before, but had occasionally corresponded on entomological subjects, and felt quite as
old friends. Mr. Edwards had some literary tastes and had
a pretty good library, so that in the intervals of work and
talk I spent many hours reading. He had lived twenty-five
years in this valley, where he had been among the first to
work the coal, and was still business manager of some of the
mines. He confirmed what Judge Holman had told me
about the Irish, who, he said, were industrious and very
intelligent and enterprising, many of them rising to high
positions. As workmen they are, in his opinion, better than
the Welsh, and equal to the Germans. And these are the
people we have for a century driven out of their native
country by despotic rule and the cruel oppression of absentee
landlordism, and still declare to be " incapable of self-government." The force of racial pride, ignorance, and impudence
can no further go.

During several drives and walks I saw a good deal of the
country and population. The villages and detached houses
were usually very poor and untidy, fences and pigsties are
built of odd bits of board, and there were hardly any gardens
or cultivation of any kind, the result probably of the people
being mostly miners and mere temporary residents. In one
village, however, where the miners owned their own cottages,
these were neat and sometimes pretty, in good repair, and
with gardens well attended to. Here, again, the magic of
property (or of permanent occupation) turns a hovel into a
home, a desert into a garden—as Arthur Young remarked
more than a century ago.

On the 13th of April at 8.30 a.m. I bade farewell to Mr.
Edwards, his daughter and son, who had made my visit a
very agreeable one, and went on to Cincinnati. The journey
was very interesting. For a long way it was through a series
of small valleys bounded by low vertical bluffs and sandstone,
and with many lateral valleys opening out of them, with

wooded slopes above. In the flat valley-bottoms the white-washed American planes were abundant, and in the villages peach trees were in blossom, but there was no sign of spring foliage in the woods. We then passed through a country of horizontal beds of rock, alternately hard and soft, looking like our Oolite, but really of Silurian age.

I remained in Cincinnati twelve days, met a good many people who were very kind to me, and saw a good deal of the very interesting country around the city. I also had the use of the Cuvier Club, where there was a nice collection of American birds, a library, reading-room, chess-room, etc., equally accessible on Sunday as during the week. Among my first visitors next morning was Mr. Charles Dury, an enthusiastic naturalist and collector, and Mr. R. H. Warder, also fond of natural history. They took me to call on Mr. J. R. Skinner, who showed me some fine arrow-heads of jade, and then took us for a drive round the beautiful suburb of Clifton, where the handsome villas are scattered about a wooded park-like country, with shrubs and wild flowers, but with no fences of any kind, either between the different properties or along the road-sides. This gives a delightfully rural aspect to the whole place, and enables every one to enjoy an uninterrupted view over the hills and valleys, and also to walk across in any direction that he may be going. Returning, Mr. Skinner asked me to dine with him, and talked about spiritualism, pyramid and Bible measures, etc., etc. For two hours he poured out Hebrew names and mystic numbers, deducing π, and all kinds of geometrical data and measures from Hebrew biblical names. He seemed to be a regular "paradoxer," and afterwards gave me many papers he had published, but I was quite unable to follow them, or to decide whether or not there was anything of value in them. In all other subjects he was a pleasant companion, interested in local antiquities, and an enthusiastic lover of native birds.

In the evening Dr. H—— and Dr. L—— called on me. The former stayed an hour and a half, a great talker, mostly about himself, his sayings and thinkings, his philosophy, his admiration of Herbert Spencer, his recollection of Sir Charles

Lyell, etc., etc. On Saturday, May 16, I went with Mr.
Skinner to meet Mr. Warder at Valley Junction, about
twenty miles below Cincinnati, and he drove us in a light
waggon a few miles to see some old Indian mounds. One
very large tumulus, about twenty-five feet high, had been
opened by a pit in the centre down to the ground level. At
a farmhouse near we found that the farmer had opened it,
had found a skeleton, two copper bracelets, several large
stone weapons and tools, some very finely worked, and a
lump of pure graphite. Mr. Skinner thought that graphite
had never been found before in the mounds. On the way
back we saw a very large elongate mound, covered with trees
and close to a village. The valley of the Ohio was here
very pleasant, with its rich fields and low wooded hills of
varied outline. Many birds were seen, the brown thrush,
red-winged blackbird, and many others, all well known to
my companions. The American Judas tree (*Cercis canadensis*) was in full flower and very abundant, and the little
spring beauty (*Claytonia virginiana*) formed sheets of pale
pink blossoms on the skirts of the woods. We saw a few
patches of virgin forest on the hills, and here and there a
rather fine tree, but these are always scarce.

The following day being very wet, our excursion to the
Madisonville Cemeteries was delayed a week. But on Sunday, the 24th, Dr. Dunn took me in his buggy, accompanied
by several other friends in a carriage, for a long drive to
the Turner group of mounds, which are very extensive, but
have been ploughed over. Near them is the cemetery, consisting of a great number of small mounds in a wood, many
of which have been opened, and bones, with numbers of
stone weapons, ornaments, etc., found in them. Circular
plates of mica are common here. On the way back we
visited a field where quantities of pottery, flints, bones, etc.,
have been found near to a small oval mound. The country
we passed through was very pleasant, and some of it quite
picturesque, with swelling hills, ridges, and valleys, often
finely wooded and park-like.

During the week preceding this excursion I had spent

four days with Mr. Dury at Avondale, where he has a small house and some land. There were some patches of the original forest near, with moist little valleys, and here I saw for the first time the American spring vegetation in its full beauty. The woods were full of an anemone-like flower (*Thalictrum anemonoides*), the curious Dutchman's breeches (*Dicentra cucullaria*) in continuous sheets, the spring beauty (*Claytonia pulchella*) equally abundant, with patches of *Phlox divaricata*, the dwarf blue *Delphinium tricorne*, the little blue-eyed Mary (*Collinsia verna*), yellow, blue, and white violets, *Jeffersonia diphylla*, and many other flowers strange to English eyes. During one walk I found a fine plant of *Mertensia virginica* in flower. But though these were wonderfully attractive to me, owing to there being so many forms of flower quite unknown in England, the actual amount of floral colour and beauty was not to be compared with our own. There was nothing to equal the sheets of bluebells, primroses, and anemones in our woods, the buttercups and early orchises of our meadows, or the marsh-marigolds of our marshes and river-banks. This subject of the comparative abundance and the striking differences between North America and Europe in this respect I have discussed somewhat fully in my *Fortnightly Review* article on " English and American Flowers," reprinted in my " Studies, Scientific and Social " (vol. i. p. 199).

One evening when at Mr. Dury's an interviewer called, and showed the most remarkable ignorance. He thought Darwin's theory was limited to the change of monkeys into men ; that Englishmen were all either Lord Dundrearys or roughs ; that the lowest Cockney talk was the " English accent," which he was much surprised that I did not possess ; and, above all, that America was the finest and the greatest country in the world, and that all who were born elsewhere were to be pitied and condoled with. But this was quite an exceptional type. All my other American interviewers were educated men and knew their business.

My friend Mr. Dury had had the rare experience of being bitten by a dead rattlesnake with very painful consequences. When in Florida he shot a very large rattle-snake, and decided to take its head only, in order to examine its dentition. He opened its mouth with a stick, and saw it had tremendous fangs, and proceeded to tie it up in a hand-kerchief, and while doing so supposes he must have touched a nerve in the cut part, for the mouth suddenly snapped, and a fang pierced his thumb. He instantly put a ligature round the base of the thumb, got a friend who was with him to lance it deeply with a penknife, and sucked it for some time. On taking off the ligature an hour afterwards, the arm swelled as well as the side of his body, and he suffered great pain. He applied water constantly, drank a good deal of whisky, and kept quiet for some days; but the thumb suppurated, and half the bone of the terminal joint came away. Then it healed, but the thumb was reduced to about half its normal size, with a correspondingly small nail; but it is quite serviceable, and being so small is for many purposes more useful than the other !

Mr. Dury had a very fine collection of land and fresh-water shells from all parts of the States, and I spent one morning looking over them. They were exceedingly numer-ous, and of curious forms, many having strange contortions of the lips, supposed to be for the purpose of protection against the smaller birds, ants, etc. The freshwater shells —mostly mussels (Unionidæ)—were wonderfully fine and varied, some curiously tubercled, some with ribs, others with long spines. They are also often finely coloured inside—white, pink, yellow, or orange—while in many of the species there is a variation of form in the two sexes. Altogether it was a most interesting collection. Mr. Dury told me he began collecting when a boy, owing to a gentleman offering him a few cents for every different kind of shell he could find, however small, and he was thus led to search for them, and to notice their forms and colours, and was surprised to find how many different kinds there were, even within a walk of his own home. He was thus induced to become a professional collector. There are about two hundred and fifty species of land-shells in temperate North America, while the fresh-water species are still more numerous, its magnificent water-system, including the great lakes and such grand rivers as the Mississippi, being richer in mollusca than any other part of the world, considerably more than a thousand species having been described.

On Friday, April 22, I returned to Cincinnati to deliver my lectures on "The Colours of Animals" for the Natural History Society. The audience was, however, a small one, and the lantern very bad, so that the slides were not shown to advantage, but the subject was evidently so new to the hearers that they were much interested. The next evening I gave the same lecture at College Hill, fifteen miles out of town. I had tea with Dr. and Mrs. Myers, who were pleasant and sympathetic people. Dr. Myers told me that he had become a sceptic through Spencer and Darwin, but is regaining belief through spiritualism. Here I had a good lamp, and everything went off well; but I only received one hundred dollars for the two lectures, out of which I had to pay fifteen dollars for the lamp and operator at the last one, so that my net receipts only paid my hotel bill. But I had a very pleasant visit, and met a number of intelligent people.

My next engagement was at Bloomington, Indiana, where I was to lecture on the Darwinian theory to the university students. I stayed with Dr. Branner, the professor of geology, who had spent many years in Brazil, so that we had a common interest. He showed me his drawings of palms, and photographs of Brazilian scenery. The university here, like all colleges and schools in the West, is open to both sexes. They meet in the classes, in lecture rooms, and in debates on a perfect equality, and Mrs. Branner thinks the results are entirely beneficial. The next morning Dr. Branner took me a long drive through the country. The rocks were of Carboniferous age, and were of limestone and sandstone in nearly horizontal strata, leading to pretty undulations of hill and valley, with abrupt slopes. We passed through some fine tracts of forest, but there were very few flowers, though the red maples in the woods, and the white *Amelanchier canadensis* were pretty.

Returning at 3 p.m., we found that my large roll of diagrams, which could not be brought as passengers' luggage, had not arrived, and I could not well give my lecture without them. There was nothing for it but to make some rough sketches from memory ; so we went to the lecture-room, got some large sheets of paper, and I sketched out the four or five diagrams (of curves of variation, lines and dots showing amounts of variation, etc.) on a small scale, and then Dr. Branner and myself, with the assistance of one of the students, set to work to enlarge them, and draw them in thick black ink, the result at a distance being almost as good as the more accurate originals, which turned up after I had left, and were sent after me. Then we had to hurry back to dinner at 6.30, to which several professors were asked to meet me, and then to the lecture at 8, which went off very well, notwithstanding the makeshift diagrams.

My next destination was Sioux City *viâ* Kansas City, but I stopped for a day at St. Louis in order to see the Trelease Botanic Garden, recently given to the town, and which I had heard highly spoken of. I travelled mostly through various kinds of prairie country, level or rolling, with occasional hilly tracts covered with wood. Everywhere some wood was in sight, and the land seemed very rich ; but the general effect was usually ragged from the ugly, rough wood fences. Crossing the fine three-arched bridge over the Mississippi to St. Louis, I went to the Laclade Hotel. After breakfast next morning I called on Dr. Trelease, who was out. I then went on to the gardens, a little outside the city. Though rather poor as a botanical garden, there were a number of fine conserva-tories and plant-houses, and plenty of seats. The many American, Rocky Mountain, and other plants I wanted to see were not to be found, ordinary South European garden plants and a few Cape and Australian species being the chief occu-pants of the garden. In the afternoon Dr. Trelease called on me. He was a youngish, pleasant man, and we had two hours' talk on natural history and other subjects. He kindly offered me plants, seeds, etc.

I left at 8.20 in a sleeping-car for Kansas City, and at

sunrise next morning saw the Missouri river on our right, from half to three-quarters of a mile wide, the opposite bank wooded. We soon left it, crossing the prairie in a nearly straight line for Kansas City, over a rich alluvial plain, with numerous clumps of trees—poplars, planes, etc. Steep bluffs, from one hundred to two hundred feet high, were frequent, either bare or wooded. As we approached the city we came near the river again, and here there were bluffs of rock of cretaceous sands or limestones—a typical rich prairie country. The Missouri here was like liquid mud, with a swift stream and numerous eddies. On reaching the city I breakfasted at the Station Hotel, bought my ticket for Sioux City, and after much trouble got my trunk and lecture diagrams checked through. We started at 11 and reached Council Bluffs, where I had to stay the night, at 6.30, the whole way along level prairie with the river always in sight. At the hotel here were pleasant female waiters instead of the usual white, brown, or black men waiters. Leaving early next morning I saw abundance of water-birds, especially thousands of grebes, scuttling off from the banks as the train passed, leaving long trails on the water. At Missouri Valley, a large village, we had to wait an hour and a half. Here the plain was several miles wide, bounded by sloping bluffs of loess, often covered with deep black mould. I walked on to some waste ground, but could find no flowers, the soil being very dry, with a little grass and a few stunted shrubs just sprouting. About twenty miles further we reached Sioux City, where the bluffs come close to the river. The city is on gentle slopes which merge into high rolling prairie inland, intersected by deep valleys; but at this time of year it was looking rather arid.

Three lectures had been arranged for me here by Mr. D. H. Talbot on behalf of the Natural History Society, and Mr. E. H. Stone had kindly offered me hospitality in his very pretty house in the suburbs. In the afternoon Mr. Talbot took me to call on a lady who made beautiful drawings in oil of native flowers. These were very skilfully executed, and almost equal to those of Miss North at Kew. I

lectured here Monday, Tuesday, and Wednesday, on " Colours
of Animals," " Mimicry," and " Oceanic Islands," and every day
had drives or excursions about the country or to Mr. Talbot's
zoological farm. On the Sunday morning after my arrival
Mr. Talbot called in a two-horsed buggy to take me to his
farm ; two other gentlemen in another ; Judge Wakefield,
Miss W. (the lady who painted flowers), and two children in a
third. We first went to a bluff near the town to see a thick
bed of loess resting on glacial drift, and this on Cretaceous
sandstone. Then up the valley of the great Sioux river, a
fine, clear stream, passing another bluff showing a thick bed
of obliquely stratified gravel with enclosed pebbles and
boulders, and about one hundred and fifty feet of loess over
it. We then turned up a thinly wooded valley to Mr. Talbot's
farm, about four miles from the city. Here we picnicked in
a rather scrubby wood with very little shade, as no leaves
were yet out, and it was very hot and dusty ; but we had
quite a luxurious feast and enjoyed ourselves thoroughly,
lighting a fire and making tea and coffee to finish with.

We then inspected our host's animals—six fine American
bisons, twelve elks, an East Indian zebu, a drove of solid-
hoofed pigs, a flock of four-horned sheep, hybrids of zebu
and cattle, a fine trotting colt, wolves, foxes, rabbits, wild
geese, and other aquatic birds, pigeons, rattlesnakes, and
other curious birds and reptiles. He has here six thousand
acres of land, wooded valleys, and prairie, where, besides
keeping all these animals in order to observe their habits,
make experiments on their instincts, etc., he carries on a
considerable business in growing agricultural seeds of choice
qualities, breeding the solid-hoofed hogs, which are said to
be superior for fattening purposes, as well as the four-horned
sheep, Angora goats, hybrid cattle, etc. He has also
patented metallic tags for identifying cattle and other farm
stock, and several agricultural implements. These animals
are all looked after by youths trained by himself—boys and
girls, who are, he finds, as soon as they take an interest in
the work, much more trustworthy than any men. He has
also a large building for a museum, or rather laboratory, of

experimental zoology. Here he showed me several hundred skins of wild geese, roughly prepared, but every one with numbered labels giving the date, hour, and exact spot where they were each shot, with the direction of their flight, while the contents of the stomach of each is preserved for examination. These have been obtained from various north-western States, and by a close study of them he hopes to trace out the exact course of their migration year by year. He hoped that in time some of his land would be included within the city limits, and would sell for a high price, in which case he would leave the rest as a zoological experimental station to the public. I made some suggestions to him as to experiments in regard to instinct, heredity, and evolution, which were much needed, and he said he would take them in hand when his affairs were more settled.

Sioux City had recently become a centre for agricultural produce, and had a large pork-curing establishment; and, as in many other Western cities, there had been " a great boom in real estate." Land two miles from the town, which was bought three or four years back for ten dollars an acre, is now selling at a hundred and fifty dollars; while in the residential parts of the city plots of one hundred and fifty feet square sell for nine thousand dollars, equal to £1800, or about £3500 per acre, and in the business part of the city twice as much.

One morning Mr. Talbot took me to see the pork-curing establishment, where, during the season, they kill a thousand hogs a day. The animals are collected in pens close to the building, with a gate opening to an inclined pathway of planks up to the top of the building. They walk up this of their own accord in a continuous procession, and at the top are caught up one after another by a chain round their hind legs, and swung on to the men who kill, scald, scrape, and cut them up; all the separate parts going through the several stages of cleaning and curing till the result is bacon, hams, barrels of pork, black puddings, sausages, and bristles, while the whole of the refuse is dried and ground up into a valuable manure. The ingenuity of the whole process is undeniable ; but to go through it all, as I was obliged to do, along narrow planks and ladders slippery with blood and water, and in the warm, close, reeking atmosphere, was utterly disgusting. My friend was, however, quite amazed at my feeling anything but admiration of the whole establishment, which was considered one of the sights and glories of the city.

On coming out I was told something that interested me more than the wholesale pork factory had done. A gentleman was standing at the door of an office close by, and in the course of conversation with him, the subject of tornadoes came up, in reference to one that had done some damage there two years before. There was a very large iron oil reservoir a few yards from the office, something like the largest-sized cylindrical steam boilers, supported on a strong wooden framework. The tornado struck this cylinder, lifted it off its support, and threw it down some yards away. Yet our friend's office and other small wooden buildings close by were absolutely untouched by it. This illustrates a peculiar feature of these storms, which, though sometimes sweeping along the surface and destroying everything in their track for miles, at other times seem to pass overhead, descending occasionally to the surface and then rising again, picking up a house or a tree at intervals. The kind of destruction a tornado often produce is well shown in the photograph of the main street of Sauk Rapids, Minnesota, after the tornado of the preceding year (April 14, 1886). This town is about two hundred miles north of Sioux City.

Leaving Sioux City in the afternoon, with several stoppages and changes I reached Kansas City at six next morning. After breakfasting there, I went on to Lawrence through a pretty country in the valley of the Kansas river, the rich alluvial land still partly covered with wood, and apparently unoccupied. Several camps of emigrants (or migrants) with waggons, etc., were passed. On the sides of the railway there were dots, clumps, and even large patches of the beautiful *Phlox divaricata*, with brilliant bluish-purple flowers. No other flower was seen, but the trees were just coming out

SAUK RAPIDS, MINNESOTA. EFFECTS OF A TORNADO.

into leaf, hardly so forward as with us at the same time of year, though twelve degrees further south.

At Lawrence, a small town of ten thousand inhabitants, is the State University, where I was to lecture on the "Colours of Animals." The buildings are on the top of a hill a little way out of the town, on a plateau of rock almost like a natural pavement. There are fine views over the plains of Kansas all round, something like the view from Blackdown over the Weald, but less woody and less cultivated. In the museum I saw a good collection of the fossil plants from West Kansas. They are found in a fine-grained iron sandstone, mostly in nodules which split open showing the leaf most beautifully, often with the stalk and articulation perfect and in one case a complete bud in the axil of a leaf. The interesting thing is, that they are mostly Dicotyledons of very peculiar forms, though the rock is of Cretaceous age. Icthyosaurus remains are also found, sometimes with portions of the skin and keeled scales.

After my lecture in the evening there was a reception of the professors and their families. I heard much of the co-education system, and, as usual, all in its favour. A lady is professor of Greek, and at Des Moines a lady is the principal, although there are pupils of both sexes up to eighteen years old. Everywhere the girls hold their own with the boys, and are often superior to them in languages. At the last high school examination here, thirteen girls and eleven boys "graduated."

Next day I went on to Manhattan, where there is a State Agricultural College, at which I was to lecture. During the journey of about one hundred miles, I passed through much rich alluvial land, with rolling prairies in the distance. Sometimes there were bluffs of horizontal strata, with frequent projecting masses of rock, many of which had broken off and lay at the foot of the slope. There were many wooded gullies with the trees nearly in full leaf, but no flowers anywhere. About the farmhouses there were usually a few trees, also some good-looking orchards and a few vineyards.

At Manhattan, which I reached early in the afternoon, it

was very hot and very dusty. At five o'clock President Geo. F. F. Fairchild called, and we had some interesting talk about the college. This, too, is open to both sexes, and one-third of the pupils are women. Some come direct from the common and high schools, others are adults. The men learn the theory and practice of agriculture, agricultural chemistry, English, mechanics, use of tools, etc. The girls and women learn horticulture, cooking, domestic economy, poultry rearing, etc. In the evening I strolled about the town; no liquor-shops, but abundance of "real estate" and loan offices, the former a common mode of gambling in Western America.

Next morning (Sunday) Professor Marlott called with Mr. Hogg, a young Englishman farming here, who had a ranch of a thousand acres twelve miles out. He offered to take me for a drive. We went a few miles round the city, by fine grassy fields on the improved prairie, but saw very few flowers. Mr. Hogg complained of the climate; the long very cold winter, often 20° below zero Fahr., and the hot dusty summer. There are only a few pleasant months in winter and spring, few nice houses, and no gardens. After dinner I took a walk alone across the river to some woods and alluvial meadows, but all very dusty and no flowers. After tea, Professor E. A. Popenhoe, a botanist, called in his buggy and took me for a drive to the top of a rocky bluff, where there were a number of interesting plants, of which a few were in flower, among them *Tradescantia virginica*, a Sisyrinchium, a yellow Baptisia, etc.

On Monday morning I went to the college to put up my diagrams, and was then, of course, taken over the buildings from top to bottom. Everybody wanted to show me everything in their departments—the clothes the girls made, the nice cupboards they kept the clothes in, the store-rooms for flour, potatoes, sugar, spices, jams, etc., the kitchen and all its arrangements. Then every class-room, and all the classes, and all the teachers. Then out-of-doors to see the sheds and stables, the cattle and the horses, and the machines; how the calves and the cows are fed; to inspect the tool- and work-shops, the gardens, the greenhouses, the tree-nursery, etc., which latter interested me most.

I spent the afternoon at the hotel writing letters, and in the evening I went to tea with President Fairchild—a regular country high tea ; cold meat, oranges, strawberries, cakes, weak green tea, etc. We talked politics, and especially prohibition. Kansas, like Iowa, is a prohibition State ; had been so seven or eight years. It had had a most beneficial effect— not one-twentieth of the noise, dirt, and bad language formerly met with. I had noticed myself how quiet was the hotel and the streets in the evenings. The feeling in favour of prohibition was increasing. Spirits were sold by druggists, with mineral waters, etc.; but it was in the open shop, and did not lead to drunkenness. And even this had been recently restricted. The President had never heard of the Gothenburg system, but thought it good in principle. Afterwards I gave my lecture on " Darwinism," which went off well, and gave much satisfaction.

The next day, after dinner, Mr. Popenhoe came in his buggy to show me some good botanizing ground, chiefly on rather dry, rocky slopes with loose stones. Here we found a fine dwarf, large-flowered form of *Baptisia australis,* besides others seen on Sunday, and a number of very interesting dwarf plants not yet in flower, including species of ruellia, houstonia, echinacea, aster, delphinium, and others, which make these banks very gay about the end of the month. We also saw a phrynosoma, one of the curious lizards commonly known as "horned toads." A Californian species which had been sent me by my brother, when irritated ejects a red, blood-like secretion from its eye. Professor Popenhoe had been in the Rocky Mountains, and told me that flowers were very abundant, and that some of the little valley-bottoms were complete flower gardens. I received a letter from Colonel Phillips, whom I met at Washington, and who invited me to stay a week with him at Salina, a new town he had himself founded, and where he was a large landowner.

Next day (May 11) I went on to Salina in the afternoon,

and to the Wittanann Hotel, where Colonel and Mrs. Phillips lived when in the country. On an elevation, called Iron Hill, Colonel Phillips was going to build a house, which would have a rather extensive view. The hill was covered with yuccas, and with the elegant tradescantia with blue or pink flowers in great abundance. I also found the fine dwarf Baptisia and *Penstemon cobœa*. As I required a lantern for my lecture, we called first on a Mr. Seitz, a druggist, who sent us to the Masonic Hall, but in vain. Then we tried the Wesleyan College and the Normal University, but both were recently established, and not the possessors of a lantern. At length we found one at a Mr. Chapman's, but it was an ordinary magic-lantern, suitable for a disc about four feet diameter, and with a common oil-lamp, giving a poor light. When the lecture was given, to add to my difficulties the lamp went out in the middle, and I had to go on talking till it was set right. There were only about a hundred people, so that there were none very far off, and they seemed fairly well satisfied.

One day we drove over to call on an old French farmer, M. Joseph Henry, who was a botanist and a student of mosses and grasses. He was out, but his wife showed us a little heap of stones near the house, in which, on the north side, he had a few very small mosses growing, one of them a new species he had discovered, named after him, *Barbula Henrici*. It was a shabby, rickety wooden farmhouse, with a few sheds in the usual style of small prairie farmers. Going back we met the owner returning home, and stayed a few minutes to talk. He had been in the country twenty years but could only speak very broken English, and when he found we could not speak any better French, he was quite indignant that a scientific man could not speak in his beautiful language—the language of the civilized world! He made me feel quite small. However, he managed to tell us that the American botanists did not know their own country. "They all say there are no mosses in Kansas. But *I* have found mosses! I have found new species of mosses! And when I send them my discoveries they will not give me the names—they will not write to me even!" So we condoled with him, and said good-bye to the unappreciated botanist of the arid plains of Kansas.

Twenty-nine years before my visit great herds of buffalo roamed over the site of Salina, and there was not a house or a hut for fifty miles around. It is now a rapidly growing town, with five railways diverging from it, and land speculation is rampant. In the business part of this small town lots twenty-five feet wide and one hundred and twenty deep, in the main street, sell at from $6000 to $10,000 (from £1200 to £2000); in the suburbs (a mile from centre of town) about $1200. Farms near the town, of good land, can be had at from £6 to £10 an acre. But the climate, the solitude, the dreariness of such a life must be a great drawback.

I left Salina in the evening of May 18 for Denver and San Francisco. We soon reached open undulating prairie, small villages or towns fifteen or twenty miles apart, and often not a house visible; very little cultivation, and rarely any trees. At five o'clock next morning, still in undulating sandy plains with very little grass but a few tufts of herbaceous plants with white composite flowers. Then low hills of horizontal strata of sandstone; and we crossed some small streams in broad sandy beds, with sometimes a few cotton-wood trees growing near them. Here I first saw some of the prairie-dog cities, as they are called—sandy mounds thrown up by these pretty rodents, one of which would be often seen sitting upright on the top of it.

We reached Denver at 8.20, and having four hours to wait here, after breakfast I called on Professor James H. Baker, Principal of the High School, to inquire if he knew of any local botanist who could give me information as to any good localities in the mountains for alpine plants. He told me that one of his lady teachers was a botanist, and took me into her class-room. As she was engaged in giving a lesson on ancient history to a class of boys and girls, we sat down and waited till it was over, when I was introduced to her, and we had an hour's talk, and she showed me dried plants she had collected on Pike's Peak. She told me that Graymount,

near Gray's Peak, was a fine spot, and I decided to visit it on
my return from California.

At 1.30 p.m. I continued my journey to Cheyenne, across
open plains of thin grass partly irrigated. Near me in the train
was a lady chewing gum ; I saw her at intervals for an hour,
her jaws going regularly all the time, just like those of a
cow when ruminating. *Not* a pleasant sight, or conducive
to beauty of expression. It must be tiring to beginners. We
had supper at Cheyenne, good, but a crush ; and then turned
west up the slope towards a pass in the Rocky Mountains.
The valley we ascended was among rounded hills, more like
our downs than mountains. Though the country was quite
wild, there were here and there lines of high posts and rails of
strong, rough timber, sometimes on one side sometimes on
the other, sometimes below and sometimes above the level of
the railway. These, I was told, were snow-guards, and were
placed just where experience showed they would check the
drifts and keep the line clear. In a few places there were
snow-sheds with one or two short tunnels, and we reached the
summit level at 8 p.m., only 8240 feet above the sea. The
next morning we were going through similar rolling, half-
desert scenery, with greasewood bushes and bare sand or
mud flats white with alkali. At Green river, one of the upper
tributaries of the great Colorado river, we got into more
picturesque scenery, with rocks standing up like castles, and
further on rocky valleys, with wind-worn rocks in strange
detached pinnacles. Fine precipices occur at Echo Cañon
and Weber's Cañon. The Devil's Slide is formed by two
vertical dykes descending a steep mountain-side only two
or three feet apart, leaving a narrow passage or " slide "
between them.

Reaching Ogden in the afternoon, I took the train to Salt
Lake City, passing the fine highly cultivated plain on the
shores of Salt Lake, the fields being all irrigated. Some of
the meadows were blue with the beautiful *Camassia esculenta*,
an easily grown garden plant with us. I spent next morning
roaming about the city and suburbs. The tabernacle is a
wonderful hall that will seat six thousand persons, and is so

shaped that a speaker at one end can be heard distinctly over the whole building when speaking in an ordinary conversational tone. To produce this effect it is a flat semi-ellipsoid, so that the regularly curved ceiling is very low for the size of the building. But the result is acoustically perfect, and such as none of our architects have equalled.

The city itself is in many respects unique and admirable. It is a kind of " Garden " city, since every house (except in the few business streets) stands in from half an acre to one acre and a half of garden. Some are pretty stone-built villas, some mere rude hovels, but all have the spacious garden. And they are real gardens, the first I have seen in America, full of flowers and fruit trees, and with abundant creepers over the houses.

The streets are about one hundred and thirty feet wide, with shady trees, and a channel of clear water on both sides of each street brought from the mountain. Every garden is thus supplied with abundance of water for irrigation, when required, by small channels under the side walks, and sluice gates to regulate the supply. Crops can thus be grown during a large part of the year. I walked a few miles into the country, and seeing a small house and pretty flower garden with some of our commonest garden flowers, roses, stocks, marigolds, etc., I spoke to a homely looking woman and found she was Welsh. A good many Welsh have become Mormons.

In the afternoon I returned to Ogden, and went on by train in the evening. All the next day (Sunday, May 22) we passed through an arid dreary country, the ground covered with saline incrustations, and almost the only vegetation the sage bush (*Artemisia spinescens*). At the stations in more fertile spots there was a little verdure and sometimes a few wild flowers—œnotheras or composites. At all the stations there were groups of Indians, usually with painted faces but with European dress, one old man only with the native blanket, boys shooting with bows and arrows, groups of men and women playing cards. The passengers give them money or buy ornaments, etc., and thus they live idly, get fat, and are thoroughly demoralized. At Reno, where we supped, the country began to get less arid, and there were some good farms in the valleys. We passed over the pass of the Sierra Nevada at night, and before sunrise were in the foothills of California, bare, except for a few second-growth pines ; then farms, orchard, and vineyards, with eucalyptus trees planted round the houses; then a low, flat country to Oakland, where huge ferry-boats cross the bay to San Francisco.

Here my brother John, whom I had not seen since I left for the Amazon in 1848, met me, and we went on to the Baldwin Hotel in San Francisco, where he had taken rooms for us, and had made arrangements for me to give two lectures on Wednesday and Friday. In the afternoon we had many callers, including Professor Holden of the Lick Observatory, Dr. Leconte, Mr. Davidson of the Geological Survey, and many others, as well as one or two interviewers. Dr. Holden kindly invited us to dinner on Thursday, where we met Professor Hilgard and Mr. Sutro. The latter gentleman invited us to breakfast with him at his beautiful cottage on the cliff, looking over the Pacific and the seal rocks, and surrounded by beautiful gardens. Mr. Sutro was a wealthy merchant and one of the magnates of San Francisco ; he gave us one of the most luxurious and pleasant breakfasts I ever enjoyed, beginning with cups of very hot, clear soup, followed by fish, cutlets, game, etc., with various delicate wines, tea and coffee, hot cakes of various kinds, and choice fruits. He entertained us also with interesting conversation, being a man of extensive knowledge and culture. My two lectures on " Darwinism " and " Colour" were fairly attended.

On Saturday Dr. Gibbons of Alameda, on the Bay of San Francisco, took me for a drive into the foothills to see the remains of the Redwood forest that once covered them, but which had all been ruthlessly destroyed to supply timber for the city and towns around. Our companion was Mr. John Muir, whose beautiful volume, " The Mountains of California," is, in its way, as fine a piece of work as Mr. Hudson's " Naturalist in La Plata." On our way we passed a dry hilly field, brilliant with hundreds of the lovely *Calochortus luteus*, which grew in a soil of stiff, hard-baked clay. We wound about among the hills and valleys, all perfectly dry, till we reached a height of fifteen hundred feet, where many clumps of young redwoods were seen, and, stopping at one of these, Dr. Gibbons took me inside a circle of young trees from twenty to thirty feet high, and showed me that they all grew on the outer edge of the huge charred trunk of an old tree that had been burnt down. This stump was thirty-four feet in diameter, or quite as large as the very largest of the more celebrated Big Trees, the *Sequoia gigantea.* The doctor has searched all over these hills, and this was the largest stump he had found, though there were numbers between twenty and thirty feet. The tree derives its botanical name, *semper-virens,* from the peculiar habit of producing young trees from the burnt or decayed roots of the old trees. These enormous trees, being too large to cut down, were burnt till sufficiently weakened to fall, and this particular tree had been so burnt about forty years before. We lunched inside this ancient mammoth tree, and saw several others on the way back. Among the few plants I saw in flower were the *Diplacus glutinosus,* a favourite in our greenhouses.

The next day we went to Stockton, where my brother lived, and found his wife, whom I had last seen as a little girl, two of his sons and his only daughter, as well as two of his grandchildren. I gave one lecture in Stockton—a combination of Darwinism and Oceanic Islands—but only had a small audience. I made the acquaintance here of Mr. Freeman, a friend of my brother, who had called on me at Godalming with his wife two or three years before, on their way round the world on a pleasure tour. He told me then that he had had good luck in his business, had made a few thousand dollars, his only daughter was just married, so he thought that he and his wife might as well see the world. On asking him how he had made the money, he said, " By handling mules," and this enigmatic profession was explained as buying them in some of the Western States, where they are largely bred, and selling them in Nevada, where there was a great demand for them at the mines, etc. Now he had taken to store-keeping, while his wife kept poultry, and as soon as they had made some more money they meant to go another tour. They had been through Central Europe and Italy, the Holy Land, India, China, Japan, and the Sandwich Islands, and had brought home many ornaments and fabrics from the East ; but what Mrs. Freeman most valued were some bottles of water which she had filled with her own hands from the River Jordan. This water she had given to some of her dearest friends to baptize their children with, a distinction of the highest kind.

While in San Francisco I had agreed to give a lecture on "Spiritualism," under the management of Mr. Albert Morton, and I went over on Sunday, June 5, and had an audience of over a thousand people in the Metropolitan Theatre. The title of my lecture was, "If a Man die, shall he live again?" The audience was most attentive, and it was not only a better audience, but the net proceeds were more than for any single scientific lecture I gave in America. I had spent the morning in the fine Golden Gate Park, where I saw some eucalyptus trees over sixty feet high, with numerous acacias and other greenhouse plants growing out-of-doors. I also had a fine view of the extensive sandhills, covered with huge clumps of blue and yellow tree-lupines, which produced a splendid effect. The interesting *séances* I had here will be described later on.

Returning to Stockton, I went with my brother and his daughter for a few days in the Yosemite Valley. The journey there—two hours by rail and two days by coach— was very interesting, but often terribly dusty. The first day we were driving for nine hours in the foothills, among old mining camps with their ruined sheds and reservoirs and great gravel heaps, now being gradually overgrown by young pines and shrubs. Here and there we passed through bits of forest with tall pines and shrubby undergrowth, but generally the country was bare of fine trees, scraggy, burnt up, and the roads insufferably dusty. At 9 p.m. we reached Priest's (two thousand five hundred feet elevation), where we had supper, bed, and breakfast.

Next day was much more enjoyable. The road was wonderfully varied, always going up or down, diving into deep wooded valleys with clear and rapid streams, then up the slope, winding round spurs, crossing ridges, and down again into valleys, but always mounting higher and higher. And as we got deeper into the sierra, the vegetation continually changed, the pines became finer both in form, size, and beauty. At about three thousand feet we first saw the beautiful Douglas fir, and the cedar (*Libocedruo decurrens*), both common in our gardens; then still higher there were silver firs and the fine *Picea nobilis*, as well as a few of the Big Trees (*Sequoia gigantea*), the road being cut right through the middle of one of these (at about five thousand eight hundred feet). Higher up still we saw the tamarisk pine (*Pinus contorta*) and the grand sugar-pine (*Pinus Lambertiana*) the resin of which is quite sugary, with very little of the turpentine taste; and among these, especially on the valley slopes, is an undergrowth of the beautiful white azalea and the handsome dogwood (*Cornus Nuttallii*), with very large white bracts. Then on the highest spur (seven thousand feet), where there were still patches of snow, we saw many of the strange snow plants (*Sarcodes sanguinea*), a thick fleshy root-parasite with a dense spike of flowers of a blood-red colour. It belongs to the heath family, and is allied to our *Monotropa hypopitys*. The sarcodes is figured in one of Miss North's pictures at Kew. From the summit we descended towards the valley, and then down a steep zigzag road, with the beautiful Bridal Veil Fall opposite, and the grand precipice of El Capitan before us, then into the valley itself with its rushing river, to the hotel in the dusk.

As both hotel and excursions were here very costly, we only stayed two clear days, and went one "excursion" to the Nevada Fall, the grandest, if not the most beautiful, in the valley. My brother and niece rode up, but I walked to enjoy the scenery, and especially the flowers and ferns and the fine glaciated rocks of the higher valley. The rest of my time I spent roaming about the valley itself and some of its lower precipices, looking after its flowers, and pondering over its strange, wild, majestic beauty and the mode of its formation. On the latter point I have given my views in an article on "Inaccessible Valleys," reprinted in my "Studies." The hotel dining-room looks out upon the Yosemite Falls, which, seen one behind the other, have the appearance of a single broken cascade of more than two thousand five hundred feet. I walked up about a thousand feet to get a nearer view of the upper fall, which, in its ever-changing vapour-streams and water-rockets, is wonderfully beautiful. To enjoy this valley and its surroundings in perfection, a small party should come with baggage-mules and tents, as early in the season as possible when the falls are at their grandest and the flowers in their spring beauty, and where, by camping at different stations in the valley and in the mountains and valleys around it, all its wonderful scenes of grandeur and beauty could be explored and enjoyed. It is one of the regrets of my American tour that I was unable to do this.

Returning from the Nevada Falls on foot, I had the advantage of passing close to the lower Vernal Fall, where a natural parapet of rock enables one to look over and almost touch the water at the brink of the fall, which shoots clear of the rock and falls four hundred feet. Here, with great skill and daring, a series of ladders has been constructed from ledge to ledge to near the foot of the fall, whence a thoroughly alpine path leads down to the main valley. Growing in clefts of the rock, and wetted by the spray of the fall, was the beautiful *Pentstemon Newberryi*—a dwarf shrub with deep-red flowers, more like those of some ericaceous plant than a pentstemon. On the return journey I noted several interesting plants. At Crockett's (where we dined), a little beyond the summit, there was a vase full of the beautiful orchis *Cypripedium montanum*, which they told me grew in the bogs near; and I also found the brilliant scarlet *Silene Californica*. Lower down, the *Calochortus venustus* was abundant and in richly varied colour, the curious *Brodiæa volubilis*, and the handsome blue *B. grandiflora*.

On our way back I turned off at the foot of the hills to visit the Calaveras Grove of big trees which my brother and niece had seen before, and I had to sleep on the way. I stayed three days, examining and measuring the trees, collecting flowers, and walking one day to the much larger south grove six miles off, where there are said to be over a thousand full-grown trees. The walk was very interesting, over hill and valley, through forest all the way, except one small clearing. At a small rocky stream I found the large *Saxifraga peltata* growing in crevices of rocks just under water, and I passed numbers of fine trees of all the chief pines, firs, and cypresses. At the grove there were numbers of very fine trees, but none quite so large as the largest in the Calaveras Grove. Many of them are named. "Agassiz" is thirty-three feet wide at base, and has an enormous hole burnt in it eighteen feet wide and the same depth, and extending upwards ninety feet like a large cavern; yet the tree is in vigorous growth. The Sequoias are here thickly scattered among other pines and firs, sometimes singly, sometimes in groups of five or six together. There are many twin trees growing as a single stem up to twenty or thirty feet, and then dividing. But the chief feature of this grove is the abundance of trees to be seen in every direction, of large or moderate size, and with clean, straight stems showing the brilliant orange-brown tint and silky or plush-like glossy surface, characteristic of the bark of this noble tree when in full health and vigorous growth. In no forest that I am acquainted with is there any tree with so beautiful a bark or with one so thick and elastic.

In the chapter on "Flowers and Forests of the Far West" (in my "Studies"), I have given a summary of the chief facts known about these trees, with particulars of their dimensions and probable age. I need not, therefore, repeat these particulars here. But of all the natural wonders I saw in America, nothing impressed me so much as these glorious trees. Like Niagara, their majesty grows upon one by living among them. The forests of which they form a part contain a number of the finest conifers in the world—trees that

in Europe or in any other Northern forest would take the
very first rank. These grand pines are often from two hundred
to two hundred and fifty feet high, and seven or eight feet in
diameter at five feet above the ground, where they spread
out to about ten feet. Looked at alone, these are noble
trees, and there is every gradation of size up to these. But
the Sequoias take a sudden leap, so that the average full-
grown trees are twice this diameter, and the largest three
times the diameter of these largest pines ; so that when first
found the accounts of the discoveries were disbelieved. My
brother told me an interesting story of this discovery. The
early miners used to keep a hunter in each camp to procure
game for them, venison, and especially bear's meat being
highly esteemed. These men used to search the forests for
ten or twenty miles round the camps while hunting. The
hunter of the highest camp on the Stanislaus river came
home one evening, and after supper told them of a big tree
he had found that beat all he had ever seen before. It had
three times as big a trunk as any tree within ten miles round.
Of course they all laughed at him, told him they were not
fools : they knew what trees were as well as he did; and
so on. Then he offered to *show* it them, but none would go;
they would not tramp ten or twelve miles to be made fools
of. So the hunter had to bide his time. A week or two
afterwards he came home one Saturday night with a small
bag of game; but he excused himself by saying that he had
got the finest and fattest bear he had ever killed, and as next
day was Sunday he thought that six or eight of them would
come with him and bring the meat home.

The next morning a large party started early, and after
a long walk the hunter brought them suddenly up to the
big tree, and, clapping his hand on it, said, "Here's my fat
bear. When I called it a tree, you wouldn't believe me.
Who's the fool now ? " This was the great pavilion tree
of the Calaveras Grove, twenty-six feet in diameter at five
feet from the ground—over eighty feet in circumference, so
that it would require fourteen tall men with arms outstretched to go round it. This tree was cut down by boring

into the trunk at six feet from the ground with long pump-augers from each side, so as to meet in the centre. The first fourteen feet was then cut into sections, and one supplied to each of the older States. The rest remains as it fell, and can be walked on to a distance of about two hundred and ten feet from the stump, and here it is still six feet in diameter. To examine this wonderful wreck of the grandest tree then living on our globe is most impressive. The rings on the stump of this tree have been very carefully counted by Professor Bradley, of the University of California, and were found to be 1240, which no doubt gives the age of the tree very accurately, as the winters are here severe, and the season of growth very well marked.

On reaching Stockton, on Saturday evening, I found a letter from Senator Leland Stanford, one of the Californian millionaires whom I had met at Washington, inviting me to visit him at his country house at Menlo Park on the following Monday. Senator Stanford's father was a large farmer near Albany, New York State, who was also the first railroad contractor in America. Up to twenty years of age he had lived and worked for his father. He then became a lawyer, and when his studies were completed, went to Wisconsin to practise. A few years later he removed to California, where he had several brothers who were merchants, and after keeping a store of his own, and thus acquiring business knowledge, he joined them. In 1861 he became Governor of California and President of the Central Pacific Railroad Company, of which he was one of the founders, and by means of which, with the large State and Union subsidies to help its construction and the enormous grants of land which became of value through the making of the railroad, he acquired his great fortune of five or six millions sterling.

When I met him and Mrs. Stanford in Washington, through the introduction of Mrs. Beecher Hooker, it was as a spiritualist, and to talk about spiritualism. Their only son, a youth of sixteen, had died three years before at Florence, and they both assured me that they had since had long-continued intercourse through several different mediums, and under circumstances that rendered doubt impossible. Senator Stanford has shown himself throughout his life a man of exceptional ability and intellectual vigour, and would hardly be imposed upon in such a matter.

Mr. Stanford met me at the station, and drove me to his house, about a mile and a half. It is a large, roomy cottage, luxuriously furnished, with very wide verandahs shaded by trees and awnings, carpeted and furnished so as to form open-air rooms, very delightful in a Californian summer. The grounds are spacious and fairly wooded with some old pines and large eucalypti, as well as many beautiful shrubs. For some distance round the house there are grass lawns, as green and smooth as any I have ever seen, with beautiful borders and flower-beds, the whole kept in the most perfect order by Chinese gardeners, with water laid on everywhere to keep up the perpetual verdure during the six or seven months of continuous heat and drought.

In the house, as in the garden, all the servants are China-men and boys, and both Mr. and Mrs. Stanford spoke of them in the highest terms. One of these boys had charge of her private rooms, and as they continually moved backward and forward between this house and their mansion at San Francisco, going and coming without notice, on her return she always found everything in the most perfect order, and has never missed the smallest article, though jewellery was often left on her dressing-table. Mr. Stanford declared that the Chinese had been the making of California, doing all kinds of domestic work, gardening, and shop-keeping when every European was rushing after gold. He had incurred much obloquy on account of his opposition to the anti-immigration laws and through his employing Chinese servants, but had now, to a large extent, lived it down.

After dinner we drove out to see some of the other millionaires' residences. The most remarkable of these was Mr. Flood's—a kind of fairy palace built entirely of wood, highly decorated with towers and pinnacles, and painted pure white throughout. There were also fine grounds and gardens, but none we saw were so exquisitely kept up as Mr. Stanford's by his thirty Chinese gardeners.

Next morning I was taken to see the site of the great university he was going to build to the memory of his son. He had here about eight thousand acres of land, in the midst of which the buildings and residences were to stand. There were large wooden offices close by, occupied by the architect and draughtsmen preparing the plans and working drawings ; and the surrounding land was already planted with shade-trees and avenues. The plans showed a central chapel in a Norman, or rather Moorish, style of architecture, surrounded by low, one-storey buildings arranged around spacious courts, about five hundred feet by two hundred and fifty feet, to be laid out in grass, trees, and flower-beds. These buildings were to comprise dwellings for professors and students, class-rooms, workshops, libraries, museums, etc., and could be almost indefinitely extended as desired. It was intended for all classes, from the poorest to the most wealthy, and to furnish a complete education from the kindergarten up to the highest departments of human knowledge, including the applications of science to industry and the arts. Arrangements would be made for the students to board themselves at the lowest possible cost. Mr. Stanford had gone into this question, and he assured me that in the best American hotels, where the rates are four or five dollars a day, the actual cost of the food, including cooking, is not more than from two or three dollars a week for each person.[1]

[1] My friend, Professor J. C. Branner, has kindly sent me the latest register of the university, together with a popular account of it, with excellent photographic illustrations and plans ; and it may interest my readers to have some particulars of this newest and in many respects most remarkable, of great educational institutions.

The whole design, of which I saw the drawings, appears to have been now carried out, and the result is very striking. The educational buildings, including a magnificent church, are arranged around a central quadrangle, five hundred and eighty feet long by two hundred and forty-six feet wide. Around this are arranged twenty-six spacious buildings, each devoted to one department of study, and these are grouped around a series of outer courts, the whole forming a quadrangle about nine hundred feet by seven hundred and seventy feet. Quite detached, at various distances around, are the boarding-houses for the students, the residences of the professors, a general library, gymnasium, workshops, and laboratories, and a magnificent museum round a central court, six hundred feet by two hundred feet.

Senator Stanford had a very high opinion of his adopted State, California, as being the richest part of the Union. He dilated on its million inhabitants producing corn enough for ten millions, of its illimitable possibilities of fruit production, and on the general well-being of the people. He expressed surprise that *we* do not federate all our English-speaking colonies, and thus form a " union " comparable in strength and extent with their own ; and it is no doubt the great and

The educational portion is massively constructed of stone or concrete, and a very striking feature, and one well adapted to the climate, is that both the inner and the outer quadrangles are surrounded by continuous arcades, supported on massive stone pillars with groined roofs and about twenty feet wide, thus affording com- munication between the whole of the buildings, with complete protection from the ardent sun of California. These magnificent cloisters aggregate a mile and a quarter in length ; and at the more important entrances the semicircular arches are highly decorated with carved ornamentation in the Mooresque style, and are supported on clustered columns.

The museum is a very fine building in a graceful Romano-Grecian style, and is full of fine works of art of all periods, as well as specimens of natural history. But ornament has been most lavishly bestowed upon the church, which is cruciform, one hundred and ninety feet long by one hundred and sixty feet wide, with a central tower, one hundred and ninety feet high. It is decorated with costly mosaic work both inside and out, and must be one of the most magnificent of modern churches.

At the present time there are more than fifteen hundred students, and nearly one hundred and fifty professors and teachers. The entire education is free for residents in California, with very moderate fees for those from other States. The entire cost of board and lodging, with incidental expenses, is about £60 a year ; but it is stated that a very considerable number of the students are able to support themselves by about three hours' daily work, either in or outside the university, more especially those who are bookbinders, printers, carpenters, or mechanics ; while many others, who can perform any domestic or manual labour thoroughly, can do the same. There are also several scholarships, which give free education and board.

The university has been endowed by Senator and Mrs. Stanford with about eighty thousand acres of land, besides the estate of Palo Alto in which it is situated (about nine thousand acres) and the Stanford mansion in San Francisco, amount- ing in all to about six millions sterling. It only remains to state the purpose for which the university was established by its founders.

" The object of the university is to qualify students for personal success and direct usefulness in life ; it purposes to promote the public welfare by exercising an influence in behalf of humanity and civilization, teaching the blessings of liberty regulated by law, and inculcating love and reverence for the great principles of government as derived from the inalienable rights of man to life, liberty, and the pursuit of happiness."

It is to be hoped that this last clause will be taught in its spirit as well as in its letter. Never, surely, has a grander memorial been raised by parents to a beloved son.

fatal mistake of our Governments not to have seen this before it has become too late, and the absurd and useless tariffs in every colony have created insuperable difficulties to what would at first have been natural and acceptable to all. His view as to the general well-being of the people was, however, fallacious. He looked at the world, just as our legislators do, from the point of view of the employer and the capitalist, not seeing that *their* prosperity to a large extent depended on the presence of a mass of workers struggling for a bare subsistence. At the very time of our interview the actual fruit-grower could hardly earn the scantiest subsistence, because he was dependent on the middlemen and railway companies to get his crop to market, and because the very abundance of the crop often so lowered prices as to make it not pay to gather and pack. Since then, year by year, the unemployed and the tramp have been increasing in California as in the Eastern States, while San Francisco reproduces all the phenomena of destitution, vice, and crime characteristic of our modern great cities. But neither capitalists nor workers yet see clearly that production for *profit* instead of for *use* necessarily leads to those results. The latter class, however, thanks to the socialists, are rapidly learning the fundamental principle of social economy. When they have learnt it, the beneficent and peaceful revolution will commence which will steadily but surely abolish those most damning results of modern (so-called) civilization—insanitary labour, degrading over-work, involuntary unemployment, misery, and starvation —among those whose labour produces that ever-increasing wealth which their employers are proud of, and which their rulers so criminally misuse.

On returning to Stockton I went with my brother to Santa Cruz, one of the health resorts on the Pacific coast south of San Francisco, and thence to the forest tract of the Coast Range, where are a few of the finest trees of the red-wood left in Southern California. We stayed the night at the hotel, and till the following afternoon, quite alone. The trees themselves are more beautiful than those of the *Sequoia gigantea*, the foliage being more like that of our yew. The largest tree is forty-seven feet round at six feet from the ground (sixty feet at the base), and only a few feet less than three hundred feet high. The forests in which they grow are not, however, either so picturesque or so full of other fine trees, shrubs, and flowers as are those of the Sierra Nevada.

While at Santa Cruz for a day, both going and returning, I saw something of the luxuriance of Californian gardens. The common scarlet geranium grew into large bushes, forming clumps six or eight feet high, a mass of dazzling colour, and in the small back garden of a lady we visited was a plant of *Tacsonia van Volxemi*, which grew all over the house, and had sent branches out to an apple tree some yards away, and covered it completely with its foliage and hundreds of its drooping crimson flowers. On the sand of the sea-beach were masses of calandrinia a yard across, covered with their gorgeous blossoms, which seemed to luxuriate in the intense heat and sun-glare.

Returning to Stockton for a week, I had the opportunity of witnessing a Fourth of July celebration. There was a great procession of all the trades and professions, firemen, army corps, volunteers, officials, etc., to the town hall. A school-boy read the Declaration of Independence, and then the "Oration" was delivered. It was pretty good in substance, but declaimed with outrageous vehemence and gesture. Then a patriotic poem was recited by a lady, but two crying infants and exploding crackers outside much interfered with the effect. All the rest of the day there were crackers all over the town, and in the evening another procession of animals, clowns, etc., crowds of people, carriages and buggies, crackers and fire-works—a kind of small and rough carnival. This over, I bade farewell to my brother and sister-in-law, my nephews and nieces, my grand-nephew and grand-niece, and left for the summit level of the Sierra Nevada on my way across the continent to Quebec, whence I was to sail for Liverpool.

CHAPTER XXXII

As my only lecture engagement on my way home was at the Michigan Agricultural College on July 29, I proposed to spend a fortnight among the alpine flowers of the Sierra Nevada and the Rocky Mountains ; and as on my way to San Francisco I had passed over the Sierra in the night, I left Stockton at 7 a.m. in order to proceed by a local mid-day train from Sacramento to the summit level, where there is a small, rough hotel, chiefly used by the men engaged in the repair of the railway.

I had three hours to wait at Sacramento, the State capital, a pleasant town, with abundance of trees and gardens in the suburbs. I bought here a very handy two-foot rule, which folded up into a length of four inches, being thus most convenient for the pocket. It was also very usefully divided in a variety of ways. The *outer* side of one face was divided into eighths of an inch, and the inner side into tenths. The other face was divided into sixteenths and twelths of an inch, while the outer edge was divided into tenths and hundredths of a foot. It was well made, would go into my waistcoat pocket, and has been very useful to me ever since. I have never seen one like it in any English tool-shop, and though it was rather dear (three shillings), it has served as a pleasant and useful memento of my American tour.

Leaving Sacramento at noon, we reached the foothills in about two hours, and soon began to see the effects of hydraulic mining in a fine valley reduced to a waste of sand, gravel, and rock heaps, the fertile surface soil broken up and buried under masses of barren and unsightly refuse, which may in time become covered with trees, but will probably never be profitably cultivable. Having passed this, at one spot I saw a group of tall golden yellow lilies, which blazed out grandly as the train passed them. When we had reached a height of forty-five hundred feet snow-sheds began, short ones at first, and at considerable intervals, but afterwards longer and closer together, and for the last fourteen miles below the summit they were almost continuous. They are formed of massive roughly-hewn or sawed logs completely enclosing the line, but with so many crevices as to let in a good deal of light; but the snow soon stops these up, and in the winter they are as dark as a bricked tunnel.

Before entering them we had fine views, looking backward, down deep valleys and lateral ravines, among the slopes and ridges of which the line wound its way at a nearly uniform incline in order to avoid tunnelling. Everywhere within sight the country had been denuded of its original growth of large timber, but there were abundance of young trees of the sugar-pine, white pine, Douglas and silver firs, and a few cedars, which, if allowed to grow, will again clothe these mountains with grandeur and beauty for a future generation. The visible rocks were either granite or talcose slaty beds and decomposing gneiss. There were also considerable tracts of white volcanic clay or ash, in which the gold-miners work, and the layers of large round pebbles here and there showed where ancient river channels had been cut across by the existing streams.

We reached the summit (seven thousand feet above the sea) at 6.13 in a large snow-shed opening into the railroad warehouses and workshops, and into the hotel. After dinner I strolled out to a small marshy lake in a hollow, and found a fine subalpine vegetation with abundance of flowers, promising me a great treat in its examination. The country immediately around consists of bare granite hills and knolls, with little lakes in the hollows. Just beyond the hotel there is a short tunnel which brings the railway out to the western slope of the Sierra, whence it winds round the southern shore of Donner Lake on a continuous descent to Truckee and the great Nevada silver-mines. The granite rocks in the pass are everywhere ground smooth by ice into great bosses and slopes, in the fissures of which nestle many curious little alpine plants.

I stayed here four days, taking walks in different directions, ascending some of the nearest mountains, exploring little hidden valleys, and everywhere finding flowers quite new to me, and of very great interest. The pentstemons were of great beauty, especially one which grew in fissures of the granite rocks, with clusters of sky-blue flowers and yellow buds, forming a most striking combination. The curious and beautiful *Pedicularis greenlandica* was common in bogs, with tall spikes of purple-red flowers, having long, strangely curved beaks, giving the appearance of some fantastic orchid. The genus Gilia was abundant in various curious modifications, one species (*G. pungens*) being like a minute furze-bush. On some of the hillsides there were sheets of the pretty butterfly-tulip (*Calochortus Nuttallii*), and in moister places the blue *Camassia esculenta*, the very dwarf *Bryanthus Breweri* like a miniature rhododendron, the pretty starlike dodecatheons, the brilliant castillejas, and a host of others. Eriogonums, allied to our polygonums, were abundant and varied, and there were many curious composites and elegant little ferns in the rock-crevices. One of the higher mountains was of volcanic rock, and having once seen their characteristic forms, it was evident that most of them were of this formation, being the sources of the great extent of Pliocene lava-streams and ash-beds which cover so much of the country in California, Nevada, and Idaho. The older rock here is a kind of gneiss, full of fragments of other rocks, both crystalline and volcanic, producing a result similar to the rocks I found in the granitic region of the Upper Rio Negro, and which I have figured in my "Amazon and Rio Negro" (p. 423, cheap ed. p. 293). The smooth, rounded forms of the rocks here are plainly due to glaciation, and have quite a different character to the globular or dome-form at the Yosemite and in Brazil, due to sub-aërial decomposition and exfoliation. Here they show the remains of what were rugged or jagged peaks worn down smooth into rounded hummocks of very varied forms. Striation is sometimes faintly visible, but under the intense climatic changes of this region, weathering has in most cases quite obliterated it.

Having read Miss Bird's account of Lake Tahoe as being superbly beautiful, I determined to see it, and if the country looked promising to stay a few days. I accordingly left by the train on Monday morning, stayed the night at a very poor hotel at Truckee, and took the stage at seven the next morning for the lake, a distance of fourteen miles. The road was up a very picturesque, winding valley, very precipitous and rocky on the east side, more sloping on the west. The bottom of the valley seemed to be granite or gneiss, but the craggy heights on the east side were all of lava, sometimes scoriaceous, sometimes almost columnar basalt, and occasionally laminated. Sometimes there were precipices, peaks, and detached pillars of scoriaceous lava, two hundred to five hundred feet high, of strange forms and highly picturesque. This valley had a rapid stream, which was the outlet of the lake. It had once probably been full of lava and ashes, when the lake would have been much deeper and larger. This was indicated by stratified deposits in places at different levels, and by layers of rock full of rounded pebbles. The lake itself, though a fine piece of water, did not come up to my expectations. The mountains around were bare and monotonous, rather higher and snow-flecked on the west, but the highest peaks visible not more than ten thousand feet. On the west side there was most wood, but the mountains were not more than two thousand to four thousand feet above the lake, and therefore not high in proportion to its size, which is thirty-five miles long and fifteen miles wide. It is really less striking than Loch Lomond or Windermere, where the mountains are more picturesque and more precipitous; while it can bear no comparison with the sub-alpine Swiss and Italian lakes.

I strolled about the shores of the lake, and into some of

the woods near, but all was very dusty and arid, and I found only a few flowers already familiar to me. The hotel looked clean and comfortable, and I had a very good dinner there, and in the afternoon sat in the verandah admiring the view over the lake, it being too hot and dry to go out. I was glad I had seen it, and especially the valley up to it, but I preferred to get on to the Rockies as soon as possible. I therefore went back to Truckee by the return of the stage in the afternoon, and went on to Reno by the evening train. While waiting at the station, two ladies addressed me, and said they had met me last autumn at the meeting of the American Association at Boston. They were both botanists, and had been camping out in the Californian mountains ; so we compared notes, and had some interesting botanical conversation. Their names were Miss J. W. Williams and Miss Sarah W. Horton, of Oakland, California.

The line from Truckee to Verdi (twenty-four miles) passes through a very interesting series of gorges in the volcanic district. The rocks and precipices exhibit all the varied characteristics of basalt, lava, and volcanic ash, with frequent intercalated layers of gravel, and glacial drifts. The lateral gorges give frequent peeps into the interior, with strange castellated cliffs and pinnacles. Sometimes the main gorge narrows, leaving barely room for the railway, with the river foaming against the black, rugged, precipice. The whole country from Gold Run, in California, to Verdi, in Nevada (eighty miles), is a region of extinct (Pliocene?) volcanoes, but at and near the summit these rocks have been denuded down to the gneiss and granite, which there exhibits the grinding power of ice as in the mountains of Europe. In this region we have the results of fire, water, and ice action well illustrating their respective shares in modelling the earth's surface. The long and deep valley of the Truckee has probably been entirely excavated through volcanic rocks since a quite recent geological period.

Leaving Reno the next morning, we passed through similar volcanic country, for about fifty miles, in the Truckee valley ; then across an arid plateau to the valley of the

Humboldt river, only reaching stratified rocks at the Humboldt mountains, towards the source of the river, in the evening ; and the next morning found ourselves near Ogden, where I changed for the Denver and Rio Grande Railway, in order to see a different portion of the mountains, two hundred miles further south, and to visit Colorado Springs and the celebrated Garden of the Gods.

I left Ogden at 10 a.m. July 14, passing Salt Lake City, about fifty miles beyond which, near Provo, we entered a fine gorge of the Wasatch Mountains, leading to an upland valley with abundant vegetation. The cliffs were of a red conglomerate with pebbles, and among the flowers I noticed *Cleome integrifolia,* yellow œnotheras, handsome thistles, a fine golden-rod, and red castillejas. When the train stopped at small stations, for water or other causes, I would jump out and gather any flowers I saw near me, keeping a sharp watch for the conductor's cry of "All aboard!" Having with me Coulter's "Flora of the Rocky Mountains," I was able to make out many of the species. Climbing up a high, open valley, we reached Soldier Summit, where there was half a mile of snow-sheds. This was the divide between the Salt Lake and the Colorado basins, and we then entered Pleasant Valley, and winding about came to the picturesque Castle Gate, where a mass of rock like the ruins of a mediæval castle rises close to the line. Passing this, we entered an almost desert region, with great bare flats of mud and clay, with occasional low ridges of gravel. During the night in this district we were stopped by a "wash out ; " a few hours' deluge of rain having fallen, turned dry channels into roaring torrents, and destroyed the track for some yards in several places. These were rapidly repaired by building up the line with sleepers laid across and across to the required level, and at eight o'clock we went on again ; but were again stopped early in the afternoon. Here I strolled about, but it was a miserable desert, with only a few stunted, ugly spiny bushes. Some of the cliffs around were splendid, in strata of red, yellow, bluish, and green. This district is between the Green and the Gunnison rivers, the latter a very turbid stream.

ON THE DENVER AND RIO GRANDE RAILWAY.

[*To face p.* 176, VOL. II.

Here were a few patches of cultivated land and little rude cabins.

Entering Clear Creek valley, the country becomes smoother, the hills more rounded and more clothed with vegetation, like parts of Wales or Scotland, with some pines and cedars. The occasional bare slopes show a covering of earth and boulders, washed from above by the melting of the winter snows. Here we wound in and out among the mountains up to the heads of all the lateral valleys, then returning on the other side so as to see the line we had come by many hundreds of feet below us. Several short snow-sheds were passed through before reaching the summit between two branches of the Gunnison river, just short of eight thousand feet above the sea. On the east side we again wound about, in and out of valleys, sometimes round such sharp curves that the train made almost a semicircle, till in the evening we reached Cimarron, where we stopped the night, as there is a fine gorge of the Upper Gunnison river through which the line passes.

Starting at 9 a.m. on July 16, we at once entered the gorge, and for fifteen miles had a succession of very fine scenery, the gneissic rocks forming grand precipices, sometimes overhanging, or in picturesque forms with towers and pin-nacles, at others widening into little basins with fine peeps of mountain summits. Pines and firs clung to the rocks, increas-ing the beauty of the scene. On emerging from the gorge, the valley became wider with moderate slopes and table-topped mountains. We reached Gunnison (7580 feet) at 11.10 a.m., situated in a rather bare open plain, with rounded hills ; then entering an open upland valley with fine-looking meadows full of flowers—a perfect garden speckled with pale and dark yellow, pink, blue, and white flowers—the most flowery valley I have seen during my American tour, and the only one that equalled the finest of the European Alps. I could distinguish great patches of dodecatheon, masses of lupins, and white and pink gilias. Then we came to patches of pines and firs, and reached Sargent, 8400 feet above the sea, and I should think a fine station for a botanist at this time of the year.

From here we entered a series of high branching valleys, up and round which we wound to ascend to Marshall Pass, the summit level of the main range of the Rocky Mountains, at an elevation of 10,850 feet. Stopping a few minutes on the summit, I saw many fine flowers, among which was a pentstemon with blossoms of a very dark vinous purple. The descent into the Upper Arkansas valley was very interesting from the way we entered and wound round the head of every lateral valley to gain distance for the descent at a practicable slope, so that in one place we could see three lines of the railway, one below the other, which we had just passed along. Salida, where we stayed to dine, is in a flat valley near the sources of the Arkansas river, and on leaving it we soon entered upon a very fine narrow valley with lofty mountains of conical or pyramidal forms, either smooth or jagged. Then we came to a granite district, with tors of strange and fantastic forms, with huge blocks, peaks, and balanced rocks, like hundreds of Dartmoor tors crowded together. Then more open rocky valleys before we reached the "Royal Gorge," where we beheld towering rocks of fantastic form and colouring closing in upon the river and hardly leaving room for the railway. In places there were vertical precipices about a thousand feet high, side cañons like narrow slits, or winding majestic ravines, often with vertical walls, or with quartz dykes running up the precipitous valley sides, and always the river roaring and raging in a tumultuous flood close alongside of us. It was a fine example of the cañons of the Rocky Mountains, and of the skill and enterprise required to build a railway through such a country. But there are many other lines which penetrate still wilder gorges, and which have overcome much greater difficulties, and I greatly regret I could not afford the time and cost of visiting these. As compared with Switzerland, the Rocky Mountains are very poor in snow-clad peaks and high alpine scenery, but are quite equal, and perhaps even superior, in the number, extent, and grandeur of its cañons or deep valley-gorges.

On leaving this gorge the country became flat and

"THE SQUATTER," GARDEN OF THE GODS, COL.

[*To face p* 179, VOL II.

uninteresting, and we reached Colorado Springs (six thousand feet above the sea) at half-past ten at night, having travelled about six hundred miles, through the most varied, grand, and interesting portion of the Rocky Mountain system. The next morning, after breakfast, I went on by the branch railway to Manitou Springs (6360 feet)—the "Soda Springs" of the old-time trappers, mentioned in some of Mayne Reid's inimitable stories. Here, where the mountains rise abruptly from the great plains, which are themselves more than six thousand feet above the sea, are a group of springs situated near together on a small plateau, yet each of different character and composition. The most interesting is the "boiling spring" or "soda spring," which is so full of gas that it *looks* as if boiling, but is really effervescing. It is as clear as crystal, and tastes just like good aërated water. The springs are surrounded by several pretty hotels, and a small number of shops, boarding-houses, and private residences. I spent the morning walking up some of the curious little valleys that open at once into the mountains, and found a few interesting plants, among which was the *Monarda fistulosa*, of a very bright lilac pink colour, some campanulas, and a few others. After dinner, it being too hot to walk, I hired a buggy to drive me round the Garden of the Gods and Glen Eyrie, a distance of about seven miles. This consists of a tract of un-dulating or hummocky land backed by a range of cliffs, and presenting scores and even hundreds of isolated rock masses of varying heights, but generally about ten or twenty feet, and worn by wind-action into the strangest forms, which have received distinctive names. They are composed of sandstone in nearly horizontal strata of varying hardness, whence has resulted their curious shapes. Some are like pillars with overhanging tops, but most of them, when seen from the right point of view, are ludicrous representations of men or animals. In one we see an old Irish peasant, in another a Scotchman with plaid and glengarry cap, and one is named the Lady of the Garden. There is a cobbler, a bear, a buffalo, a Punch and Judy, and the Squatter;—the last is here reproduced from a photograph.

But even more remarkable than these are the wonderful group of isolated rocks, forming what is called the gateway to the garden. Here are two enormous walls or slabs of red sandstone rising abruptly out of the smooth grassy surface to a height of three hundred and fifty feet, and leaving about the same distance between them, in the centre of which is a smaller similar rock. Through this opening is seen the fine rocky mass of Pike's Peak, snow-clad in spring and flecked with snow in summer, contrasting with the rich red of the sand-stone gateway and the flower-specked sward, so as to produce a landscape which for singularity and beauty I have never seen equalled. In nature, as in the view here reproduced, the precipices forming the gateway have the appearance of rocky hills pierced by a chasm, and it is only when one goes through the gate and looks back, and then walks completely round them, that one sees that they are mere vertical slabs of sand-stone, quite comparable with those which form the fantastic groups and pillars already described, but of much greater dimensions. Looked at from another point of view, the upper ridge is seen to be worn into strange shapes with open-ings and pinnacles, the central mass having excellent representations of a seal and a bear, while on the left is seen the figure of an Indian in his robes. From yet another point the same masses, when seen edgeways, appear as a wonderful group of lofty rock-pinnacles, which are appropriately named the Cathedral Spires. Glen Eyrie, a little way further north, is a small valley terminating in a narrow gorge full of isolated columnar masses of various forms and overhanging; often mushroom-like tops, as shown in the photograph, have quite a distinct character, but have not the varied beauty of the " Garden."

The next day (Monday, July 18) I went on to Denver, and arranged with Miss Eastwood, whom I had met in May, to go to Graymount, the nearest station to Gray's Peak, for a few days' botanizing. Starting at eight the next morning, we went up very picturesque valleys to the mining settlement of Georgetown (eight thousand five hundred feet), and thence on to Graymount, eight miles further, in which distance we

GATEWAY TO GARDEN OF THE GODS : WITH PIKE'S PEAK.

[*To face p.* 180, VOL. II.

THE SEAL AND BEAR, GARDEN OF THE GODS.

"CATHEDRAL SPIRES," GARDEN OF THE GODS.

[*To face p.* 180, VOL. II.

ascended 1170 feet. On the way we had recourse to a loop,
the line crossing the valley winding up its side, then crossing
back again by a lofty viaduct and thus overcoming the greatest
abrupt rise in the valley, making, in fact, an aërial instead of a
subterranean corkscrew as they so often do under similar cir-
cumtances in the Alps. At Graymount we found a tolerable
hotel, where we stayed a few days to explore. There were
two valleys from here, the most northerly and larger, called
Grizzly Gulch, penetrated further into the mountains to the
north of Gray's Peak ; the smaller and steeper leading to a
small collection of miners' huts called Kelso's Cabin, and then
along a wide, upland valley just above timber-line up to the
very foot of Gray's Peak, whence a winding mule-track led
to its summit.

On the second day, when going up Grizzly Gulch, we
came to a miner's cabin, and two men we saw there asked us
to have dinner with them. They gave us some good soup,
pork, and peas, with hot coffee. They told us that a little
higher up there was a fine place for flowers, and that they
were going by there to their work. So we went with them,
and about a quarter of a mile up we came to some patches of
snow at the foot of a fine alpine, rocky slope, and all around
it was a complete flower-garden. We remained here some
hours to botanize, and gathered thirty-five species of alpine
plants in flower. Some, as *Mertensia alpina, Parnassia
fimbriata, Phacelia sericea*, and *Primula angustifolia*, were
among the gems of the Rocky Mountain flora.   Others
were European, as *Anemone narcissiflora, Ranunculus nivalis,*
Astragalus alpina, and *Androsace septentrionalis ;* while others,
again, were British, as *Silene acaulis, Dryas octopetala,* and the
rare *Swertia perennis*, which here dotted the grass with its
curious slaty-blue flowers. The scenery was just like many
a Swiss Alp, where snow-peaks were not in sight, and the
flowers, if not quite so brilliant or so numerous in species,
were especially charming to me from the curious mixture of
European and American species.

On our second visit to Kelso's Cabin we were overtaken
by Mr. Thomas West, an English mining engineer, in whose

house in Grizzly Gulch we had dined the day before, and he asked us to make use of his hut high up the valley, so as to have plenty of time for our visit to Gray's Peak. We took our lunch in a miner's hut, and I was greatly pleased with the little chipmunks—a very small ground-squirrel—which came round the door to pick up crumbs, and after a little time entered the house and ate whatever we gave them without any fear. The miners are as fond of these little creatures as we are of robins, and thus they become quite pets about houses in the wilds where they abound. In the evening we made our way to the cabin, said to be the highest house in the States (about thirteen thousand feet), where it freezes at night nearly all the year round. Some of Mr. West's men had brought up stores, the house being used for prospecting purposes and trial-workings. They made us quite welcome, and we had supper together.

The next morning we walked up to the top of Gray's Peak (14,340 feet), one of the highest in the Rocky Mountains. On this side the ascent was very easy, over grassy slopes inter-spersed with streams of loose stone fragments, everywhere dotted with interesting alpine plants. The summit was a nearly level plateau, with precipices on the north-west, and with a magnificent view all round, only limited by the yellow haze which cuts off the horizon. We had, however, a view of the celebrated Holy Cross mountain, about thirty-five miles to the south-west, below the summit of which some deep gorges preserve perpetual snow in the shape of a cross. Over an area of about three hundred miles from north to south, and two hundred from east to west, there are said to be over thirty summits which reach fourteen thousand feet, and many more above thirteen thousand—a clear indication of this whole region having been once a nearly level plateau, which, during the process of elevation, has been cut into innumerable valleys and cañons by sub-aërial denudation. This is the more remarkable, as the geological structure of the region is very complex, consisting of ancient rocks, and has probably once been covered by the Secondary and Tertiary deposits which now everywhere surround it, as illustrated by the belt of Triassic sandstones of the Garden of the Gods.

We luxuriated here in plants which were altogether new to me. By the side of the road up were great clumps of the common *Silene acaulis*, embedded in which were little tufts of the exquisite blue *Omphalodes nana*, var. *aretioides*, closely allied to a rare alpine species. In damp, shady spots was a curious alpine form of columbine (*Aquilegia brevistyla*), while minute saxifrages, potentillas, trifoliums, and many dwarf composites starred the grassy slopes with beauty. In the afternoon we crossed over a low pass and descended through a precipitous forest into Grizzly Gulch, and then up to Mr. West's house and laboratory, where he did a good deal of work as an assayer of minerals for the numerous prospectors in the district. In the boggy parts of the wood we found great masses of the fine purple *Primula Parryi*.

We spent Sunday with Mr. West and his son, who were working a mine here in partnership with several other men, and these invited us to dine with them. After a morning among the flowers on the way up, we reached the mine tunnelled into the face of the mountain. After going in a few feet the whole surface of the tunnel becomes a mass of ice-crystals as white as snow, showing that the mean temperature of the earth at a few feet deep is below the freezing-point. This continues for a distance of about five hundred feet, when the increase of temperature with depth becomes just sufficient to prevent freezing, and with every twenty or thirty yards further an increase of warmth is felt. We dined with about a dozen men in a large, rough cabin with sleeping-bunks all round. Our table and benches were of rough planks, but they were covered with a clean table-cloth, and our hosts gave us a most excellent dinner of soup, stew, fruit, and cheese, with very good coffee. In these camps they always get a good cook.

In the afternoon we walked up the main gulch into a high, upland valley, with Gray's Peak on our left. Here I found *Bryanthus empetriformis*, a pretty dwarf, heath-like plant new to the flora of Colorado. The grassy slopes here were

wonderfully flowery, with the beautiful *Aquilegia cærulea* and
the scarlet Castillejas, and higher up was a little moraine lake
where *Primula Parryi* and *Arnica cordifolia* were abundant.
Some account of the relations of the American and European
alpine plants is given in my chapter on "Flowers and Forests
of the Far West," in my "Studies" (vol. i. p. 217).

The next morning, after gathering a few more choice
plants to send home to England, we bade farewell to our
kind friends the miners, walked down to Graymount and
took the train to Denver, noticing many fine plants on the
way, as well as the grand precipices of Clear Creek cañon,
where the strata are seen to have been "twisted and tortured
into indescribable forms," as I noted in my journal. In the
evening I had a visit from Mr. and Mrs. Eastwood, and the
next day, at 8.30 a.m., left Denver for Chicago. For some
time I had the sleeping-car, eighty feet long, all to myself,
there being three alternative lines to Chicago, all starting at
the same time. I was now going by the northernmost, so as
to see the prairie country along a new line, about two hundred
miles north of that by which I had come.

For some time after starting I had a fine view of the
range of the Rockies, Long's Peak, to the north-west, being
the most conspicuous object. At Julesberg, two hundred
miles from Denver, we stopped to allow the train for
California to pass us, and I took a short walk out on the
prairie. All around was a boundless expanse of slightly
undulating country, covered irregularly with short wiry grass,
with a few patches of weeds here and there, a purple and a
yellow cleome, and a dwarf entire-leaved golden rod. There
was also a yellow-flowered prickly solanum and a small white-
flowered asclepiad, with linear crowded leaves, like a mare's
tail. The soil was mainly gravel, composed of small crystalline
pebbles, not much rounded. The smallest of these, about the
size of very small peas, were gathered into many anthills
about a foot high. Coming near the North Platte river, the
fine blue *Iris missouriensis* was seen in the marshes. There
was good grass here, and plenty of cattle grazing. The river
was about a mile wide, but shallow and full of mud-banks.

The next morning (Wednesday, July 27) we were near Omaha, in a flat but fertile and cultivated country of un-dulating prairie, with meadows, and even hedges! the hay-stacks, horses, and cattle near the farmhouses having a more homely aspect than the usual half-desert waste of prairie. After crossing the Missouri, and leaving Council Bluffs, the country became more undulating, with fields of maize and rather more flowers. Among these were yellow œnotheras showy rudbeckias, orange marigolds, and the white euphorbia. Further on, the country was almost all cultivated, wheat being all cut, maize growing vigorously, grass all closely cropped off, and no flowers. Every engine that has passed us has poured out a column of smoke of intense blackness, the result of bad coal, careless stoking, and total disregard of the comfort of passengers. We passed the Mississippi about midnight, and in the morning found ourselves near Chicago. For miles before reaching it there are grass-grown streets laid out in the bare open country, with a house here and there—indications of a land " boom," such as are continually got up by speculators.

Having six hours to wait for my train to Michigan, I took a bus and some walks after breakfast to see the town. The chief impression was of endless vistas of long parallel streets ending in the lake-shore, the whole enveloped in a smoky mist worthy of London itself. Like all new American cities, there were great incongruities in the buildings, small two-storey wood houses next door to handsome shops or palatial warehouses seven or eight stories high. This extreme irregu-larity is the more an eyesore from the contrast of wood and granite, or other fine building stone; but, of course, this will gradually disappear. The great lake, which might have given the city a grandeur and dignity of its own, has been spoilt by the railroad companies, for though there is a belt of park and promenade, the shore-front itself is given up to eight parallel lines of railways with ugly iron railings, and a notice that the public will cross this at its own risk. There is here a great area of black dust or mud, screeching engines pouring out dense volumes of the blackest smoke, and at this time of year the grass is dried up, and the trees all blacker than in London. The Dearborn Railway Station is a fine building, but the restaurant attached to it is very poor—ragged table-cloths or bare tables, and a general air of shabbiness pervades it. I did not regret having no business to keep me in Chicago.

Leaving at noon, I passed through a nearly level country of prairie and wood, with luxuriant meadows near the streams or burnt-up pastures, and reached Trowbridge Station at 5.30, where I was met by Mr. Cook, with whom I was to stay. We had supper, of tea, fruit, etc., and I afterwards tried their stereopticon lantern, which was very poor. The next evening (Friday) I gave my lecture on "Darwinism," and offered to give that on " Colours of Animals " on Monday evening as a return for their hospitality. The next morning Professor Beal took me to a fine bit of original swamp forest, with features which were quite new to me. Throughout my wanderings in the Sierra and the Rockies I had never met with any sphagnum moss, which I should often have been glad of to pack my plants in. In this bit of forest, however, there were acres of such sphagnum as I had never seen before, forming a continuous carpet more than a foot thick, and in this congenial rooting medium there were numbers of very interesting plants. American pitcher-plants (sarracenia) were abundant, but what pleased me more were quantities of the elegant orchis (*Habenaria ciliaris*), with curious fringed flowers, making quite a sheet of yellow in places, its tubers not in the soil, but embedded in the sphagnum a few inches below the surface. There was also a curious little plant called gold-thread, allied to the hellebore, and a number of ferns. Among the shrubs were tall vacciniums, and the beautiful red-berried *Nemopanthes canadensis*, allied to the holly, but deciduous.

On Sunday I saw the botanical garden attached to the college, the library, and the insect collections, which latter were very fine as compared with our English species. Of moths of the genus Catocala, instead of our four species there were about twenty, many of them much larger and more gorgeously coloured, while the Saturnias and other groups

were in equal proportion. After giving my lecture on
" Colour " in the evening, I had to hurry off to catch the
train, in which I slept, and reached Kingston the next day
early in the afternoon. Here I had been invited to spend a few
days in a delightful old country house on the shores of Lake
Ontario, in the refined and very congenial society of Mr. and
Mrs. Allen, and their two daughters. I much enjoyed this
visit, and my genuine admiration of the writings of their only
son, Grant Allen, was a bond of sympathy. The house is a
roomy old-world mansion, situated in a small park with grand
old trees, and fruit, flower, and kitchen garden sloping down
to the water. Mr. Allen himself worked at his flowers, and
had a magnificent collection of gladioli now in full bloom.
But what interested me even more was to see rows of vines
in the open ground laden with as fine fruit as we grow in a
vinery, though the winters are far longer and more severe
than ours. But the higher temperature due to the more
southern latitude, combined with a clearer atmosphere and
greater amount of sunshine, are far more favourable to
all fruit and flowers which are uninjured by low winter
temperatures.

One afternoon I went to visit a relative of the Allens at
Gananoque, where they have a small cottage on the rocky
bank of the St. Lawrence, looking on to the celebrated
Thousand Islands. There is an acre of wild ground, with
a little woody ravine bounded by granite rocks, where interesting wild plants are found. The next morning I was taken
among the nearer islands in a small yacht, landing on some
to collect ferns. They are all ice-ground, often mere bosses
rising a few feet above the water, some of the larger ones
having pretty villas and gardens on them. A description of
this place has been made the subject of one of Grant Allen's
bright magazine articles.

One evening Mr. Allen took me to tea at Sir Richard
Cartright's, one of the Canadian ministers, at his fine country
house in a spacious park, a few miles in the country. One of
the sons took me to a wood where trilliums were in flower ;
afterwards we had tea in a spacious hall. There were several

visitors, and the conversation was chiefly about Ireland. Here, as elsewhere in America, our conduct in persistently refusing self-government to Ireland is hardly intelligible, and is almost universally condemned.

On Sunday morning, August 7, I took leave of my very kind hosts, and went by a small steamer through the Thousand Islands to Alexandria Bay on the American shore, and stayed the night at the Thousand Islands Hotel. The trip of about thirty-five miles was most interesting among the countless islands, varying from mere granite rocks to others several miles long. The hotel has a broad verandah out of the dining-room on the first floor, affording a magnificent view up the river, of varied and beautiful combinations of rock, wood, and water hardly to be surpassed. After dinner at 3 p.m., I walked a few miles into the country, consisting of cultivated fields alternating with rock-masses or ridges. These were all rounded, furrowed, and smoothed by ice, and on some of them, where hard quartzose sandstones occurred, the striæ, furrows, and deep scooping were perfectly developed, all following the general direction of the St. Lawrence valley, whatever their shape or aspect. This is the most conclusive indication of ice-action as opposed to other causes. In the evening the scene from the hotel was charming. In addition to the natural beauties of the surface, there were many pretty or elegant villas on the larger islands, with fine lawns and masses of bright flowers, while many pretty yachts were sailing about or lying at anchor. American wealth had here displayed itself to some advantage in a tract of country of such a nature as hardly to admit of any serious deterioration of its natural beauty.

The next morning at seven I went on by steamer to Montreal, passing many picturesque islands, and with occasional distant views of the Adirondacks. We also passed down the whole series of rapids, not very remarkable as compared with those of the Rio Negro, except the two named the " Coteau " and the "Lachine." These rush and boil, and form waves as in a chopping sea, with occasional eddies and whirls where the vessel had to pass between reefs and

rock-ledges, requiring good steering ; but there is nowhere any
perceptible *fall* of the water, and on the whole the scenery of
the St. Lawrence was somewhat monotonous. We passed
under a fine girder bridge and the great Victoria Tubular
Bridge before reaching Montreal, the appearance of which is
much spoilt by factory chimneys and the usual but quite
unnecessary pall of smoke. For all this unsightliness in
almost every city in the world, land monopoly and competition are responsible. If each city owned its own land, it
would be no one's interest to destroy its beauty and healthi-
ness with smoke and impure water ; and if every parish,
district, or county owned its own land, factories would only
be permitted away from centres of population, and would be
so regulated as to prevent all injury or even inconvenience to
those who worked in them.

I had been kindly invited by Mr. Iles, the manager of the
Windsor Hotel, to stay there a day or two as his guest. He
was a great admirer of Herbert Spencer, who had visited
him when in America, and through him I obtained a fine
photograph of our great philosopher, the very best I have
seen, both for likeness and expression. The next morning
he took me for a drive round the city, and up to the top
of Mount Royal, whence there is a magnificent view of the
sloping plain below, on which the city stands, with its
abundance of churches and of trees, which give it a characteristic aspect. It is curious to see all public notices in French
and English, even in this comparatively English part of
Canada. Mr. Iles is a literary man as well as a hotel
manager. He lent me an article of his on " Mathematics
and Evolution," in which he made use of the theory of per-
mutations and combinations to illustrate Spencer's principle
of " multiplication of effects," applied especially to sociology
—an ingenious and well-written paper. He is also a student
of Emerson and Darwin, and he entertained Butler, the
author of " Erewhon," a few years before, and gave me a
copy of the inimitably humorous rhapsody on Montreal,
which I have quoted in Chapter XXVIII.

In the evening at 9.30 I went on board the steamer

Vancouver for Liverpool, and we reached Quebec at 3.30 the next afternoon. As the ship stayed here the night to coal, I determined to sleep on shore and see this celebrated city. Taking my bag in my hand, I walked to the town. On my way I saw a gardener at work—an Irishman—and inquired for a quiet place for a night's lodging. He directed me to a small private hotel—the other hotels, he said, were too noisy and too dear. Securing a room and leaving my bag, I walked to Dyffryn Terrace, where is the monument to Wolfe and Montcalm. Then up to the ramparts of the citadel, from which there is a grand view of the river and the country round, and where the strength of the position can be well seen. For dinner they gave me beef-steak pie, quite English, the first real homely pie I have met with on the American continent. I then strolled into the town and bought a few trifles in the shops. Everywhere they were talking French. The terraces and gardens with electric lights were very pretty.

Next morning I went out at 7 a.m., called on the Irish gardener again, and asked the way to the best part of the town. He offered to show me : went along St. Louis Street and the Grande Allée by the new Parliamentary Buildings, which are very large and handsome ; a new Drill Hall, fantastic Mooresque ; then to| the open down and the Plains of Abraham. The gardener said there were many Irish and Scotch in Quebec, but more French than all the others. He thought they could not become independent, because they could not pay their share of the Canadian Debt. I suggested that perhaps France would help pay it in order to get back their old colony. Yes, he thought they might some day ; but he did not think the French people wanted that. He told me he had been in Quebec forty-six years, and the winters were not nearly so cold as they used to be. He is sure of it. Noses and ears were often frozen and lost then ; now one never hears of such a thing.

I got back to breakfast soon after eight, and then descended to the lower town by the elevator, and to the wharf, where a tender took us on board in a drizzling rain and very cold wind; and at 10 a.m. we started down the St. Lawrence. Fortunately, I had a cabin to myself, as I was very unwell during the whole voyage, with chest oppression, and asthma for the first time in my life.

Having now left North America, I may say just a few words of my general impressions as to the country and the people. In my journal I find this note: "During more than ten months in America, taking every opportunity of exploring woods and forests, plains and mountains, deserts and gardens, between the Atlantic and Pacific coasts, and extending over ten degrees of latitude, I never once saw either a humming-bird or a rattlesnake, or even any living snake of any kind. In many places I was told that humming-birds were usually common in their gardens, but they hadn't seen any this year! This was my luck. And as to the rattlesnakes, I was always on the look out in likely places, and there are plenty still, but they are local. I was told of a considerable tract of land not far from Niagara which is so infested with them that it is absolutely useless. The reason is that it is very rocky, with so many large masses lying about overgrown with shrubs and briars as to afford them unlimited hiding-places, and the labour of thoroughly clearing it would be more costly than the land would be worth."

The general impression left upon my mind as to the country itself is the almost total absence of that simple rural beauty which has resulted, in our own country and in some other parts of Europe, from the very gradual occupation of the land as it was required to supply food for the inhabitants, together with our mild winters allowing of continuous cultivation, and the use in building of local materials adapted to the purposes required by handwork, instead of those fashioned by machinery. This slow development of agriculture and of settlement has produced almost every feature which renders our country picturesque or beautiful: the narrow winding lanes, following the contours of the ground; the ever-varying size of the enclosures, and their naturally curved boundaries; the ditch and bank and the surmounting hedgerow, with its rows of elm, ash, or oak, giving variety and sylvan beauty to the surroundings of almost every village or hamlet, most of which go back to Saxon times ; the farms or cottages built of brick, or stone, or clay, or of rude but strong oak framework filled in with clay or lath and roughcast, and with thatched or tiled roofs, varying according to the natural conditions, and in all show-ing the slight curves and irregularities due to the materials used and the hand of the worker;—the whole, worn and coloured by age and surrounded by nature's grandest adornment of self-sown trees in hedgerow or pasture, combine together to produce that charming and indescribable effect we term picturesque. And when we add to these the numerous foot-paths which enable us to escape the dust of high-roads and to enjoy the glory of wild flowers which the innumerable hedgerows and moist ditches have preserved for us, the breezy downs, the gorse-clad commons and the heath-clad moors still unenclosed, we are, in some favoured districts at least, still able thoroughly to enjoy all the varied aspects of beauty which our country affords us, but which are, alas! under the combined influences of capitalism and landlordism, fast disappearing.

But in America, except in a few parts of the north-eastern States, none of these favourable conditions have prevailed. Over by far the greater part of the country there has been no natural development of lanes and tracks and roads as they were needed for communication between villages and towns that had grown up in places best adapted for early settlement ; but the whole country has been marked out into sections and quarter-sections (of a mile, and a quarter of a mile square), with a right of way of a certain width along each section-line to give access to every quarter-section of one hundred and sixty acres, to one of which, under the home-stead law, every citizen had, or was supposed to have, a right of cultivation and possession. Hence, in all the newer States there are no roads or paths whatever beyond the limits of the townships, and the only lines of communication for foot or horsemen or vehicles of any kind are along these rectangular section-lines, often going up and down hill, over bog or stream, and almost always compelling the traveller to go a much greater distance than the form of the surface rendered necessary.

Then again, owing to the necessity for rapidly and securely fencing in these quarter-sections, and to the fact that the greater part of the States first settled were largely forest-clad, it became the custom to build rough, strong fences of split-trees, which utilized the timber as it was cut and involved no expenditure of cash by the settler. Again, to avoid the labour of putting posts in the ground the fence was at first usually built of rails or logs laid zigzag on each other to the height required, so as to be self-supporting, the upper pairs only being fastened together by a spike through them, the waste of material in such a fence being compensated by the reduction of the labour, since the timber itself was often looked upon as a nuisance to be got rid of before cultivation was possible. And yet again, this fact of timber being in the way of cultivation and of no use till cut down, led to the very general clearing away of all the trees from about the house, so that it is a comparatively rare thing, except in the eastern towns and villages, to find any old trees that have been left standing for shade or for beauty.

For these and for similar causes acting through the greater part of North America, there results a monotonous and unnatural ruggedness, a want of harmony between man and nature, the absence of all those softening effects of human labour and human occupation carried on for genera-tion after generation in the same simple way, and in its slow and gradual utilization of natural forces allowing the renovat-ing agency of vegetable and animal life to conceal all harsh-ness of colour or form, and clothe the whole landscape in a garment of perennial beauty.

Over the larger part of America everything is raw and bare and ugly, with the same kind of ugliness with which we also are defacing our land and destroying its rural beauty. The ugliness of new rows of cottages built to let to the poor, the ugliness of the mean streets of our towns, the ugliness of

our "black countries" and our polluted streams. Both
countries are creating ugliness, both are destroying beauty;
but in America it is done on a larger scale and with a more
hideous monotony. The more refined among the Americans
see this themselves as clearly as we see it. One of them has
said, "A whole huge continent has been so touched by human
hands that, over a large part of its surface it has been reduced
to a state of unkempt, sordid ugliness; and it can be brought
back into a state of beauty only by further touches of the
same hands more intelligently applied." [1]

Turning now from the land to the people, what can we
say of our American cousins as a race and as a nation?
The great thing to keep in mind is, that they are, largely
and primarily, of the same blood and of the same nature
as ourselves, with characters and habits formed in part by
the evil traditions inherited from us, in part by the influence
of the new environment to which they have been exposed.
Just as we owe our good and bad qualities to the inter-
mixture and struggle of somewhat dissimilar peoples, so
do they. Briton and Roman, Saxon and Dane, Norsemen
and Norman-French, Scotch and Irish Celts—all have intermingled in various proportions, and helped to create that
energetic amalgam known the world over as Englishmen.
So North America has been largely settled by the English,
partly by Dutch, French, and Spanish, whose territories were
soon absorbed by conquest or purchase; while, during the
last century, a continuous stream of immigrants—Germans,
Irish, Highland and Lowland Scotch, Scandinavians, Italians,
Russians—has flowed in, and is slowly but surely becoming
amalgamated into one great Anglo-American people.

Most of the evil influences under which the United States
have grown to their present condition of leaders in civilization,
and a great power among the nations of the world, they
received from us. We gave them the example of religious
intolerance and priestly rule, which they have now happily
thrown off more completely than we have done. We gave

---

[1] *The Century*, June, 1887.

them slavery, both white and black—a curse from the effects of which they still suffer, and out of which a wholly satisfactory escape seems as remote as ever. But even more insidious and more widespread in its evil results than both of these, we gave them our bad and iniquitous feudal land system; first by enormous grants from the Crown to individuals or to companies, but also—what has produced even worse effects— the ingrained belief that *land*—the first essential of life, the source of all things necessary or useful to mankind, by labour upon which all wealth arises—may yet, justly and equitably, be owned by individuals, be monopolized by capitalists or by companies, leaving the great bulk of the people as absolutely dependent on these monopolists for permission to work and to live as ever were the negro slaves of the south before emancipation.

The result of acting upon this false conception is, that the Government has already parted with the whole of the accessible and cultivable land, and though large areas still remain for any citizen who will settle upon it by the mere payment of very moderate fees, this privilege is absolutely worthless to those who most want it—the very poor. And throughout the western half of the Union one sees everywhere the strange anomaly of building lots in small remote towns, surrounded by thousands of uncultivated acres (and perhaps ten years before sold for eight or ten shillings an acre), now selling at the rate of from £1000 to £20,000 an acre! It is not an uncommon thing for town lots in new places to double their value in a month, while a fourfold increase in a year is quite common. Hence land speculation has become a vast organized business over all the Western States, and is considered to be a proper and natural mode of getting rich. It is what the Stock Exchange is to the great cities. And this wealth, thus gained by individuals, initiates that process which culminates in railroad and mining kings, in oil and beef trusts, and in the thousand millionaires and multi-millionaires whose vast accumulated incomes are, every penny of them, paid by the toiling workers, including the five million of farmers whose lives of constant toil only result for the most part in a bare livelihood, while the railroad magnates and corn speculators absorb the larger portion of the produce of their labour.

What a terrible object-lesson is this as to the fundamental wrong in modern societies which leads to such a result! Here is a country more than twenty-five times the area of the British Islands, with a vast extent of fertile soil, grand navigable waterways, enormous forests, a superabounding wealth of minerals—everything necessary for the support of a population twenty-five times that of ours—about fifteen hundred millions—which has yet, in little more than a century, destroyed nearly all its forests, is rapidly exhausting its marvellous stores of natural oil and gas, as well as those of the precious metals ; and as the result of all this reckless exploiting of nature's accumulated treasures has brought about overcrowded cities reeking with disease and vice, and a population which, though only one-half greater than our own, exhibits all the pitiable phenomena of women and children working long hours in factories and workshops, garrets and cellars, for a wage which will not give them the essentials of mere healthy animal existence ; while about the same proportion of its workers, as with us, endure lives of excessive labour for a bare livelihood, or constitute that crying disgrace of modern civilization—willing men seeking in vain for honest work, and forming a great army of the unemployed.

What a demonstration is this of the utter folly and stupidity of those blind leaders of the blind who impute all the evils of *our* social system, all *our* poverty and starvation, to over-population! Ireland, with half the population of fifty years ago, is still poor to the verge of famine, and is therefore still overpeopled. And for England and Scotland as well, the cry is still, "Emigrate! emigrate! We are over-peopled!" But what of America, with twenty-five times as much land as we have, with even greater natural resources, and with a population even more ingenious, more energetic, and more hard-working than ours? Are they over-populated with only twenty people to the square mile? There is only one rational solution of this terrible problem. The system that allows the land and the minerals, the means of communication, and all other public services, to be monopolized for the aggrandisement of the few—for the creation of millionaires—necessarily leads to the poverty, the degradation, the misery of the many.

There never has been, in the whole history of the human race, a people with such grand opportunities for establishing a society and a nation in which the products of the general labour should be so distributed as to produce general well-being. It wanted but a recognition of the fundamental principle of "equality of opportunity," tacitly implied in the Declaration of Independence. It wanted but such social arrangements as would ensure to every child the best nurture, the best training of all its faculties, and the fullest opportunity for utilizing those faculties for its own happiness and for the common benefit. Not only equality before the law, but equality of opportunity, is the great fundamental principle of social justice. This is the teaching of Herbert Spencer, but he did not carry it out to its logical consequence—the inequity, and therefore the social immorality of wealth-inheritance. To secure equality of opportunity there must be no inequality of initial wealth. To allow one child to be born a millionaire and another a pauper is a crime against humanity, and, for those who believe in a deity, a crime against God.[1]

It is universally admitted that very great individual wealth, whether inherited or acquired, is beneficial neither to the individual nor to society. In the former case it is injurious, and often morally ruinous to the possessor ; in the latter it confers little or no happiness to the acquirer of it, and is a positive injury to his heirs and a danger to the State. Yet its fascinations are so great that, under conditions of society in which the yawning gulf of poverty is ever open beside us, the amassing of wealth at first seems a duty, then becomes a habit, and, ultimately, the gambler's excitement without which he cannot live. The struggle for wealth and power is always exciting, and to many is irresistible. But it is essentially a degrading struggle, because

[1] I have discussed this subject in my " Studies," vol. ii. chap. xxviii.

the few only can succeed while the many must fail; and where *all* are doing their best in their several ways, with their special capacities and their unequal opportunities, the result is very much of a lottery, and there is usually no real merit, no specially high intellectual or moral quality in those that succeed.

It is the misfortune of the Americans that they had such a vast continent to occupy. Had it ended at the line of the Mississippi, agricultural development might have gone on more slowly and naturally, from east to west, as increase of population required. So again, if they had had another century for development before railways were invented, expansion would necessarily have gone on more slowly, the need for good roads would have shown that the rectangular system of dividing up new lands was a mistake, and some of that charm of rural scenery which we possess would probably have arisen.

But with the conditions that actually existed we can hardly wonder at the result. A nation formed by emigrants from several of the most energetic and intellectual nations of the old world, for the most part driven from their homes by religious persecution or political oppression, including from the very first all ranks and conditions of life—farmers and mechanics, traders and manufacturers, students and teachers, rich and poor—the very circumstances which drove them to emigrate led to a natural selection of the *most* energetic, the *most* independent, in many respects the *best* of their several nations. Such a people, further tried and hardened by two centuries of struggle against the forces of nature and a savage population, and finally by a war of emancipation from the tyranny of the mother country, would almost necessarily develop both the virtues, the prejudices, and even the vices of the parent stock in an exceptionally high degree. Hence, when the march of invention and of science (to which they contributed their share) gave them the steamship and the railroad ; when California gave them gold and Nevada silver, with the prospect of wealth to the lucky beyond the dreams of avarice ; when the great prairies of the West gave them illimitable acres of marvel-lously fertile soil ;—it is not surprising that these conditions with such a people should have resulted in that mad race for wealth in which they have beaten the record, and have produced a greater number of multi-millionaires than all the rest of the world combined, with the disastrous results already briefly indicated.

But this is only one side of the American character. Everywhere there are indications of a deep love of nature, a devotion to science and to literature fully proportionate to that of the older countries ; while in inventiveness and in the applications of science to human needs they have long been in the first rank. But what is more important, there is also rapidly developing among them a full recognition of the failings of our common social system, and a determination to remedy it. As in Germany, in France, and in England, the socialists are becoming a power in America. They already influence public opinion, and will soon influence the legislatures. The glaring fact is now being widely recog-nized that with them, as with all the old nations of Europe, an increase in wealth and in command over the powers of nature such as the world has never before seen, has *not* added to the true well-being of any part of society. It is also indisputable that, as regards the enormous masses of the labouring and industrial population, it has greatly increased the numbers of those whose lives are "below the margin of poverty," while, as John Stuart Mill declared many years ago, it has not reduced the labour of any human being.

An American (Mr. Bellamy) gave us the books that first opened the eyes of great numbers of educated readers to the practicability, the simplicity, and the beauty of Socialism. It is to America that the world looks to lead the way towards a just and peaceful modification of the social organism, based upon a recognition of the principle of Equality of Oppor-tunity, and by means of the Organization of the Labour of all for the Equal Good of all.

CHAPTER XXXIII

LITERARY WORK, ETC., 1887–1905

LEAVING Quebec early in the morning of Friday, August 12, after a week of cold and dull weather, we anchored at 6 a.m. on the 17th off Portrush, on the north coast of Ireland, to leave mails and passengers for Londonderry. Here and all along this coast I gazed upon the intensely vivid green of the grassy slopes, and for the first time understood the appropriateness of " Emerald Isle " as a name for Ireland ; for the colour is altogether unique, and such as I have never seen elsewhere. Two hours later we passed the grand range of basaltic cliffs above the Giant's Causeway, and here, too, all the grassy patches and slopes were of the same vivid tint. Then the Mull of Cantire, in Scotland, came into view, and later Port Patrick and the Mull of Galloway, just catching sight of Ailsa Craig between them. In the afternoon we passed south of the Isle of Man, and reached Liverpool late at night, having thus seen a portion of the British Isles that was quite new to me. Between 6 and 10 a.m. I managed to get all my baggage through the Custom House and taken to the station, had a good breakfast at the hotel, and was off by the 11.5 a.m. express to London, then to Waterloo, and home to Godalming at 5.30.

On my way from Godalming old station to Frith Hill in a fly, an extraordinary event happened. Suddenly I perceived that the driver's coat was on fire behind—actually in flames ! I called out to him. He looked round, beat it with his hands, said, " All right, sir ! " and went on. After a few minutes it began smoking again. I called out louder, it flamed again; both overcoat, trousers, and cushion were burning. Then he got down, took off his overcoat, trampled on it, and beat out the rest. We went on. A third time it burst out in smoke and flame. Again I shouted, and passers-by called out and stopped to look. And then at last, with their help, he finally extinguished the conflagration. A cab-man on fire! No more curious incident occurred during my six thousand miles of travel in America. It originated, no doubt, from his having put a lighted pipe in his pocket, or perhaps from a loose phosphorous match. But he did not seem to mind it much, even when in a blaze.

The rest of the year 1887 was occupied at home in over-taking my correspondence, looking after my garden, and making up for lost time in scientific and literary reading, and in considering what work I should next occupy myself with. Many of my correspondents, as well as persons I met in America, told me that they could not understand Darwin's "Origin of Species," but they did understand my lecture on "Darwinism;" and it therefore occurred to me that a popular exposition of the subject might be useful, not only as en-abling the general reader to understand Darwin, but also to serve as an answer to the many articles and books pro-fessing to disprove the theory of natural selection. During the whole of the year 1888 I was engaged in writing this book, which, though largely following the lines of Darwin's work, contained a great many new features, and dwelt especially with those parts of the subject which had been most generally misunderstood.

The spring of 1889 was occupied in passing it through the press, and it was published in May, while a few corrections were made for a second edition in the following October. During this time, however, I gave several of my American lectures in various parts of the country—at Newcastle and Darlington in the spring of 1888; in the autumn at Altrincham and Darwen; and in 1889 at Newcastle, York, Darlington, and Liverpool.

In the autumn of this year the University of Oxford did

me the honour of giving me the honorary degree of D.C.L., which I went to receive in November, when I enjoyed the hospitality of my friend Professor E. B. Poulton. The Latin speech of the Public Orator on the occasion has been translated for me by my friend Mr. Comerford Casey, and I here give a copy of the translation.

Addressing the Vice-Chancellor and Proctors of the University of Oxford, the Public Orator spoke words to this effect:—

"In that department of natural science which is concerned with the accurate study of animals and plants, be well assured that no living man has laboured more diligently and with happier results than Alfred Russel Wallace.

"For having wandered long in early life through the forests of Brazil, and among those islands which lie beyond the Golden Chersonese, and beneath a burning sun, he thought out and explained with wonderful insight the law according to which (as learned men now believe) new species of animals arise : namely, that a stronger and more vigorous offspring is left behind by those individuals whom nature has, in some way or other, best fitted to endure the vicissitudes of life. Thus, in the course of ages, scions are produced which differ more and more widely from the original stock.

"When this law was discovered, almost simultaneously by the distinguished naturalist, Charles Darwin, neither begrudged to the other his meed of praise; and so high-minded were they both that each was more desirous of discovering new truths than of gaining credit for himself.

"I need not enumerate the many and learned works which Alfred Russel Wallace has published, since the facts which I have related give him sufficient claim to the honorary degree of Doctor of Civil Law which this University is about to confer upon him."

Finding my house at Godalming in an unsatisfactory situation, with a view almost confined to the small garden, the south sun shut off by a house and by several oak trees, while exposed to north and east winds, and wishing for a generally milder climate, I spent some weeks in exploring the country between Godalming and Portsmouth, and then westward to Bournemouth and Poole. I had let my house from Lady Day, and had moved temporarily into another, and therefore wished to decide quickly. We were directed by some friends to Parkstone as a very pretty and sheltered place, and here we found a small house to be let, which suited us tolerably well, with the option of purchase at a moderate price. The place attracted us because we saw abundance of great bushes of the evergreen purple veronicas, which must have been a dozen or twenty years old, and also large specimens of eucalyptus; while we were told that there had been no skating there for twenty years. We accordingly took the house, and purchased it in the following year; and by adding later a new kitchen and bedroom, and enlarging the drawing-room, converted it from a cramped, though very pretty cottage, into a convenient, though still small house. The garden on the south side was in a hollow on the level of the basement, while on the north it was from ten to thirty feet higher, there being on the east a high bank, with oak trees and pines, producing a very pretty effect. This bank, as well as the lower part of the garden, was peat or peaty sand, and as I knew this was good for rhododendrons and heaths, I was much pleased to be able to grow these plants. I did not then know, however, that this peaty soil was quite unsuited to a great many other plants, and only learnt this by the long experience which every gardener has to go through.

During the eight years I had lived at Godalming, I had greatly enjoyed my garden, and had grown, more or less successfully, an immense number of hardy and half-hardy plants in about half an acre of ground. The soil was of the Lower Greensand formation, with a thin layer of leaf-mould, the whole district having been originally woodland and copse. On the whole this soil was the best for gardening purposes I have ever had, being easy to work, and well suited to a great variety of herbaceous plants and shrubs, and especially to

bulbs.   Here, without any special trouble, I was able to grow
on a raised bank *Iris susiana* and *I. durica* for several years
in succession, and the lovely jalap plant, *Exogonium purga*,
grew most luxuriantly over a low trellis at the back of the
same bed.   Here, too, I had the magnificent *Eremurus
robustus* on a raised bank, with *Leonotis leonurus*, and many
other tender shrubs in the borders.   I received contributions
of uncommon plants from many friends, and ransacked all
the nurserymen's catalogues for rarities and curiosities, and I
find that I attempted the cultivation in this garden, or in a
very small greenhouse and verandah, of about fifteen hundred
species of plants, some of which, of course, never reached
flowering size, others survived only a few years ; but the
delight of watching the growth of these, to me, new forms of
vegetable life, and seeing them flower even once or twice, was
so great that no trouble was spared to obtain it.

My gardening has always been to me pure enjoyment.
I have never made any experiments with my plants, never
attempted to study their minute structure or to write about
them ; the mere seeing them grow, noting the infinite diversities of their forms and habits, their likes and dislikes, all
made the more interesting by the researches of Darwin,
Kerner, H. Müller, Grant Allen, Lubbock, and others, on the
uses of each infinitely varied detail of stem and leaf, of bract
and flower—all this was to me a delight in itself, and gave
me that general knowledge of the outward forms and inward
peculiarities of plants, and of the exquisite beauty and almost
infinite variety of the vegetable kingdom, which enabled me
better to appreciate the marvel and mystery of plant life,
whether in itself or in its complex relations to the higher
attributes of man.

When I came to Parkstone (in June, 1889) I had a
smaller garden, but one which I thought would prove better
adapted to a variety of species which I had not hitherto
succeeded with.   I thought my peat-bank facing the south-
west might grow some of the beautiful Cape heaths which I
had always so greatly admired, so I obtained in the spring

of 1890 a dozen choice species, as well as a considerable number of Sikhim rhododendrons (seedlings and young plants) from different dealers. But although I protected them with fern, ashes, etc., every heath was killed the first winter, while most of the rhododendrons lived and have now grown into large bushes, of which two or three have flowered and others I still hope to see flower. That winter (1890–91) was the first of a series of five severe winters; while the first of them for duration of hard frost and the last for extreme low temperature were the worst known, at all events in the south of England, for about sixty years. What I regretted even more than the heaths was a fine young plant of the celebrated blue Puya, a present from my kind friend Miss North, who had raised it from seed she brought from Chile. Not having had time to get well rooted in the soil it died, like the heaths, the first winter, although when once well established it will bear a considerable amount of frost.

I made a little pond here to grow water-lilies and other aquatic plants, and here again I met with one of the commonest difficulties of the amateur who grows more than he can properly attend to, the presence of what are now termed "dangerous plants." I got a small bit of the fine red Swedish water-lily from Ware, and after the first year or two it grew well and formed one of the greatest attractions of my garden; but I also had at one side of the pond the fine native plant, *Ranunculus lingua*, and this, if left alone, would in a few years have monopolized the whole pond and destroyed the more valuable plants. Another of these rapid growers is the very pretty *Villarsia nymphœoides*, which sends out runners in all directions, and so becomes a danger to all less vigorous plants. The same thing happens with alpine plants. Many, indeed most of them, are quite easy to grow with a suitable position and soil, but they require constant protection against stronger-growing plants and weeds. The amateur must therefore either make them his chief care or else limit his rockery to small dimensions and grow only a few of the best kinds. In stocking my garden at Parkstone I received valuable contributions from many kind friends, among whom were the late Miss Owen, Mr. H. J. Elwes, Miss Jekyll, and Sir W. T. Thistelton Dyer of Kew, and many others. Among the plants which I grew here with some success were the fine blue, purple, and yellow Himalayan poppies, the curious *Periploca græca*, which produced masses of its strange blossoms, the beautiful *Akebia quinata* with its wire-coloured flowers, a very large *Solanum crispus*, and the strange Chilian climber, *Mutisia decurrens*, which we called the "glory dandelion," from its very large stellate flowers of intense orange. Even Sir Thomas Hanbury, who paid me a visit here, had not before seen this plant in flower. An unusually clear blue hydrangea on a shady bank was also one of the glories of my Parkstone garden.

As already stated, from my very schoolboy days and my early youth orchids had a fascination for me, from the strangeness of their growth and habits and their fantastic and beautiful flowers. In the parts of the tropics I visited they were comparatively few in number, while their limited flowering period made the finding of any of the more showy species in flower a rare event. It was only after my return home that at flower shows, and especially at Mr. William Bull's annual exhibition of orchids at Chelsea, I became really acquainted with their inexhaustible variety, extreme interest, and marvellous beauty. There was no exhibition in London that was at once so enjoyable and satisfying as these orchid shows, which I generally managed to visit every year.

Being, as I thought, settled for life at Parkstone, I determined at last that I would try and grow some orchids myself, and accordingly built a small house in three divisions so as to get different temperatures, and for about four or five years persevered in the attempt, with a great deal of labour and enjoyment to myself, though with only a limited amount of success. As I was always longing for new species, I did not content myself with a few of the most showy and most easily managed, but endeavoured to get examples of almost all the chief forms. Some I bought at sales, a few from dealers, and I had a nice lot of Jamaica orchids sent me by Mr. W.

Fawcett, among which was the handsome *Broughtonia san-guinea*, which flowered for several years. I also received a large case of fine Indian orchids from the Botanic Gardens at Calcutta. At last I got together more than a hundred species, most of which I had the pleasure of seeing flower once, though many refused to do so a second time.

Owing to the entrance to the orchid house being on a different floor from my study, the constant attention orchids require in shading, ventilating, and keeping up a moist atmosphere, involved such an amount of running up and down stairs, or up and down steps or slopes in the garden, that I found it seriously affected my health, as I was at that time subject to palpitations and to attacks of asthma, which were brought on by any sudden exertion. I was therefore obliged to give up growing them, as I found it impossible to keep them in a satisfactory condition. This was partly owing to the position of my houses, which were exposed to an almost constant wind or draught of air, which rendered it quite impossible to keep up the continuously moist atmosphere and uniform temperature which are essential conditions for successful orchid-growing. One of my friends who began growing orchids soon after I did, having a well-sheltered position and better aspect, succeeded far better, although he was able to give them much less attention and often did not enter the house for days together, having a boy to keep up the fire, shade from the sun, and moisten the floors twice a day. It is a well-known fact that, even under the same gardener, orchids will grow well in one house, while in another, perhaps only twenty yards distant, it is almost impossible to keep them in health.

It was in the early part of my residence at Parkstone that I received a visit from the great French Geographer, Elisée Reclus, who had, I think, come to England to receive the gold medal of the Royal Geographical Society. He was a rather small and very delicate-looking man, highly intellectual, but very quiet in speech and manner. I really did not know that it was *he* with whose name I had been familiar for twenty years as the greatest of geographers, thinking it must have been his father or elder brother; and I was surprised when, on asking him, he said that it was himself. However, we did not talk of geography during the afternoon we spent together, but of Anarchism, of which he was one of the most convinced advocates, and I was very anxious to ascertain his exact views, which I found were really not very different from my own. We agreed that almost all social evils—all poverty, misery, and crime—were the creation of governments and of bad social systems; and that under a law of absolute justice, involving equality of opportunity and the best training for all, each local community would organize itself for mutual aid, and no great central governments would be needed, except as they grew up from the voluntary association of their parts for general and national purposes.

On asking him if he thought force was needed to bring about such a great reform, and if he approved of the killing by bombs or otherwise of bad rulers, he replied, very quietly, that in extreme cases, like that of Russia, he thought there was no other way to force upon the rulers' notice the determination of the people to be free from their tyrants; but under representative governments it was not needed, and was not justifiable. Few would think to look at this frail man that he was not only in the very first rank among the students and writers of the nineteenth century, but that he had fought for his country against the foreign invader, as well as against the despotism of enthroned officialdom which succeeded it.

He has now passed away (1905), having completed one of the greatest (if not the very greatest) literary works of the past century. But he will also be remembered as a true and noble lover of humanity—a firm believer in the goodness, the dignity, and the perfectibility of man.

During the first half of my residence at Parkstone (1889-96), I did not write any new books, having, as I thought, said all that I had to say on the great subjects that chiefly interested me; but I contributed a number of articles to

reviews, wrote many notices of books, with letters to *Nature*
on various matters of scientific interest. A short account of
the more important of these will show that I was not altogether inactive as regards literary work.

In the spring of 1890 I lectured at Sheffield and at Liverpool, and have since declined all invitations to lecture, partly
from disinclination and considerations of health, but also
because I believed that I could do more good with my pen
than with my voice. During the year I prepared a new
edition of my " Malay Archipelago," bringing the parts dealing with natural history up to date.

In the same year I contributed to the *Fortnightly Review*
an article on " Human Selection," which is, I consider, though
very short, the most important contribution I have made to
the science of sociology and the cause of human progress.
The article was written with two objects in view. The first
and most important was to show that the various proposals
of Grant Allen, Mr. Francis Galton, and some American
writers, to attempt the direct improvement of the human
race by forms of artificial elimination and selection, are both
unscientific and unnecessary ; I also wished to show that the
great bugbear of the opponents of social reform—too rapid
increase of population—is entirely imaginary, and that the
very same agencies which, under improved social conditions,
will bring about a real and effective selection of the physically,
mentally, and morally best, will also tend towards a diminution of the rate of increase of the population. The facts and
arguments I adduce are, I believe, conclusive against the two
classes of writers here referred to.

A year later I contributed a paper to the Boston *Arena*,
dealing more especially with the laws of heredity and the
influence of education as determining human progress, showing that such progress is at present very slow, and is due
almost entirely to one mode of action of natural selection,
which still eliminates *some* of the most unfit. And I pointed
out that a more real and effective progress will only be made
when the social environment is so greatly improved as to
give to women a real choice in marriage, and thus lead both

to the more rapid elimination of the lower, and more rapid increase of the higher types of humanity.

Shortly afterwards I was interviewed for the *Daily Chronicle* on this subject, in which I gave a condensed sketch of these two articles, and this drew attention to them, and brought me a very kind and appreciative letter from the late Frances Willard, who was then in England.

In 1891 I wrote the two articles on the American flora already referred to, and prepared a new edition of my two books on "Natural Selection" and "Tropical Nature," now forming one volume, but from which some of the more technical portions were omitted, while two new chapters were added —"The Antiquity of Man in North America," and "The Debt of Science to Darwin." I also wrote two articles on "Apparitions" for the Boston *Arena*, which are included in the later editions of my "Miracles and Modern Spiritualism;" and I reviewed a few books in *Nature*, among which was the important work of Professor Lloyd Morgan on "Animal Life and Intelligence."

In 1892 I wrote four review articles, three of which are reprinted in my "Studies," and I reviewed (in *Nature*) Mr. W. H. Hudson's delightful volume, "The Naturalist in La Plata."

In the year 1893 I was pretty fully occupied with literary work. I prepared for Mr. Stanford a new edition of the Australian volume of his "Compendium of Geography," involving a large amount of new matter; I contributed five articles to reviews or books, two of which, on "The Ice Age and its Work," gave an entirely new argument in favour of the ice-origin of valley-lakes in glaciated regions; and I also reviewed two books and wrote a number of letters to *Nature* on biological and physical problems. In the summer of this year I went with my wife to the lake district—our first visit; we ascended two of the mountains, and I paid particular attention to the phenomena of glaciation, which are every-where prominent in rounded rocks, glacial striæ, and abund-ance of moraines.

In the year 1894 I read a paper to the Cambridge
Natural Science Club on the question, " What are Zoological
Regions?" which was printed in *Nature* (April 26). But
my conclusion—that the six regions first defined by Dr. P. L.
Sclater are, for all practical purposes of the study of distribution, the most convenient and those which best illustrate the
actual facts of nature—was contested by my friend, Professor
Alfred Newton as regards the Nearctic and Palæarctic
regions, which he contended formed but one natural region.
I therefore thought it necessary to go into the subject in
more detail, and contributed a paper to *Natural Science* in
the following June, entitled " The Palæarctic and Nearctic
Regions compared as regards the Families and Genera of
their Mammalia and Birds."

The first of these papers was for the purpose of showing
that, to be of any practical use to naturalists, zoological regions
must be so defined as to serve to elucidate the distribution
of *all* land animals.   This will be evident if we consider the
results of the contrary view, that many classes, orders, and
even families, require a special set of regions to exhibit their
distribution with any approach to accuracy.   Now as there are
some hundreds of these groups in the animal kingdom, we
should, perhaps, require fifty or a hundred sets of zoological
regions—each set differing in the number of regions and in
the boundaries of each, involving a different set of names in
each case.   The result would be that each specialist would
have his own set of regions, with different names and
different boundaries ; and as no one could be familiar with
all these, the conclusions of each could be unintelligible and
useless to others.   With one set of regions, on the other
hand, the distribution in every case can be described in terms
which would be intelligible to all ; and the comparison of the
distribution of groups differing in powers of dispersal and in
other ways, would often lead to an explanation of the differences of distribution, which is the whole aim and end of the
study, and which, so far as I can see, can be arrived at in no
other way.

The second and more technical paper was for the purpose

of showing the great importance of the *absence* of extensive groups from one region that are present in the adjacent region, even though these groups are not peculiar to either. Thus, the fact that both the bear and the deer families are absent from Africa south of the Sahara, though abundant throughout all Asia and North America, marks out the Ethiopian region as distinctly as does the presence of giraffes and hippopotami, which are now peculiar to it.

But I show that, in mammals, about one-third of the families in the Palæarctic and the Nearctic regions respectively are *not* found in the other ; while in birds, one-third of the families found in the Palæarctic region are not found in the Nearctic, and one-fourth of those in the Nearctic are not found in the Palæarctic region. These facts prove, I maintain, a radical dissimilarity, although, owing to the fact that temperate Europe and Asia are continuous with tropical Africa and Asia, and temperate with tropical America, neither of the regions we are considering have any important families of birds altogether peculiar to them. Any of my readers who are interested in the problems here stated should read the two articles above referred to.

Other articles were, "A Representative House of Lords," in the *Contemporary Review* (June), and "A Suggestion to Sabbath-Keepers," in the *Nineteenth Century* (October), both which articles attracted notice in the Press. I also wrote a paper criticizing the Rev. George Henslow's views as to the origin of irregular flowers, and of spines and prickles, in *Natural Science* (September), the three articles being included in my "Studies." I also reviewed James Hutchinson Stirling's "Darwinianism" in *Nature* (February 8), and Mr. Benjamin Kidd's "Social Evolution" in the same paper (April 12), as well as an anonymous volume, entitled "Nature's Method in the Evolution of Life," by a writer who suggests vague theories, less intelligible even than those of Lucretius, as a substitute for the luminous work of Darwin.

In the next year (1895) I wrote an important article on "The Method of Organic Evolution" (*Fortnightly Review*,

February-March), which was chiefly devoted to showing that the views of Mr. Francis Galton, and of Mr. Bateson in his book on "Discontinuous Variations," are erroneous; and that such variations, which are usually termed "sports," and in extreme cases "monstrosities," do *not* indicate the method of evolution. Darwin gave special attention to this view, and finally rejected it; and I think I have shown *why* it is not effective in nature. It is a view which is continually cropping up as if it were a new discovery, and a Dutch botanist, De Vries, has recently written a large work claiming that new species are produced in this manner, through what he terms "mutations." It was therefore important to show that all such methods are fallacious, and that owing to the constancy, universality, and extreme severity of elimination through survival of the fittest, such large and abrupt variations, except through some extraordinary and almost impossible concurrence of favourable conditions, can never permanently maintain themselves.

Another article (in the October issue of the same Review) on "The Expressiveness of Speech" develops a new principle in the origin of language, and brought me a holograph (and partly unintelligible) letter from Mr. Gladstone, expressing his concurrence with it. I also brought out a new edition of my "Miracles and Modern Spiritualism," containing two new chapters, and a new preface giving a sketch of the changes of opinion on the subject during the preceding half century.

In July I went with my friend Mr. William Mitten for a short botanizing tour in Switzerland. We walked a good deal of the time, and I thus had a further opportunity of examining glacial phenomena. We went to Lucerne, whence we ascended the Stanzerhorn by the electric railway, and found a very interesting flora on the summit. Then to the head of the lake, and to Gœschenen, whence we walked to Andermatt; then over the Furca pass to the Rhone glacier, staying two days at the hotel; then over the Grimsel pass, where we greatly enjoyed both the flowers and the wonderful indications of glacial action, especially on the slope down to and around the Hotel Grimsel, where we stayed the night. The valley down to Meiringen was excessively interesting, being ice-worn everywhere. We stayed an hour at the fine Handeck cascade, and then, with the help of a chaise, into which two ladies hospitably received us, got on to Meiringen. Here we stayed two days, exploring the gorge of the Aar and the wonderful rock-barrier of the Kirchet, visited the Reichenbach falls, and had an excursion to Brunig, where, in some hilly beech woods, we were greatly pleased to find the beautiful *Cephalanthera rubra* in fair numbers and in full flower. This is one of the rarest of British orchises, having been found only at long intervals in Gloucestershire and Somersetshire. I remember, I think about fifty years ago, seeing a newly gathered specimen exhibited at the Linnean Society. Other orchises which occur at similar long intervals are the beautiful ladies slipper (*Cypripedium calceolus*) in some Yorkshire woods, and the strange goat-orchis (*O. hircina*) in copses in Kent and Suffolk. In all these cases, no doubt, the plant persists in the respective localities, but is accidentally prevented from flowering, or requires some specially favourable seasons which only recur at long intervals. We then went on to Lauterbrunnen and the Wengern Alp, where we stayed two days, botanizing chiefly among the woods and slopes near the Trummetthal. We were, however, so dreadfully persecuted by swarms of blood-sucking flies, which filled the air and covered us in thousands, piercing through our thin clothing, that we returned home some days earlier than we had intended.

In 1896 I wrote three articles. "How best to model the Earth," in the *Contemporary Review* (May), was a discussion of the proposal by Elisée Reclus to erect an enormous model of the globe, about four hundred and twenty feet in diameter, giving a scale about one-third smaller than our ordnance maps of one inch to a mile. It was to be modelled in minute detail on the convex side, and would therefore require to be completely covered in by a building nearly six hundred feet high, and would need an elaborate system of platforms and staircases in order to see it, while only a very small portion of it could be seen at once, and accurate photographs could only be taken of very small areas.

My proposal was to adopt the plan of Wyld's great globe in Leicester Square, many years ago, giving all the *detailed* features on the inside surface, while the outside could be boldly modelled in some indestructible material to show all the chief physical features, which might also be coloured in fresco as naturally as possible, and would then be a grand object seen either near or at a distance, while a captive balloon would afford a splendid view of the polar regions and of all parts of the northern hemisphere. The numerous advantages of this plan are explained in some detail, and I have little doubt that it will be realized (perhaps on half the scale) some time during the present century. The article is contained in the second volume of my "Studies."

I also wrote an article on "The Gorge of the Aar and its Teachings," as serving to enforce my papers on the "Ice Age and its Work" three years before. But my most important scientific essay this year was a paper I read to the Linnean Society on "The Problem of Utility." My purpose was to enforce the view that all specific and generic characters must be (or once have been) useful to their possessor, or, owing to the complex laws of growth, be correlated with useful characters. It was necessary to discuss this point, because Mr. Romanes had unreservedly denied it, and Professor Mivart, the Rev. Mr. Henslow, Mr. Bateson, and others, had taken the same view. I endeavoured to show that the problem is a fundamental one, that utility is the basic principle of natural selection, and that without natural selection it has not been shown how *specific* characters can arise. By *specific* is, of course, meant characters which, either separately or in combination, distinguish a species from all others, and which are found in all, or in the great bulk, of the individuals composing the species; and I have shown that it is for want of clear thinking and accurate reasoning on the entire process of species formation that the idea of useless specific characters has arisen (see "Studies," vol. i.).

I also reviewed Copes' "Primary Factors of Evolution" and Dr. G. Archdall Reid's "Present Evolution of Man" in *Nature* (April 16), and wrote a long letter in *Nature*

(January 9) on "The Cause of the Ice Age," pointing out
the extreme complexity of the subject, and the fallacy
of discussing the problem as if it were merely one of the
amount of sun-heat received in different latitudes under
differing degrees of eccentricity, as several eminent mathematicians had done. In the same issue Sir Robert Ball
pointed out the same fallacy ; and this affords a good illustration of the fact that specialists are usually not well fitted to
arrive at the true explanation of great natural phenomena
which are highly complex in their nature, and which require
the consideration of a great variety of physical forces and
laws in order to arrive at their causes. It is for this reason
that Mr. Croll's theory is so much more satisfactory than any
of the modern substitutes for it. His views were, however,
spread over many different periodicals, and are often rather
obscure and disconnected, while few of his recent critics
appear to have studied the whole of them. I venture to
think that my chapter viii. of "Island Life" gives the
best connected and systematic statement of Croll's views
which are to be found, and that the further explanations
of essential points, and some modifications in detail, render
it the completest and most rational theory which has yet
been set forth. Being myself a mere outsider, neither a
geologist nor a mathematician, and only an amateur physicist,
none of the writers on the subject appear to have read my
chapter, since I have never seen it referred to. Yet it
appeals throughout to astronomical, physical, geographical,
and meteorological facts, showing their actions and reactions
on each other, and how they co-operated to produce the
glacial epoch, as they now co-operate to bring about the
strikingly contrasted climates of the eastern and western
shores of the North Atlantic, and the still more striking
contrasts of the Arctic and Antarctic regions.

During this summer I was invited by Dr. H. S. Lunn to
go with him and his party to Davos for a week early in
September, and to give them a lecture on Scientific Progress
in the Nineteenth Century. As I had never been in this part

of Switzerland I accepted the invitation, and had a very
pleasant time. My companion on the first part of the journey
was Mr. Le Gallienne, and at Basle we were joined by Dr.
and Mrs. Lunn and others. At Davos we were a large party
in one of the best hotels, and our special party, who sat
together at meals, included the Rev. Hugh Price Hughes and
the Rev. H. R. Haweis, both talented and witty men, whose
presence was enough to render almost any party a brilliant
success. Mr. Price Hughes was, I think, without exception,
the most witty man and one of the best companions I ever
met. At breakfast and dinner he was especially amusing
and brilliant, ranging from pure chaff with his old friend Dr.
Lunn to genial wit and admirably narrated anecdotes. He
often literally kept the table in a roar of laughter. But this
was only one side of his character. He was a Christian and
a humanitarian in the best sense of the words. I saw a good
deal of him in private, and we often walked out together, at
which times we discussed the more serious social problems
of the day ; and he gave me details of his rescue work in
London which were in the highest degree instructive, showing that even those who are considered to be the most
degraded and irreclaimable can be reached through their
affections. Their degradation has usually been brought
about by society, and has been intensified into hate and
despair by the utterly unsympathetic and cruel treatment of
our workhouses and prisons. Mr. Price Hughes gave me an
account of one of these cases—a woman who had reached
the uttermost depths of drunkenness and vice, and who was
besides so violent that it was dangerous to approach her.
Knowing her case, a lady who was one of Mr. Hughes' chief
helpers in his rescue work went to the prison to receive her
on her discharge, and begged to be allowed to go to her cell
and take her with her. She was assured it was not safe,
that she would be instantly attacked, and perhaps seriously
injured. But the lady insisted, and at length was allowed to
try, with several of the strongest female warders at hand to
assist or rescue her from one whom they described as an
utterly irreclaimable wild beast. Mrs. —— entered without

the least fear, opened her arms, kissed the poor woman with every indication of compassion and love, and spoke to her as if she were an unfortunate and ill-used daughter or sister. The woman was utterly disarmed by the *reality* of the affection showed her, and burst into tears. She was taken to the home of which the lady was the head, and at the time Mr. Hughes was speaking had been there several years, and was one of his most useful and earnest helpers. This woman had not, for years, received a single word of real sympathy or love. A similar marvellous effect was produced by Mrs. Fry on the female prisoners in Newgate by her intense sympathy and affection for them; yet we still go on with our crude, harsh system of prison discipline, which inevitably degrades and brutalizes the great majority of those subject to it. And we dare call ourselves enlightened, humane, civilized, and even Christian!

I also had some pleasant intercourse with Mr. Haweis, and one day we spent a whole afternoon in a private room, talking chiefly about spiritualism, of which he had a considerable practical knowledge. He was one of the few clergymen of the Church of England who not only acknowledged his belief, but preached the doctrines of spiritualism openly from his London pulpit.

Dr. Lunn arranged for his party some amusement for several evenings in each week, either a concert, lecture, or conversazione. Mr. Le Gallienne gave a very interesting lecture on "English Minor Poets," reading selections from their works to illustrate their style. Among these he included Grant Allen, better known as a delightful writer on nature-study and a novelist, but who was also gifted with the true poetic power; and, the lecturer thought, had he devoted himself to developing his power he might have become a major instead of a minor poet. As an example of his work, a very agnostic and even atheistic poem was quoted.

I cannot find this, as I remember it, in his little volume of verse, "The Lower Slopes;" but there is one which

expresses the same idea, and which perhaps may be it—
A Prayer—as follows :—

> " A crowned Caprice is god of this world ;
> On his stony breast are his white wings furled.
> No ear to listen, no eye to see,
> No heart to feel for a man hath he.
> But his pitiless arm is swift to smite,
> And his mute lips utter one word of might ;
> ' Mid the clash of gentler souls and rougher,
> Wrong must thou do, or wrong must suffer.'
> Then grant, oh, dumb, blind God, at least that we
> Rather the sufferers than the doers be."

The lecturer stated that, however extreme and even outrageous these views would appear to many of his audience,
he could assure them from personal knowledge, that they
represented the opinions of almost all of the poets of whom
he had spoken.

After the lecture Dr. Lunn protested against the idea that
poets were generally agnostic or even irreligious, referring to
Milton, Browning, Tennyson, and many others ; but Mr. Le
Gallienne had said nothing about these—the major poets—
and he assured me afterwards that he was well acquainted
with all the poets he had referred to, and that every one of
them were more or less pronounced agnostics. This seems
to me an interesting fact.

My own lecture was mainly devoted to a sketch of the
chief great advances of science during the century, but I
added to it a kind of set-off in discoveries which had been
rejected and errors which had been upheld, referring to
phrenology as one of the first class, and vaccination as one
of the second. There were, of course, in such a place as
Davos, many doctors among the audience, and they signified
their disapproval in the usual way ; but I assured them that
some of them would certainly live to see the time when the
whole medical profession would acknowledge vaccination to
be a great delusion.

Although Davos has no grand alpine scenery immediately
around it, there are many delightful walks through woods full
of flowers and ferns, alpine meadows with gentians and

primulas, and stony passes from which the snow had just retreated. On the Strela pass, about eight thousand feet, I found some charming little alpines I had not seen before, among them the very dwarf *Viola alpina*, growing among stones, the leaves hardly visible and the comparatively large flat flowers of a very deep blue-purple, with a large orange-yellow eye. This is peculiar to the Eastern Alps, and seems difficult to cultivate, as few dealers have it in their lists. I sent home a few plants, but could not succeed in keeping them alive.

On leaving Davos, I made my way across to Adelboden, where my wife and daughter, with some friends, were staying. This is surrounded with fine alpine peaks and snow-fields, and though the weather was unsettled we spent a pleasant week here—probably the last visit I shall make to ever-delightful Switzerland—the sanatorium and alpine garden of overworked Englishmen.

From this time onwards I did not write many articles or reviews, the more important being " The Problem of Instinct," in 1897, in which I gave an attempted solution of bird migration, though the article was really a review of Professor Lloyd-Morgan's " Habit and Instinct ; " an article on the question whether "White men can work in the Tropics," which most English writers declare to be impossible without thinking it necessary to adduce evidence, but which, I affirm, is proved by experience to be quite easy. Both these are reprinted in my " Studies," as is also a short essay on " The Causes of War and the Remedies," written for *L'Humanité Nouvelle*. I also wrote letters to the *Daily Chronicle* on America, Cuba, and the Philippines ; and a protest against the Transvaal War in the *Manchester Guardian*.

In the year 1900 I wrote an article for the *New York Journal* on "Social Evolution in the Twentieth Century— An Anticipation," for which I received a very complimentary letter from the editor. During the next two years I was engaged in preparing new editions of my books on " Darwinism " and " Island Life," and I also wrote several letters on political and social subjects, such as an " Appreciation of the Past

Century" (in 1901, in the *Morning Leader*), and (in 1903) an article on "Anticipations and Hopes for the Immediate Future," which was written for a German paper (the *Berliner Local Anzieger*), but which was too plain-spoken for the editor to publish, and which I accordingly sent to the *Clarion*. As it gives my latest views, expressed in the plainest words, on some of the most important problems of the day, I give it here for the consideration of a wider circle of readers.

ANTICIPATIONS AND HOPES FOR THE IMMEDIATE FUTURE.

I am looking to the coming year with no expectation of any great change, political or social, but with a hope and belief that the great movement among the workers in favour of a more rational and more equitable system of government, and of social organization, will continue to grow as it has been growing during the last few years. I trust that, in the more advanced countries—especially in Germany and France—it may become sufficiently powerful, even within the coming year, to exercise a decided control over the reactionary party, and even be able to initiate, and perhaps to secure, some important legislation for the extension of individual freedom, and for checking military expenditure.

As to the future (limiting ourselves here to the twentieth century), I look forward to the same movement as destined to produce great and beneficent results.

The events of the past few years must have convinced all advanced thinkers that it is hopeless to expect any real improvement from the existing governments of the great civilized nations, supported and controlled as they are by the ever-increasing power of vast military and official organizations.

These organizations are a permanent menace to liberty, to national morality, and to all real progress towards a rational social evolution. It is these which have given us during the first years of this new century examples of national hypocrisy and crimes against liberty and humanity —to say nothing of Christianity—almost unequalled in the whole course of modern history.

Scarcely was the ink dry of the signatures of their representatives at the Hague Conference, where they had expressed the most humane and elevated ideas as to the necessity for reduction of armaments, for the amelioration of the horrors of war, and for the principle of arbitration in the settlement of national difficulties, than we find all the chief signatories engaged in destroying the liberties of weaker peoples, without any rational cause, and often in opposition to the principles of their own constitutions, or to solemn promises by their representatives, or in actual treaties.

England carried fire and sword into South Africa, and has robbed two

Republics of the independence guaranteed to them after a former unjust annexation ; a crime aggravated by hypocrisy in the pretence that British subjects were treated as " helots ; " whereas their own committee of inquiry into the war has now demonstrated that it was a pure war of conquest in order to secure territory and gold-mines, determined on years before, and only waiting a favourable opportunity to carry into effect.

The United States, against their own " Declaration of Independence " and the fundamental principles of their constitution, have taken away the liberties of two communities, the one—Porto Rico—by mere overwhelming power, the other—the Philippinos—after a bloody war against a people fighting for their independence, the only excuse being that they had been purchased—land and people—from their former conquerors and oppressors.

Russia itself, the originator of the Peace Conference, forthwith persecutes Jews and Doukhobors on account of their religion, and takes away their solemnly guaranteed liberties from the Finns—a people more really civilized than their persecutors.

All three of these governments, as well as Germany and France, invaded China, and committed barbarities of slaughter, with reckless devastation and plunder, which will degrade them for all time in the pages of history.

Such are the doings of the official and military rulers of nations which claim to be in the first rank of civilization and religion ! And there is really no sign of any improvement. But, for the first time in the history of the world, the workers—the real sources of all wealth and of all civilization—are becoming educated, are organizing themselves, and are obtaining a voice in municipal and national governments. So soon as they realize their power and can agree upon their aims, the dawn of the new era will have begun.

The first thing for them to do is, to strengthen themselves by unity of action, and then to weaken and ultimately to abolish militarism. The second aim should be to limit the bureaucracy, and make it the people's servant instead of its master. The third, to reorganize and simplify the entire legal profession, and the whole system of law, criminal and civil ; to make justice free for all, to abolish all legal recovery of debts, and all advocacy paid by the parties concerned. The fourth, and greatest of all, will be to organize labour, to abolish inheritance, and thus give equality of opportunity to every one alike. This alone will establish, first, true individualism (which cannot exist under present social conditions), and this being obtained, will inevitably lead to voluntary association for all the purposes of life, and bring about a social state adapted to the stage of development of each nation, and of each successive age.

This, in my opinion, is the ideal which the workers (manual and intellectual workers alike) of every civilized country should keep in view. For the first time in human history, these workers are throwing aside international jealousies and hatreds ; the peoples of all nations are becoming brothers, and are appreciating the good qualities inherent in each and all of them. They will therefore be guilty of folly as well as crime if they much longer permit their rulers to drill them into armies, and force them to invade, and rob, and kill each other.

The people are always better than their rulers. But the rulers have power, wealth, tradition, and the insatiable love of conquest and of governing others against their will. It is, then, in the People alone that I have any hope for the future of Humanity.

ALFRED R. WALLACE.

In 1904 I wrote a short letter on the "Inefficiency of Strikes" for the *Labour Annual,* and a rather long one to the *Clarion,* suggesting a policy for socialists in opposition to continued military expenditure as advocated by Robert Blatchford ; but this was, I fear, too much advanced even for the readers of this very advanced paper, since no one came forward in my support. I feel sure, however, that there are many who, when it is clearly put before them, will approve of the policy I have sketched out, since it is merely one of justice and consideration for nations as well as for individuals —of adopting the same rules of right and wrong in the one case as in the other. The letter may be termed—

A SUBSTITUTE FOR MILITARISM.

I will first say a few words on the, to me, extraordinary statement that, though fifty years of continuously increasing expenditure on our national defences has resulted in " an inadequate and imperfect " outcome, and what a military writer in the July *Nineteenth Century* called " our pitiable military situation," yet, only give to our rulers unlimited money and conscription, and our defences will instantly become "adequate and efficient." With all respect, this seems to me nothing less than pure delusion. One Government after another has had a free hand to reform our military and naval forces, and all have utterly failed. They have wasted countless millions with no adequate result. And now we are asked to give them more millions to waste, and the very same body of official rulers and organizers and titled officers will suddenly be imbued with wisdom, unselfishness, and economy, and all will be well. Our defences, as by a miracle, will become " adequate and efficient." For what has to be done must be done at once. Germany, we are told, is ready ; we are not. Therefore the money and the men must be given to the Government *now*. To any such proposal I venture to hope that, by an overwhelming majority, the Socialist and Labour parties will reply in the now historic words : " Never again."

But this is only preliminary. We will now come to the real issue.

Robert Blatchford proceeds to ask a number of questions, and to offer a
number of alternatives, as if they were exhaustive and there was nothing
more to be said or done. Shall we leave the Empire defenceless? Shall
we abandon our country and our colonies to the invasion of any power
that cared to take them? Russia covets India. We must either defend
India or surrender it to Russia. If we made India a self-governing
nation, the result would be civil war and a Russian conquest. More than
one foreign power envies us our possessions. And so on, and so on; with
the one conclusion : We must increase army, navy, and home defences,
and be prepared to fight all the world. Not one word about there being
any alternative to all this blood-and-iron bluster and defiance; not one
syllable to show that the writer is a great Socialist teacher, a believer in
the goodness of human nature and the brotherhood of man. "But," he
replies by his heading, "this is very good in theory, and very true, but it
is not practical politics. The danger is urgent. Tell us, ye Labour
leaders, what you propose to do *now* ?"

I am not a Labour leader, but I hope I am a true friend of Labour
and a true Socialist ; and I will now state the case as it appears to me,
and suggest what, in my opinion, is the only course of action worthy of
Socialism or politic for Labour, and, besides, the only course which has
the slightest chance of succeeding *in the long run :* in one word, the only
RIGHT course.

It is a notorious and undeniable fact that we—that is, our Governments—are, with few exceptions, hated and feared by almost all other
Governments, especially those of the Great Powers. Is there no cause for
this? Surely we know there is ample cause. We have either annexed
or conquered a larger portion of the world than any other Power. We
have long claimed the sovereignty of the sea. We hold islands and forts
and small territories offensively near the territories of other Powers. We
still continue grabbing all we can. In disputes with the powerful we
often give way ; with the weak and helpless, or those we think so, we are
—allowing for advance in civilization—bloody, bold, and ruthless as any
conqueror of the Middle Ages. And with it all we are sanctimonious.
We profess religion. We claim to be more moral than other nations,
and to conquer, and govern, and tax, and plunder weaker peoples for
their good ! While robbing them we actually claim to be benefactors !
And then we wonder, or profess to wonder, why other Governments hate
us ! Are they not fully justified in hating us ? Is it surprising that they
seek every means to annoy us, that they struggle to get navies to compete with us, and look forward to a time when some two or three of them
may combine together and thoroughly humble and cripple us ? And who
can deny that any just being, looking at all the nations of the earth with
impartiality and thorough knowledge, would decide that we deserve to be
humbled, and that it might do us good?

Now the course I recommend as the only true one is, openly and
honestly, without compulsion and without vainglory, to do away with
many of the offences to other peoples, and to treat all subject peoples and

all foreign Powers on exactly the same principles of equity, of morality, and of sympathy, as we treat our friends, acquaintances, and neighbours with whom we wish to live on friendly terms.

And, to begin with, and to show that our intentions are genuine, I would propose to evacuate Gibraltar, dismantle the fortress, and give it over to Spain ; Crete and Cyprus should be free to join Greece ; Malta, in like manner, would be given the choice of absolute self-government under the protection of Britain, or union with Italy. But the effect of these would be as nothing compared with our giving absolute internal self-government to Ireland, with protection from attack by any foreign Power ; and the same to the Transvaal and Orange Free State ; and this last we should do " in sackcloth and ashes," with full acknowledgment of our heinous offences against liberty and our plighted word.

Now we come to India, which our friend Blatchford seems to consider the test case. And so it is ; for if ever there was an example of a just punishment for evil deeds, it is in the fact that, after a century of absolute power, we are still no nearer peace and plenty and rational self-government in India than we were half a century ago, when we took over the government from the " Company " with the promise to introduce home-rule as soon as possible. And now we have a country in which plague and famine are chronic—a country which we rule and plunder for the benefit of our aristocracy and wealthy classes, and which we are therefore in continual dread of losing to Russia.

If we had honestly kept our word, if we had ruled India with the one purpose of benefiting its people, had introduced home-rule throughout its numerous provinces, states, and nations, settling disputes between them, and guarding them from all foreign attack, we should by now have won the hearts of its teeming populations, and no foreign Power would have ventured to invade a group of nations so united and so protected. Such a position as we might have now held in India—that of the adviser, the reconciler, and the powerful protector of a federation of self-governing Native States—would be a position of dignity and true glory very far above anything we can claim to-day.

But, it will be replied, all this is foolish talk ; it will be a century before the British people will be persuaded to give up its possessions and its power ; and, in the mean time, if we do not defend ourselves we shall not have the opportunity of being so generous, hardly shall we keep our own liberties. I have not so low an opinion of my countrymen as to believe that they really *wish* to keep other peoples subject to them against their will ; that they are really *determined* to go on denying that freedom to others which is so dear a possession to themselves. And if there is not now a majority who would agree to act at once as I suggest, I am pretty confident that there is, even now, a majority who would acknowledge that such action is theoretically just, and that they would be willing to do it by degrees, and as soon as it is safe, to look forward to it, in fact, as an ideal to be realized at some future time.

Now, what I wish to urge is, that it is of the most vital importance to us, now, that all who agree with me that there can be no national honour or glory apart from justice and mercy, and that to take away people's liberty and force our rule upon them against their will is the greatest of all national crimes, should take every opportunity of making their voices heard. If, for instance, every Socialist in our land, and I hope a very large proportion of workers and advanced thinkers who may not be Socialists, would agree to maintain this as one of their fundamental principles, to be continually brought before the people through the Press and on the platform, to be urged on the Government at every opportunity, and to be made a condition of our support of every advanced Parliamentary candidate, we should create a body of ethical opinion and feeling that would not only be of the highest educational value at home, but which would influence the whole world in their estimate of us. It would show them that though our Government is bad—as all Governments are—yet the people at heart are honest and true, and that it will not be very long before the people will force their Governments to be honest also.

This, I submit, *would* be really " practical politics." At the present day we *have* got so far as this—that none of the Great Powers wages a war of aggression and conquest against another Power without some quarrel or some colourable pretence of injury. But surely the fact of there being such a party as I have outlined, and especially if it would (as I think it certainly could) compel the next Government to make some of the smaller concessions here indicated and adopt the general principle of respecting the liberties of even the smallest nationalities, would so reduce the amount of envy and hatred with which we are now regarded as to considerably diminish the danger of combined aggression upon us.

I should have liked to say something about Russia, and the fact that *we* are answerable for the present war in the Far East, by so long upholding Turkey, and preventing Russia from acquiring free egress into the Mediterranean, in exchange for which concession she would (after the Russo-Turkish War) have willingly agreed to the neutralizing of Constantinople as a free port under the guarantee of the Powers. We had at that time a preponderance of power in Europe, as shown by what occurred at the Peace Congress ; but Lord Beaconsfield used that power for a bad purpose, as Lord Salisbury afterwards admitted.

I greatly regret being obliged to differ so radically from a man I admire and respect so much as I do Robert Blatchford ; but, as I am known to be a Socialist and a constant reader of the *Clarion*, it might be thought that my silence would imply some degree of agreement. The present letter is merely for the purpose of making my views clear on this vitally important question, and with the hope that others who agree with me will not longer keep silence.

ALFRED R. WALLACE.

About the year 1899 our house at Parkstone became no longer suitable owing to the fact that building had been going on all around us and what had been pretty open country when we came there had become streets of villas, and in every direction we had to walk a mile or more to get into any open country. I therefore began to search about various parts of the southern counties for a suitable house, and as this was almost impossible to obtain, I endeavoured to induce a sufficient number of friends to join together to buy a small estate which we could divide between us, so as to secure the benefit of pleasant society and picturesque sur-roundings—to create, in fact, a kind of very limited garden-city, or rather garden settlement. With one or two friends interested in the project, I spent a good deal of time examin-ing estates within thirty or forty miles of London, but though we found several that were in most ways suitable, it was found impossible to find any that exactly fulfilled the re-quirements of the parties most interested or to raise the necessary funds for the purchase. We then returned to the search for a house or land for ourselves, and after almost giving up the attempt in despair, we accidentally found a spot within four miles of our Parkstone home and about half a mile from a station, with such a charming distant view and pleasant surroundings that we determined, if we could get two or three acres at a moderate price, to build a small house upon it.

After a rather long negotiation I obtained three acres of land, partly wood, at the end of the year 1901 ; sold my cottage at Godalming at a fair price, began at once making a new garden and shrubbery, decided on plans, and began building early in the new year. The main charm of the site was a small neglected orchard with old much-gnarled apple, pear, and plum trees, in a little grassy hollow sloping to the south-east, with a view over moors and fields towards Poole harbour, beyond which were the Purbeck hills to the right, and a glimpse of the open sea to the left. In the foreground were clumps of gorse and broom, with some old picturesque trees, while the orchard was sheltered on both sides by patches of woodland. The house was nearly finished in about a year, and we got into it at Christmas, 1902, when we decided to call it Old Orchard.

Being so near to our former house, I was able to bring all our choicer plants to the new ground, and there was, fortu-nately, a sale of the whole stock of a small nursery near Poole in the winter, at which I bought about a thousand shrubs and trees at very low prices, which enabled us at once to plant some shrubberies and flower borders, and thus to secure something like a well-stocked garden by the time we got into the house. Since that time it has been an ever-increas-ing pleasure, and I have been able to satisfy my craving for enjoying new forms of plant-life every year, partly by raising numbers of seeds of hardy and greenhouse plants, always trying some of the latter in sheltered places out-of-doors, and partly by exchanges or by gifts from friends, so that every year I have the great pleasure of watching the opening of some of nature's gems which were altogether new to me, or of others which increase year by year in beauty. In one end of my greenhouse I have a large warmed tank in which I grow blue, pink, and yellow water-lilies, which flower the greater part of the year, as well as a few other beautiful or curious aquatic plants, while the back wall of the house is covered with choice climbers.

In this hasty sketch of my occupations and literary work during the last nine years, I have purposely omitted the more important portion of the latter, because the circumstances that led me on to undertake three separate works, involving a considerable amount of labour, were very curious, and to me very suggestive, and I will now give a connected account of them.

When in 1896 I was invited by Dr. Lunn to give a lecture to his friends at Davos, I firmly believed that my scientific and literary work was concluded.   I had been for some years in weak health, and had no expectation of living much longer. Shortly after returning from America I had another very severe attack of asthma in 1890, and a year or two after it recurred and became chronic, together with violent palpita-tions on the least sudden exertion, and frequent colds almost invariably followed by bronchitis. Any attempt at continuous work was therefore very far from my thoughts, though at times I was able to do a fair amount of writing. My friend and neighbour, Professor Allman, had suffered from the same affliction during a large part of his life, and only found very partial relief from it by the usual fumigations and cigarettes, with occasional changes of air, and it was often quite painful to witness his sufferings, which continued till his death in 1898. As he was himself a medical man, and had had the best advice attainable, I had little hope of anything but a continuance and probably an increase of the disease.

But the very next year I obtained relief (and up to the present time an almost complete cure) in an altogether accidental way, if there are any " accidents " in our lives.   Mr. A. Bruce-Joy, the well-known sculptor (a perfect stranger to me), had called on me to complete the modelling of a medallion which he had begun from photographs, and I apologized for not looking well, as I was then suffering from one of my frequent spells of asthma, which often prevented me from getting any sleep at night.   He thereupon told me that if I would follow his directions I could soon cure myself.   Of course, I was altogether incredulous; but when he told me that he had himself been cured of a complication of allied diseases—gout, rheumatism, and bronchitis—of many years' standing, which no English doctors were able to even alleviate, by an American physician, Dr. Salisbury; that it was effected solely by a change of diet, and that it was no theory or empirical treatment, but the result of thirty years' experiment on the effects of various articles of diet upon men and animals, by the only scientific method of studying each food separately and exclusively, I determined to try it.   The result was, that in a week I felt much better, in a month I felt quite well, and during the six years that have elapsed no attack of asthma or of severe palpitation has recurred, and I have been able to do my literary work as well as before I became subject to the malady.

I may say that I have long been, and am still, *in principle*, a vegetarian, and believe that, for many reasons, it will certainly be the diet of the future.   But for want of adequate knowledge, and even more from the deficiencies of ordinary vegetable cookery, it often produces bad effects. Dr. Salisbury proved by experiment that it was the consumption of too much starch foods that produces the set of diseases which he especially cures; and when these diseases have become chronic, the only cure is the almost complete abstention from starchy substances, especially potatoes, bread, and most watery vegetables, and, in place of them, to substitute the most easily digestible well-cooked meat, with fruits and nuts in moderation, and eggs, milk, etc., whenever they can be digested. Great sufferers find immediate relief from an exclusive diet of the lean of beef. I myself live upon well-cooked beef with a fair proportion of fat (which I can digest easily), a very small proportion of bread or vegetables, fruit, eggs, and light milk puddings. The curious thing is that most English doctors declare that a meat diet is to be avoided in all these diseases, and many order complete abstinence from meat, but, so far as I can learn, on no really scientific grounds. Dr. Salisbury, however, has experimentally proved that this class of ailments are all due to malnutrition, and that this malnutrition is most frequently caused by the consumption of too much of starch foods at all meals, which overload the stomach and prevent proper digestion and assimilation. My case and that of Mr. Bruce-Joy certainly show that Dr. Salisbury has found, for the first time in the history of medicine, a *cure*—not merely an *alleviation*—for these painful and distressing maladies. This personal detail as to my health is, I think, of general interest in view of the large number of sufferers who are pronounced incurable by English doctors, and it was here an essential preliminary to the facts I have now to relate, which would probably not have occurred as they did had my health not been so strikingly renovated.[1]

[1] In addition to the foregoing, I have suffered at intervals from diseases contracted abroad, which have recurred in acute paroxysms, and sometimes threatened to become serious. For years together they have given me much anxiety and required constant care and attention. Since my general health has improved, however, they have so much diminished as no longer to give me much trouble. I have also suffered twice from severe eye troubles. My sight has always been

The lecture which I gave at Davos on the science of the nineteenth century (a subject suggested by Dr. Lunn) led me to think that an instructive and popular book might be made of the subject, as I found there were so many interesting points I could not treat adequately or even refer to in a lecture. I therefore devoted most of my spare time during the next year to getting together materials and writing the volume, which I finished in the spring of 1898, and it was published in June. The work had a pretty good sale, and at the request of my publishers I prepared from it a School Reader, with a considerable number of illustrations, which was published in 1901. This suggested the idea of a much enlarged and illustrated edition of the original work, which was, as regards many of the more important sciences and arts, a mere outline sketch. Almost all the year 1902 and part of 1903 was occupied in getting together materials for this new work, as it really was, and it was not published till the autumn of the latter year.

But while I was writing three new chapters on the wonderful astronomical progress of the latter half of the century, the startling fact was impressed upon me that we were situated very nearly at the centre of the entire stellar universe. This fact, though it had been noted by many of the greatest astronomical writers, together with the indications

myopic, though otherwise strong, but in 1883 I did a great deal of work at night, requiring a continual reference to several books of different-sized print, and this brought on rather severe inflammation of the retina, which necessitated a darkened room for some weeks, and no reading or writing for several months—a tremendous trial to me, so that I was able to do no literary work in 1884. The occulist I consulted told me that with care in two or three years my eyes would be as strong as ever ; and they very gradually became so, and I had no further trouble till 1891, when some irritating substance got into my left eye and could not be got out, causing severe inflammation for some weeks, which, however, passed away without immediate bad results. From that time, however, there began a loss of the power of adjustment of the two eyes, so that I saw distant objects double, and this has increased so that I now see everything double, even at the other side of a room ; but this does not much inconvenience me except to produce a general indistinctness of objects. Two persons walking together on the other side of the street seem to me to be three or four persons, according to the angle of sight, and I often have to shut one eye in order to be sure how many there are. The divergence has now, I think, got to the worst, as I perceive no difference during the last few years.

that led to the conclusion that our universe was finite, and that we could almost, if not quite, see to its very limits, were seldom commented on as more than isolated phenomena —curiosities, as it were, of star distribution—but of no special significance. To me, however, it seemed that they probably *had* a meaning ; and when I further came to examine the numerous facts which led to the conclusion that no other planet in the solar system than our earth was habitable, there flashed upon me the idea that it was only near the centre of this vast material universe that conditions prevailed rendering the development of life, culminating in man, possible. I did not, however, dwell upon this idea, but merely suggested it in a single paragraph on pp. 329–330 of my work, and I might probably never have pursued the subject further but for another circumstance which kept my attention fixed upon it.

While I was still hard at work upon the book, the London agent of the New York *Independent* wrote to ask me to write them an article on any scientific subject I chose. I at first declined, as having no subject which I thought suitable, and not wishing to interrupt my work. But when he urged me again, and told me to name my own fee, the idea struck me that these astronomical facts, with the conclusion to which they seemed to me to point, might form a very interesting, and even new and attractive article. As the subject was fresh in my mind, and I had the authorities at hand, it did not take me very long to sketch out and write a paper of the required length, which appeared simultaneously in the *Independent* and in the *Fortnightly Review*, and, to my great surprise, created quite a sensation, and, still more to my surprise, a considerable amount of antagonism and rather contemptuous criticism by astronomers and physicists, to which I replied in a subsequent article.

But as soon as my agent, Mr. Curtis Brown, read the MSS. he suggested that I should write a volume on the subject, which he was sure would be very attractive and popular, and for which he undertook to make arrangements both in England and America, and secure me liberal terms. After a little consideration I thought I could do so, and terms were arranged for the book before the article itself was published. This enabled me to get together all the necessary materials and to begin work at once, and after six months of the stiffest reading and study I ever undertook, the book was completed in September, and published in November of the same year. In November of 1904 a cheaper edition was published, with an additional chapter in an Appendix. This chapter contained an entirely new argument, founded on the theory of organic evolution, which I had not time to introduce into the first edition. This argument is itself so powerful that, when compounded with the arguments founded on astronomical, physical, and physiological phenomena, it renders the improbability of there having been two independent developments of organic life culminating in man, so great as to be absolutely inconceivable.

The success of this volume, and the entirely new circle of readers it brought me, caused my publishers to urge me to prepare the present work, which I should otherwise have not written at all, or only on a very much smaller scale for the information of my family as to my early life.

Now it seems to me a very suggestive fact that my literary work during the last ten years should have been so completely determined by two circumstances which must be considered, in the ordinary sense of the term, and in relation to my own volition, matters of chance. If Dr. Lunn had not invited me to Davos, and if he had suggested "Darwinism" or any other of my special subjects instead of the "Science of the Nineteenth Century," I should not have written my "Wonderful Century;" I should not have had my attention so specially directed to great astronomical problems; I should not, when asked for an article, have chosen the subject of our sun's central position; and I should certainly never have undertaken such a piece of work as my book on "Man's Place in the Universe," or the present autobiography. And further, without the accident of a perfect stranger calling upon me for reasons of his own, and that stranger happening to be a man who had been so marvellously cured by Dr.

Salisbury as to induce me to adopt the same treatment, with similar results, I should never have had the energy required to undertake the two later and more important works. Of course, it may be that these are only examples of those "happy chances" which are not uncommon in men's lives; but, on the other hand, it may be true that, "there's a divinity that shapes our ends, rough-hew them as we will;" and those who have reason to know that spiritual beings can and do influence our thoughts and actions, will see in such directive incidents as these examples of such influence.

Although I have now brought the narrative of my literary and home life up to the time of writing this Autobiography, there are a number of special subjects, which, for the sake of clearness, I have either wholly omitted, or only just mentioned, but which have either formed important episodes in my life, or have brought me into communication or friendly intercourse with a number of interesting people, and which therefore require to be narrated consecutively in separate chapters. These will now follow, and will, I think, be not the least interesting or instructive portions of my work.

NOTE.

The ADDENDUM at the end of this volume should have followed here, and had better be read before the remaining chapters.—A. R. W.

CHAPTER XXXIV

LAND NATIONALIZATION TO SOCIALISM, AND THE FRIENDS THEY BROUGHT ME

SOON after I returned from the Amazon (about 1853), I read Herbert Spencer's " Social Statics," a work for which I had a great admiration, and which seemed to me so important in relation to political and social reform, that I thought of inviting a few friends to read and discuss it at weekly meetings. This fell through for want of support, but the whole work, and more especially the chapter on " The Right to the Use of the Earth," made a permanent impression on me, and ultimately led to my becoming, almost against my will, President of the Land Nationalization Society, which has now been just a quarter of a century in existence. In connection with this movement, I have made the acquaintance of a considerable number of persons of more or less eminence, and my relations with some of these will form the subject of the present chapter.

The publication of my " Malay Archipelago " in 1869, procured me the acquaintance of John Stuart Mill, who on reading the concluding pages, in which I condemn our " civilization " as but a form of " barbarism," and refer, among other examples, to our permitting private property in land, wrote to me from Avignon on May 19, 1870, enclosing the programme of his proposed Land Tenure Reform Association, and asking me to become a member of the General Committee. Its object was to claim the future " unearned increment " of land values for the State, to which purpose it was to be strictly limited. I accepted the offer, but proposed a new clause, giving the State power of resuming possession of any land on payment of its net value at the time, because, as I pointed out, the greatest evil was the *monopoly* of land, not the money lost by the community. This he himself supported, but suggested giving not the current value only, but something additional as compensation; and I think this was done. Later I proposed another addition to the programme, which he also agreed to, as shown by the following extract from a letter I received from him in July, referring to a general meeting of the Association at the Freemason's Tavern :—

"I hope that you will be able to attend, and that you will propose, as an addition to the programme, the important point which you suggested in your letter to me, viz., the right of the State to take possession (with a view to their preservation) of all natural objects or artificial constructions which are of historical or artistic interest. If you will propose this I will support it, and I think there will be no difficulty in getting it put into the programme, where undoubtedly I think it ought to be."

He then asked me to dine with him at Blackheath Park on the following Sunday at five o'clock, which I of course accepted. The only other persons present were his step-daughter Miss Helen Taylor, Mr. George Grote the historian, and the Hon. Auberon Herbert. We had a very pleasant dinner and some very interesting and instructive conversation afterwards, only one portion of which I recollect, as it referred to a subject on which I differed from Mill, and thought his views, for such an undoubtedly great and clear thinker, somewhat hasty and ill-considered. The conversation turned somehow upon the existence and nature of God. Mr. Grote seemed inclined to accept the ordinary idea of an eternal omniscient and benevolent existence, because anything else was almost unthinkable. To which Mill replied, that whoever considered the folly, misery, and badness of the bulk of mankind, such a belief was unthinkable, because it would imply that God *could* have made man good and happy, have abolished evil, and has not done so. I ventured to suggest that what we call evil may be essential to the ultimate development of the highest good for all; but he would not listen to it or argue the question at all, but repeated, dogmatically, that an omnipotent God might have made man wise, good, and happy, and as He had not chosen to do so it was absurd for us to believe in such a being and call *Him* almighty and good. He then turned the conversation as if he did not wish to discuss the matter further.

There is one point in connection with this problem which I do not think has ever been much considered or discussed. It is, the undoubted benefit to all the members of a society of *the greatest possible diversity of character*, as a means both towards the greatest enjoyment and interest of association, and to the highest ultimate development of the race. If we are to suppose that man might have been created or developed with none of those extremes of character which now often result in what we call wickedness, vice, or crime, there would certainly have been a greater monotony in human nature which would, perhaps, have led to less beneficial results than the variety which actually exists may lead to. We are more and more getting to see that very much, perhaps all, the vice, crime, and misery that exists in the world is the result, not of the wickedness of individuals, but of the entire absence of sympathetic training from infancy onwards. So far as I have heard, the only example of the effects of such a training on a large scale, was that initiated by Robert Owen at New Lanark, which, with most unpromising materials, produced such marvellous results on the character and conduct of the children, as to seem almost incredible to the numerous persons who came to see and often critically to examine them. There must have been all kinds of characters in his schools, yet *none* were found to be incorrigible, *none* beyond control, *none* who did not respond to the love and sympathetic instruction of their teachers. It is therefore quite possible that *all* the evil in the world is directly due to man, not to God, and that when we once realize this to its full extent we shall be able, not only to eliminate almost completely what we now term evil, but shall then clearly perceive that all those propensities and passions that under bad conditions of society inevitably led to it, will under good conditions add to the variety and the capacities of human nature, the enjoyment of life by all, and at the same time greatly increase the possibilities of development of the whole race. I myself feel confident that this is really the case, and that such considerations, when followed out to their ultimate issues, afford a complete solution of the great problem of the ages—the origin of evil.

The last letter I had from Mill was in April, 1871, when a great public meeting of the Association was to be held on May 3, as to which he said, " It would be very useful to the Association, and a great pleasure to myself, if you would consent to be one of the speakers at the meeting. There is the more reason why you should do so, as you are the author of one very valuable article of the programme. Were you to explain and defend that article, it would be a service which no one is so well qualified to render as yourself." I had then recently visited the stone circles and bridges of Dartmoor, and also Stonehenge, and urged the importance of preserving them. At that time there would probably have been no question of paying more than the actual selling value of the land, and we should have been spared the disgrace of having our grandest ancient monument, after centuries of neglect and deterioration, claimed to be private property, and having an exorbitant price demanded for it. But Mill's death soon afterwards put an end to the association, and we had to wait many years for the present very imperfect legislation on the subject.

The question of land nationalization continued at intervals to occupy my mind, but having become strongly impressed by the teachings of Spencer, Mill, and other writers as to the necessity for restricting rather than extending State agency, and by their constant reference to the inevitable jobbery and favouritism that would result from placing the management of the whole land of the country in the hands of the executive, that I did not attempt to write further upon the subject. But when the topic of Irish landlordism became very prominent in the year 1879–80, an idea occurred to me which seemed to entirely obviate all the practical difficulties which were constantly adduced as insuperable, and I at once took the opportunity of the controversy on the question to set forth my views in some detail. I did this especially because the Irish Land League proposed that the Government should buy out the Irish landlords, and convert their existing tenants into peasant-proprietors, who were to redeem their holdings by payments extending over thirty-five years. This seemed to me to be unsound in principle, and entirely useless except as a temporary expedient, since it would leave the whole land of Ireland in the possession of a privileged class, and would thus disinherit all the rest of the population from their native soil.

In my essay I based my whole argument upon a great principle of equity as regards the right of succession to landed property, a principle which I have since further extended to all property.[1] But the suggestion which rendered land nationalization practicable was, that while, under certain conditions stated, all land would gradually revert to the State, what is termed in Ireland the *tenant-right*, and in England the *improvements*, or increased value given to the land by the owner or his predecessors, such as buildings, drains, plantations, etc., would remain his property, and be paid for by the new state-tenants at a fair valuation. The selling value of land was thus divided into two parts: the *inherent value* or ground-rent value, which is quite independent of any expenditure by owners, but is due solely to nature and society; and the *improvements*, which are due solely to expenditure by the owners or occupiers, and which are essentially temporary in nature. My experience in surveying and land-valuation assured me that these two values can be easily separated. It follows that land as owned by the State would need no "management" whatever, the rent being merely a ground-rent, which could be collected

[1] See my "Studies, Scientific and Social," vol. ii. chap. xxviii.

just as the house-tax and the land-tax are collected, the state-
tenant being left as completely free as is the "freeholder"
now (who is in law a state-tenant), or as are the holders of
perpetual feus in Scotland.

This article appeared in the *Contemporary Review* of
November, 1880, and it immediately attracted the attention
of Mr. A. C. Swinton, Dr. G. B. Clark, Mr. Roland Estcourt,
and a few others, who had long been seeking a mode of
applying Herbert Spencer's great principle of the inequity of
private property in land, and who found it in the suggestions
and principles I had laid down. They accordingly communicated with me; several meetings were held at the
invitation of Mr. Swinton, who was the initiator of the
movement, and after much discussion as to a definite programme, the "Land Nationalization Society" was formed,
and, much against my wishes, I was chosen to be president.
Notwithstanding the scanty means of the majority of the
founders and members, the society has struggled on for a
quarter of a century. Its lecturers and its yellow vans have
pervaded the country, and it has effected the great work
of convincing the highest and best-organized among the
manual workers as represented by their Trades Unions, that
the abolition of land-monopoly, which is the necessary result
of its private ownership, is at the very root of all social
reform. Hence the future is with them and us, and though
the capitalists and the official Liberals are still against us, we
wait patiently, and continue to educate the masses in the
certainty of a future and not distant success.

Although Herbert Spencer was the first eminent Englishman of science to establish the doctrine of land nationalization
upon the firm basis of social justice, he had several fore-
runners who saw the principle as clearly as he did, declared
it as boldly, but, being far in advance of their age, were
treated with scorn, persecution, or neglect. The earliest was
Thomas Spence, a poor schoolmaster of Newcastle-on-Tyne,
who in 1775 delivered a lecture before the Philosophical
Society of that town, for which he was immediately expelled

from the society, and soon after obliged to leave the town. This lecture was reprinted by Mr. H. M. Hyndham in 1882, and a single sentence will indicate its scope and purpose:—

" Hence it is plain that the land or earth, in any country or neighbourhood, belongs at all times to the living inhabitants of the said country or neighbourhood in an equal manner. For, as I said before, there is no living but on land and its productions, consequently what we cannot live without we have the same property in as our lives."

Spence further opposed centralized government as much as any individualist of our day, and advocated a system of free communal home-rule, every parish owning its own land and managing its own affairs.

A few years later, in 1782, Professor Ogilvie published anonymously, " An Essay on the Right of Property in Land " (a volume of 120 pages). He lays down the principle that no right to property in land can justly arise except through occupancy and labour upon it, and even this must be limited by the equal rights of every other individual. And after discussing the various laws and circumstances of modern civilized communities, he shows how the laws can be amended so as to bring about a just distribution of land. This is a thoughtful, well-reasoned, and clearly written work, yet it remained almost unknown to successive generations of reformers.

A few years later than H. Spencer (in 1856), but apparently quite independently of him, a very remarkable work was published in London, under the title "On the Evils, Impolicy, and Anomaly of Individuals being Landlords and Nations Tenants," by Robert Dick, M.D. This was a very comprehensive work, anticipating the main thesis of Henry George, as shown by the following passage from the introductory chapter: " My design, in short, is to show that wealth, accumulated in individuals and classes, necessarily implies poverty elsewhere, in like manner as exemption from labour by some men and classes, of necessity implies double, treble, quadruple labour in others." He then lays down a number of fundamental propositions, which are so brief, clear, and forcible, and go so directly to the root of all those social problems which demand solution to-day even more peremptorily than they did a century ago, that I will give the more important of them here—

"*Prop. I.* The use of earth in the form of food is equally necessary for human life as the use of air—the privation of one kills in a few minutes, of the other in a few days or weeks."

"*Prop. II.* Hence the man who controls land, controls human life—excluding life that might be, holding at his mercy life that is."

"*Prop. III.* As God made a free gift to each man of life, He equally intended for a free gift the necessary condition of life—a portion of the soil."

"*Prop. V.* Hence a portion of the soil is each man's congenital and inalienable patrimony."

"*Prop. VII.* The nationalizing of the soil should have been the primary, the fundamental step in human association."

"*Prop. X.* The culture of a portion of the soil (as a man's own) has this advantage over all other labour, that it gives him directly, and at first cost, those very necessaries which he is obliged, indirectly, to seek, in manufactures, trade, and commerce, namely, *home, food, fuel,* etc., all which must otherwise be purchased at more than natural cost in labour or money."

"*Prop. XIII.* It is out of the pauper and floating masses who have been separated from the land, and have consequently no option between starvation and selling their labour unconditionally, that capital is originally formed, and is, thereafter, enabled absolutely to dictate to the very labour that creates it, and to defraud that labour of those surplusses which ought to remain wholly with the latter."

In these two last propositions is comprised the whole philosophy of social reform, the last anticipating the main thesis of Marx. And to show how well this fine writer and thinker appreciated the more human, esthetic, and ennobling aspects of the question, I will give one more short quotation, on the overcrowding and housing questions, still *talked* about

as if they could be remedied piecemeal while the fundamental cause of these and a thousand other evils remains
untouched—

"Is it to be credited that this crowding together of men
in houses dovetailed into each other, with everything of
nature—winds, flowers, verdure, the healthy smell of earth—
shut out and replaced by a thousand miasmas—is it, I say,
to be credited that this is the normal condition of beings
born with natural cravings for activity and pure air, with an
intelligent eye for nature's manifold beauties, with bodies
requiring to be exercised no less than heads ? The very
necessity for drains tells against us. All manure was meant
directly to nourish the land it accumulates on—not to pollute
our streams and rivers. Cities as they now are, are the
abscesses of nature. The soil and terrestrial space are not
meant for the rearing of food only, but to be dwelt and
moved about on—to be daily enjoyed in all the variety of
agreeable sights, sounds, and odours they afford us."

This important, thoughtful, and suggestive work, published
in our own time, and dealing most thoroughly with the ever-
growing evils of our social economy, has remained almost
absolutely unknown. With the exception of the works of
one or two land-nationalizers, I have never seen it even
referred to by the host of political writers, who weakly and
ignorantly dabble in the great questions affecting the real
well-being of our whole population.

Our Society being established, it seemed necessary to
prepare something in the form of a handbook or introduction
to the great problem of the land ; and I accordingly devoted
my attention to the subject, studying voluminous reports on
agriculture, on Irish famines, on Highland Crofters, and
numbers of special treatises dealing with the various aspects
of this vast and far-reaching question. My book was
published in March, 1882, under the title "Land Nationalization : its Necessity and its Aims," and gave, in a compact
form, the only general account of the evils of our land system
as it exists in England, Ireland, and Scotland ; a comparison

with other countries or places in which a better system prevails, together with a solution of the problem of how to replace it by the only just system, without any confiscation of property or injury to any living individual. The book has had a large circulation, and, in a revised edition, is still on sale ; and, together with numerous tracts issued by the Society, has done much to educate public opinion on this most vital of all political or social questions.

As, however, it was quite certain that it would take a very long time before even the first steps towards land nationalization would be taken, I took every opportunity of advocating such other fundamental reforms as seemed to me demanded by equity and to be essential to social well-being. One of the earliest was on the subject of *interest*, about which there was much difference of opinion among advanced thinkers. A discussion having arisen in *The Christian Socialist*, I developed my views at some length in an article which appeared in the issue of March, 1884. As it still appears to me to be logically unassailable, and is upon a subject of the very highest social importance, I give it here—

THE MORALITY OF INTEREST—THE TYRANNY OF CAPITAL.

By Alfred Russel Wallace.

Having read Professor Newman's defence of interest and your remarks thereon, I wish to make a few observations on the general question.

Your position, and also that of Mr. Ruskin, appears to be that money should be lent only as an act of benevolence or charity, and that lending it in any other way is not only, in most cases, economically and socially, injurious, but is also morally wrong. With the first part of this proposition I am very much inclined to agree, but not with the second. Looked at broadly, I believe that the *power* of obtaining interest on capital, however great, with the corresponding *desire* of the owner of capital to obtain interest on it, is, next to the private monopoly of land, the great cause of the poverty and famine that prevail in all the most advanced and most wealthy communities. To prove this would occupy too much space ; but I may just notice that bankruptcies, with the widespread misery they inflict ; the speculations of promoters and financiers often bringing ruin on hundreds or thousands of deluded investors ; and the vast loans to foreign despots, which can only be paid by the sweat and blood of their unfortunate subjects, are the direct, and in the present state of society, the necessary results of the interest-system. Professor Newman says that if it were to cease, business would be lessened by one-third. But only rotten and speculative business would be stopped; commercial men would then *be* what they only now appear to be, and no really necessary business would cease to be carried on. The late William Chambers has stated (in his " Life of Robert Chambers ") that their vast bookselling, printing, and publishing business was established and carried on from first to last without one penny of borrowed capital; and that, as a result, panics and financial crises which brought ruin to some of their competitors, only caused them a little temporary inconvenience. I believe, therefore, that it would be for the benefit of the community if loans of money, or advances of goods on credit, were not recognized by the law, but were made wholly at the risk of the lender; but I do not see that it follows that he who lends, even under these circumstances, and takes interest for his loan, is doing what is wrong. For I cannot perceive any essential difference in principle between lending on interest, and selling at a profit. If I buy a shipload of drugs or any other goods at wholesale price, warehouse them, and sell them in the course of a year at the current market rate, making a profit of, say, 15 per cent. on my money, am I doing that which is morally wrong? Of the amount gained by me, we may put perhaps 1 or 2 per cent. for my personal trouble in the matter, 2 or 3 per cent. for risk of loss, 5 per cent. for interest on capital, and the other 5 per cent. for surplus profit. Is this 10 per cent. illegitimate gain? and am I morally bound to sell my goods at so much below the market rate as to leave me only fair payment for my time and risk? If it is wrong to take interest for the money which, when lent to another man enables him to do this, surely it is wrong to take a larger share in the shape of profit; and this really means that all trade is immoral which returns more than payment for personal labour, and insurance of the capital employed. But if so, it should be so stated, and the question should not be confined to interest on money loans only, and, in fact, Mr. Ruskin does not so confine it. The quotation you make from Mr. Ruskin does not, however, seem to me at all to the point. You freely lend your friend an umbrella in his need, and you would even do the same to the merest acquaintance or neighbour, but if your neighbour called every day for your umbrella on his way to the city, and other neighbours followed his example, so that you ceased to have the use of your own umbrellas, you would soon have either to refuse to lend, or to charge a rent for the use of them, and if this were convenient to your neighbours, and they were willing to pay you sufficient to cover the wear and tear of umbrellas, your time and trouble in looking after them, and interest on your capital invested in them, it will require arguments very different from any yet advanced to satisfy me that you would be morally wrong in doing so. In like manner, though you lend your friend or neighbour cab-money, or give him a bed for a night on rare occasions when he urgently requires such aid, you would give none of these things repeatedly to a mere acquaintance. Yet, if circumstances rendered such accommodation very useful to a considerable number of persons, and you or some one else found it profitable to supply such accommodation, you would charge rent for your beds, and interest for your loans, and the transaction would differ nothing in principle from that of every tradesman who sells goods at a profit, of the innkeeper who charges beds in his bill, or of the jobmaster who charges for the use of his horses or his carriages. Nothing deserving the name of proof has yet been given that either of these things are immoral. Whether it is a good and healthy state of society in which large numbers of persons get their living by such means, is another matter altogether.

The difference of opinion on this question of usury arises mainly from the different standpoints of the disputants. Seeing that it is bound up with many of the evils of modern society, and believing that it should have no place in a system of true Socialism, you and Mr. Ruskin denounce it as immoral. Professor Newman, on the other hand, looks at it as a question of modern society, and finds nothing *in its essential nature* contrary to justice, and here he seems to me to have the best of the argument. No doubt, in a more perfect state of society, in which private accumulations of capital were comparatively small, and the land and its products were freely open to the use of all, usury would have little place, because loans of money would rarely be needed; but when they were needed, I cannot see any grounds for maintaining that it would be morally wrong to lend money on interest. On the contrary, such loans would then retain their use without the evils their wide extension now brings. There would be no great capitalists, and if one man lent to another it would be a convenience to the borrower, and certainly some loss to the lender, because, as Professor Newman well puts it, £100 paid one year or ten years hence is *not* as valuable as £100 paid to-day. To say that it is so is really to say that it has *no value* to-day, for if its payment can be delayed one year without loss it can two, or three, or ten, or a hundred, or a thousand! Where are we to stop? If we suppose a perfect social state, we suppose all men to be producers, and as capital is an aid to production, no man can give up the use of his capital to another without loss. The true solution of the problem is, I believe, to be found in the proposition that all loans should be *personal*, and, there-fore, *temporary;* and that, as a corollary, the repayment of the capital should be provided for in the annual payments agreed to be made by the borrower, either for a fixed period (if he live so long), or for the term of his life. This would abolish the idea of perpetual interest, which is as impossible in fact as it is wrong in principle, while it would avoid the injustice of compelling one man, or set of men, to pay the debts of a preceding generation from which they may have received no real benefit.

This question of interest thus becomes involved in the wider question of the tyranny of capital over labour, and its remedy. At present, civilized Governments act on the presumption that great accumulations of capital are beneficial, and even necessary, to the well-being of the community, and all legislation favours such accumulations. When the people are once convinced that the reverse is the case, and legislation is directed to favour small holders of capital, and to check its inordinate accumulation, most of the evils complained of will cease. To this end the first step would be to get rid of all Government funds, guaranteed loans, railway stocks, etc., which are the main agents and tools by which capital is accumulated and money is made to breed money. This could be done in every case by making such stocks non-transferable after a certain date, and then declaring the payments to be terminable at the death of the holders and their living heirs, just as I propose to do in the case of landlords. The railways should be taken by the State, existing shareholders receiving annuities of the amount of their average dividends, payable in like manner to themselves and their living heirs. The Limited Liability Act should be repealed, because it has served only to foster the worst and most iniquitous speculations, and has deluded the public into the idea that they could safely share in the profits of commercial enterprises of the nature and management of which they are profoundly ignorant. There would remain no safe investments for money, except in some branch of agriculture, manufactures, or commerce in which either the investor or some relation or friend was personally interested, and thus would be brought about the diminution and practical abolition of usury as a system, and of whole classes living idle lives on the interest of money derived from the accumulations of previous generations.

Of course, it will be said that the plan here proposed is wholesale confiscation and repudiation ; but a little consideration will show that it is nothing of the kind, and that it is really the best thing that can happen even to the individual holders of the stocks dealt with. In the case of the National Debt, for example, fundholders are now threatened with a reduction of interest of a quarter per cent., and later on of a half per cent. ; and *they will be forced to accept it*, because the interest on the public debt regulates that of all other good investments, which will inevitably rise in price enormously if any considerable portion of the amount now invested in the funds seeks other investments. The offer to pay off fundholders at par, will, therefore, be illusory, and the vast class who *live* upon their dividends will inevitably have their incomes reduced one-twelfth or one-sixth, while the cost of living goes on continually increasing. Would they not be far better off to have their present incomes secured to themselves and their living heirs? And when they fully realize their position, will they not choose the latter alternative if offered them? If the series of changes here sketched out were effected, the reign of capital as the tyrant and enemy of labour would be at an end. When the tools with which the financier and the speculator work no longer exist, the piling up of great fortunes will be impossible, and much personal care and attention will be required in order to make capital produce a steady return. Industry and commerce

will be the sole means of acquiring wealth, and by these means alone—
under the new conditions of society—very great wealth can never be
accumulated by one man. For the land being nationalized, and the
use of some portion of it obtainable by all, the minimum of wages will
rise far above the starvation point which now prevails, and every village
or other community, however small, will consist of small capitalists, who
will be ever ready to unite for the safe employment of their capital.
Then will arise a variety of industries on a scale adapted to the size and
wealth of the district, and calculated to utilize the surplus labour and
spare time of the surrounding population ; and these small industries
will compete successfully with the establishments of individual capitalists,
because they will have an ample and cheap supply of labour, and because
most of the labourers, or their relations, will be shareholders, and will
be thus working for themselves. The individual capitalist will then find
himself paralyzed for want of labour, unless he offers great temptations
in the form of high wages and participation in the profits. For when
a large proportion of the population are settled upon the land, and are
able to devote their savings and their spare time to local industries, they
will not, as now, be forced to become parts of a huge manufacturing
machine in the success of which they have little personal interest.

By the methods here sketched out the labourer will receive, as Karl
Marx and other social reformers maintain that he should do, the whole
produce of his labour, and he will obtain this general result without any
aid from Government, except what consists in remedying injustice, and
removing the restrictions on freedom which now hamper him. Without
any laws against usury, usury will practically cease to exist. Without
any direct restrictions on wealth, those vast and injurious accumulations
of wealth which now prevail will be impossible. The " stealers " and the
" beggars " who now, as Mr. Girdlestone has shown, are so numerous
among us, will steadily give place to " workers," and just in proportion
as that happens, poverty will diminish, and will ultimately disappear.
Now, a large portion of the working population are employed in the
production of useless and often tasteless luxuries and trifles, the direct
consequence of the large number of persons who have surplus money to
spend after all their reasonable wants and comforts are fully satisfied.
It is this, much more than the mere number of idle people, that is the
dead weight which keeps thousands starving in the midst of so much
wealth. When mere extravagant luxuries are less in demand great
masses of labourers will be set free to produce the necessaries and
comforts of life ; and these will be more abundant and cheaper (whatever their money price may be), and if all those who are now idle aid in
the production of these necessaries and comforts, it is evident that, with
free exchange, none can want.

I would particularly call attention to the fact that the results here
indicated would be all brought about by carrying out the true system of
laissez-faire now so much abused as if it had failed, when really it has
never been tried. Labour, the sole source of all wealth and well-being,

has been fettered in all her limbs, and harassed in all her actions, and then because she often stumbles or faints by the way, they cry, " See, she cannot do without help ! " But first unloose your bonds, and cease to hamper her with your legal meshes, and then see if she will not achieve a glorious success. Let Government do its duty, and no more. Let it secure peace from external foes, and safety from internal violence ; let it give free and speedy justice between man and man ; let it secure to all alike free access to the land and all natural powers ; let it abolish every monopoly of individuals and classes—either the local or central authority having the management of all institutions or industries which are essential to the public welfare, but which in private hands tend to become monopolies ; and let it enact that all debts contracted by individuals shall be payable by those individuals only, and those con-tracted by the municipality or State be payable by the generation which contracts them, so that they may never remain a burden on the succeed-ing generation. When it has done all this, then alone will labour be really free, and, being free, it will work out the well-being of the whole community without any Government interference whatever. This is the true *laissez-faire ;* and this, I believe, will enable us to realize the best social state which, *in its present phase of development,* humanity is capable of. The distant future will take care of itself; let us try to improve the future that is immediately before us. I have here very briefly and imperfectly sketched out a series of measures which I believe are best calculated to promote this object, and they have the great and inestimable advantage that they all tend to the diminution of govern-mental interference with labour and industry, instead of that indefinite increase of it which the German Socialists advocate, and which, as the greatest political thinkers maintain, and as all experience shows, must inevitably fail, while in the present condition of civilization it will probably lead to evils not less grave than those it attempts to cure.

At this time I was in correspondence with Mr. Robert Miller, a wealthy gentleman of Edinburgh, who had read my book and had given a donation to our society, but who wished to give or bequeath a large part of his fortune for the benefit of the community at large. He was, however, much disturbed by the conflicting views of writers on the subject, and though he was much inclined to land nationalization, he found it to be so strongly opposed by all the recognized authorities in political economy, as well as by most public writers and politicians, that he could not make up his mind what to do. In this uncertain frame of mind he was persuaded by some of his friends that the best thing he could do would be to have a conference of all the leading politicians and advanced thinkers to discuss the question, "What are the best means, consistent with justice and equity, for bringing about a more equal division of the accumulated wealth of the country, and a more equal division of the daily products of industry between Capital and Labour, so that it may become possible for all to enjoy a fair share of material comfort and intellectual culture, possible for all to lead a dignified life, and less difficult for all to lead a good life ? "

He proposed to devote £1000 for the expenses of the conference, and the following gentlemen agreed to act as trustees : Sir Thomas Brassey, Mr. John Burnett, Mr. Thomas Burt, the Earl of Dalhousie, Professor Foxwell, Mr. Robert Giffen, and Mr. Frederic Harrison.

But these gentlemen did not adopt the very clear statement of the problem Mr. Miller wished to be enlightened upon, nor the highly humane and moral object he had in view, as shown by his own words given above. Instead of it they adopted a comparatively hard and colourless statement in the following terms:—"*Is the present system or manner whereby the products of industry are distributed between the various persons and classes of the community satisfactory ? Or, if not, are there any means by which that system could be improved?*" And this was again rendered still more bald and systematic by being stated under five heads and ten subdivisions, in the approved manner of the political economists, so as to limit the questions discussed to utilities, while excluding as much as possible all questions of justice or equity, of moral or intellectual advancement.

The conference lasted three days, with morning and afternoon sittings ; about one hundred and fifty delegates representing the chief labour associations of the kingdom attended ; and twenty representatives of political and social science, including myself, were invited to read papers. These papers, with some valuable statistics in appendices, and a report of the discussions on the chief papers, were published by Cassell and Co. in a thick 8vo volume of over 500 pages, entitled " Industrial Remuneration Conference Report." In the

paper which I prepared, I endeavoured to go to the very
heart of the question propounded by Mr. Miller, " How to
cause Wealth to be more equably distributed." It occupies
twenty-four pages of the Report, but I give here an abstract
of it prepared for the newspapers.

HOW TO CAUSE WEALTH TO BE MORE EQUABLY DISTRIBUTED.

(ABSTRACT.)

As the bulk of the community live on wages, the only means by which
they can obtain a larger proportion of the wealth they produce is by
wages becoming generally higher, and by work being more constant ;
and in order that this change may be permanent, and be commensurate
with the evil to be remedied, it must be brought about, not by any form
of charity, or of local or individual action, but by social rearrangements
which will be self-acting and self-sufficing. The fundamental objection
often made that a general rise of wages would interfere with our foreign
commerce was shown to be unsound, and it has been refuted by Mill,
Fawcett, and other political economists.

The cause of low wages was next discussed, and was shown to be due,
not to a superabundance of labourers, but to the fact that the majority
of labourers have nothing but daily wages between themselves and
starvation, under which conditions wages are necessarily driven down to
the minimum on which life can be supported. This absolute dependence
of labourers on daily earnings is at a maximum in great cities where
access to land and to natural products is completely cut off, and it is here
that these earnings sink to their minimum, and at the same time that the
wages of highly skilled labour is at a maximum, the latter phenomenon
being that which is chiefly dwelt on by economists. Illustrative cases
of these low wages were given, and they were shown to be intimately
connected with the existence and continued growth of our great cities.

The diminution of the population of the purely agricultural districts
was next dwelt upon, and it was traced back to the circumstance that the
natural growth and extension of village communities is checked by the
direct action of landlords. Evidence of this fact was adduced from
the writings of Sir George Grey, Mr. Francis Heath, John Bright, Mr.
Thomas Hardy, and the Rev. Stopford Brooke, and its deplorable results
were shown to be a great diminution of food produced in the country,
overcrowding, and intense competition, with incalculable vice and misery
in towns.

The beneficial results of allowing labourers to have land were next
detailed. Evidence was adduced showing the great amount of produce
which is obtained by labourers from allotments, although these are,

comparatively, disadvantageous both to the labourers themselves and to society; and they are altogether condemned by John Stuart Mill as being bad in principle. But in every case in which labourers have been allowed a few acres of land at a fair rent, *and attached to their cottages,* the effects have been most beneficial. Not only have they obtained a large increase to their means by utilizing labour before wasted, but they and their families have acquired habits of temperance, industry, and thrift, so that pauperism and drunkenness have been greatly diminished, and the population has been elevated, both socially and morally.

In order to extend these beneficial results to the whole community, the labourer asks for neither charity nor loans, but fair opportunity and equal justice. It was urged that the necessary capital will be saved by the more industrious and thrifty labourers when they have before them the certainty of procuring that dream of their lives, " a homestead of their very own;" while nothing would so certainly lead to failure as any extensive system of loans to enable those who have not these essential virtues to obtain the needful land, stock, and houses without them.

The scheme suggested as most beneficial to labourers and to the community at large is as follows :—

(1) In each rural parish four *land-assessors* to be chosen by the ratepayers, two to be farmers and two labourers.

(2) Any labourer or mechanic wanting a plot of land shall have it allotted to him by two of the assessors, one named by himself and one by the existing occupier of the land, after the parties have met together on the ground and stated their wishes and objections.

(3) The rent of the land thus allotted to be fairly valued by the assessors, who will also determine the sum to be paid for improvements, unexhausted manures, etc., on the land, which last sum must be paid before obtaining possession.

(4) The rents of the plots thus allotted to be collected by the local rate-collector, and the amounts, less a percentage for collection, to be paid to the landlord.

(5) The tenure of the plots to be secure so long as they are personally occupied, and to be saleable or transferable ; while the rents are to be fixed for long periods, and only raised by a new general valuation in case the value of the land itself has risen irrespective of all improvements, which last remain the absolute property of the tenant.

By the method thus sketched out no attack is made on private property, and no new principle in dealing with land is introduced, since many industrial enterprises calculated chiefly to benefit individuals often obtain from Parliament the right to take land. It is now only asked that the same power may be given to the people at large, under strict limitations, and in order to benefit the whole community by bringing about a more natural distribution of population, and a greater and more varied production of food and other useful products.

Various popular objections to labourers having land were then answered, and it was shown that none of them has any force as applied to the proposed scheme, the claims and merits of which were summed up as follows :—

(1) That it goes to the very root of the matter, since, by rendering a large number of labourers less dependent on daily wages as their only means of obtaining food, it would immediately and necessarily raise the standard of wages ; and this is absolutely the only means by which the labouring classes may at once be enabled more fully to share in the products of industry.

(2) It does this in the simplest conceivable way, by throwing down the barriers which now prevent labour from spreading over the land.

(3) It would enable every labourer, by industry and thrift, to realize his highest aspiration—"a homestead of his own."

(4) It would largely increase the food-supply of the country, especially in dairy produce, poultry, fruit, and vegetables, now to the amount of thirty-eight millions annually imported from abroad.

(5) It would, by a self-acting, gradual process, withdrawing the congested population of the towns back to the rural districts from which they have so largely come in recent times, and would at the same time benefit all who remained by both raising their wages and lowering their rents.

(6) It would completely settle both the Irish and the Highland land questions by satisfying the just claims of the labourers and cottiers in one country, and the crofters in the other, and would open up to human industry extensive areas of both countries, once cultivated, but now devoted exclusively to cattle, sheep, or game.

(7) It would also bring about a great moral reform, since all experience proves that the possession of land on a secure tenure is the best incentive to sobriety, industry, and thrift.

(8) And, lastly, all this can be effected without any financial operation or increased taxation, and with no greater interference with landed property than is allowed to many of the speculations of capitalists of far less general utility, and often of none whatever.

Whether the originator of the conference obtained anything worth the thousand pounds expended is doubtful. There was no independent and judicial summing-up of the evidence adduced, and the opinions expressed, and the great variety and contradictory nature of these opinions, often quite unsupported by any facts, must have left his mind in a state of greater confusion and uncertainty than before. At all events, I believe he did not leave any large sum to be devoted to helping on the cause he had so much at heart. At the meeting devoted to the land question, at which my paper and one by Professor Francis W. Newman were read, the discussion, instead of being kept to the subject of the two papers, consisted mainly of a declamatory battle between the socialists and individualists, both declaring that our proposals were useless, because they were not in accordance with those of either party. Mr. A. J. Balfour, however, did criticize my proposals, declaring, without adducing any evidence, that if labourers all had from one to five acres of land on a secure tenure, they could not live on it, and would therefore be quite as much dependent on the farmers and obtain as low wages as when they were quite landless. This amazing statement was made in the face of the almost life-long experience of Lords Tollemache and Carrington, in four counties, and of facts adduced in the reports of the latest Royal Commission on Agriculture, which an M.P. and prospective Prime Minister ought to have known something about. Professor J. Shield Nicolson also sent a "Note on Dr. Wallace's Paper," the chief points being that five-acre lots would not alleviate agrarian distress in the Highlands—which I knew quite as well as he did—and more especially that my simple method of valuation would utterly break down and satisfy no one, and contending that "nothing but an eloborate system of law and judicial machinery could make such a plan tolerable"! He had himself read a paper, with suggestions for a number of mild ameliorations of the present system, which, in its essentials, was to remain untouched.

The result was, I think, to show that a conference of opposing parties, each looking at the question from an absolutely different standpoint, and with no possibility of agreement as to fundamental principles, cannot lead to any definite conclusion. The method adopted by the Land Nationalization Society was the only one calculated to produce any definite results, viz. to lay down certain fundamental principles, capable of logical demonstration, and by means of an association for the purpose to educate the public on the subject, both by argument and by a constant appeal to all facts or experiments which serve to illustrate the evils of the present system and the benefits of that which we propose to substitute for it. This has been done both by land nationalizers and socialists with, on the whole, most satisfactory results. On the one hand, the socialists are agreed that, as a first step, free access to land, with a view to its future nationalization, is vitally important ; while on the other hand, the workers no longer say, as they did at the congress, "Land nationalization will be of no use to us." This is an important advance in the short space of twenty years.

Among the few eminent men who joined our movement was Professor F. W. Newman, and I had the pleasure of meeting him several times at the house of my friend Mr. A. C. Swinton, and I also had some correspondence with him ; but there is little in the few letters I have worth quoting. The following is the concluding paragraph of a letter dated June 6, 1882 : "Our duty is to do what we can, in detail ; but the longer I live the less hope I have of justice, without changes so great in the persons who hold power that it will be called a revolution. I mean justice, not as to land tenure only, but as to many other things equally sacred, perhaps more vital. Until popular indignation rises, I expect no result ; and when it rises it may seem easier to make a clean sweep than carry a quarter measure.

"Be assured that I look up to you with gratitude.

"F. W. NEWMAN."

Soon after our society was started, Henry George, author of that remarkable work, "Progress and Poverty," came to England, and I had the pleasure of making his acquaintance. He spoke at several of our meetings and elsewhere in London, as well as in various parts of England and Ireland. He was a very impressive speaker, and always held his audience. His delivery was slow and deliberate ; so much so as to appear sometimes as if he had broken down, but he was always cool and collected, and when the next sentence came one saw that the pause was made either for the purpose of choosing the right phrase or of producing a greater effect. The following passages, in a letter written from Dublin in November, 1881, soon after his arrival, show

how a well-educated and thoughtful American was impressed by the English rule of Ireland—

"I had not intended to speak in public before coming to England; but I feel so much sympathy with the Irish people in their resistance to the degrading tyranny now rampant here, that it seems to me cowardly to refuse any little assistance I might give, and I have told some gentlemen who have been urging me that I will lecture this week for the benefit of the Political Prisoners' Aid Society, of which Miss Helen Taylor is President.

"I had the pleasure of meeting that lady here, and the pleasure of listening to her address to the ladies of the Land League—a speech that I wished could have rung through the length and breadth of England. When will the great English party to whom the future will be given raise its head? I long for its advent. If this is Liberalism which I see here, what Toryism may be I can with difficulty imagine.

"I have had the pleasure, too, of meeting an Irish Catholic bishop who is with us entirely—Bishop Nulty, of Meath—a prelate who does not hesitate to declare that private property in land is an injurious blasphemy. He is fettered to some extent, of course, but he wields great influence, and we shall hear from him before this thing is over."

The lady above referred to, step-daughter of John Stuart Mill, was an earnest land-nationalizer and a valued supporter of our society. She was always ready to speak at our meetings; she supported us liberally by donations and subscriptions, and she gave to our public proceedings that tone of sympathy, humanity, and idealistic enthusiasm which was of great importance to us.

Among my early correspondents on the land question were Mr. Jesse Collings and Mr. J. Boyd-Kinnear, both afterwards M.P.'s. The latter gentleman was so much interested in my writings on this and allied questions, that he invited me and my wife to visit him at Guernsey, where he was then living. We spent a delightful week on that beautiful island, either driving or walking over it. Mr. Boyd-Kinnear was a practical farmer and agricultural chemist, and had a small farm close to the town of St. Pierre, with the usual large vinery under glass. While here, we thoroughly discussed the land and other questions, and though I could not quite convert him, we agreed generally in our political and social heresies.

Among the most esteemed of the friends I owed to " Land Nationalization," were two eminent Scotchmen, both poets, and both ardent lovers of justice and humanity—Professor J. Stuart Blackie and Charles Mackay. The former wrote to me in July, 1882, saying that he had just finished the " careful study " of my " Land Nationalization," and that he was " happy to find it so much in accordance with my oldest and most mature speculations, and—what is of more importance—observations on the subject." He sent me a copy of his small volume, " Alteriora," with a chapter on the " Sutherland Clearances," and he concluded, " As to your remedy for the gigantic evils which our present system of land laws entail, they recommend themselves strongly to every consistent thinker."

Both he and I suffered some inconvenience from having mentioned the name of the agent who carried out the terrible Sutherland evictions in the first two decades of the nineteenth century, as it is given in all the early narratives, as well as in the report of the trial of the agent for arson and murder, when, of course, he was acquitted. His sons were at that time alive, and protested against the publication. Both our publishers were frightened. Professor Blackie withdrew his book, and published a second edition much cut down. I placed mine in the hands of a new publisher, and I promised that in a new edition I would omit the name of the agent, but refused to make any alterations in the statements of facts.

Three years later (in December, 1885), when I was lecturing in Edinburgh, I had the great pleasure of meeting Professor Blackie. I was staying with the late Mr. Robert Cox, at whose house the professor was an intimate. He called soon after I arrived, and on hearing my name, he cordially embraced me (in the continental fashion) as one with whom he was in complete sympathy, and then threw himself upon

the rug to talk to Mrs. Cox. Afterwards I had a long conversation with him on all the subjects that interested us most,
and was delighted with his geniality no less than with his
intense human sympathy, especially in the case of the cruelly
disinherited Highlanders.

Although I had for many years been a great admirer of
Charles Mackay's Songs and Poems, and that I was quite
near him while we lived at Dorking, from August, 1876, to
March, 1878, I did not make his acquaintance till some years
afterwards, as, owing to my constitutional shyness, I do not
think I ever made the first overtures to any man, or even
called upon any one without some previous correspondence
or introduction. But several years later I sent him a copy
of my "Land Nationalization" (I think probably on the
suggestion of some one who knew him), with a letter begging
his acceptance of it. This brought me three letters in rapid
succession—one acknowledging it, saying he had been very
ill for six months, but adding that he had been an adherent
of our cause for forty years, and referring me to his poem,
"Lament of Cona for the Unpeopling of the Highlands."
Five days later he wrote again, saying—

"I have read every line of your admirable volume on
'Land Nationalization' with the greatest interest and profit.
I agree with every one of your arguments, which are all
incontrovertible, and not only lucidly, but triumphantly
placed before the reader. They must convince and make
converts of every unprejudiced person who will attentively
study them with the sole view of arriving at the truth." He
then refers to his own writings in the same direction of forty
years before, naming "The Cry of the People"—and there
are many others—concluding, "I am afraid that age and ill-
health will not allow me to labour much further in the cause;
but what I *can* do, I *will* do. If my name is of any use to
your society, you are free to it.

"Believe me, with the highest esteem and regard,
　　　　　　　　　　"Yours most cordially,
　　　　　　　　　　　　　"CHARLES MACKAY."

The next day he wrote me again, and as this contains matter of wide public interest, and points to a legal public right which has been, and may still be enforced, I here give it—

"I omitted in my letter of yesterday to mention a fact, which, if you are unaware of it, may possibly be of interest to you. It is recalled to my mind by the remarks in your book (pages 128, 129) on the closing of large tracts of country, by selfish and tyrannical Highland proprietors, for the purpose of creating solitudes for the cultivation and pre-servation of deer. The practice is clearly illegal, in contra-vention of an old, and unrepealed Scottish law, entitled 'Free Foot in the Wilderness.' Many years ago, when I was editor of the *Glasgow Argus*, I fought the Duke of Athol in its columns, and appealed to the law, not without success, in the famous Glen Tilt case. I wrote some stinging verses about his grace on the occasion, entitled 'Baron Braemar,' which had a considerable spurt of popularity—which the Queen read, and of which she expressed her approval (and agreement) to her physician, Sir James Clark, an old friend of mine, who told me about it. The Glen was thrown open for a time, but, I believe, has been closed up again with as much rigour as ever."

In the following year he removed to London for good medical attendance, and wrote me a very flattering letter after reading my "Malay Archipelago." The next year (1886) I was able to call on him when in London for a day, at his apartments in Longridge Road, South Kensington, when we had a long talk, and he afterwards wrote to me as "My dear friend and philosopher." On the occasion of this visit he introduced me to his step-daughter, Miss Marie Corelli, a very pleasant young lady, whose future eminence as a writer I did not divine.

Charles Mackay is, apparently, hardly classed as a poet, since in Chambers's "Biographical Dictionary" he is spoken of as a song-writer; and a modern poet to whom I once mentioned him was ignorant of his existence. Some know him only by his "Emigrant" songs, which were set to music

by Henry Russell, and are often thought to have been
composed by him. These songs have a charm and a music
in the sentiments and the rhythm, which owe nothing to the
music. What can be more inspiring than the last lines of
" Cheer, Boys ! Cheer ! "—

> " Here we had toil and little to reward it,
>   But there shall plenty smile upon our pain,
> And ours shall be the mountain and the forest,
>   And boundless prairies ripe with golden grain."

Or the first verse of " To the West "—

> " To the West ! to the West ! to the land of the free,
> Where mighty Missouri rolls down to the sea,
> Where a man is a man if he's willing to toil,
> And the humblest may gather the fruits of the soil.
> Where children are blessings, and he who hath most,
> Hath aid for his fortune and riches to boast ;
> Where the young may exult, and the aged may rest,
> Away, far away, to the Land of the West ! "

Every fact, every hope in these songs were literally true
when they were written, as contemporary American literature
clearly shows, but the growth of capitalism and land
monopoly during the last thirty years has rendered them
almost a mockery.

Mackay had not the magic of words and phrases, or the
deep idealism which characterize the highest poetry, and
could not therefore rank with Tennyson, William Watson,
Lowell, or Edwin Markham ; but he was the equal of Longfellow or Scott, and perhaps superior to both in the infinitely
varied *matter* of his verse. His ballads and stories were
unsurpassed for vigour and originality of treatment—such as
his " Invasion of Scotland by the Northmen," his "Thor's
Hammer," and his " Lament of Cona for the Unpeopling of
the Highlands ; " while " The Man in the Dead Sea," " The
Interview," " The Building of the House," " We are Wiser
than we Know," and " Eternal Justice " deal with some of
those grand problems in the elucidation of which the poet
is so often the seer. At my request he wrote for us some

verses on the land question, which, as they are not included
in any edition of his poems, I give here.

Free Land for a Free People.

" Thank God for the Sunshine, the Air, and the Sea,
 For the Rain and the Dew, ever free to the free !
 No landlords can parcel *them* out, or conspire
 To sell them, or tax them, or let them on hire ;
 And close up with barriers what Nature design'd,
 In mercy and love for the needs of Mankind !

" There's a break in the clouds—there's a gleam in the sky,
 There's a beautiful star shining brightly on high,
 That heralds the dawn of the long-promised day,
 When Right shall be Might, and shall flourish for aye ;
 When Man in the strength of his manhood shall stand,
 To enjoy, and possess, and replenish the land.

"With our faces to heaven and our feet on the sod,
 We swear by the faith that we cherish in God—
 By the breeze of the sky, by the light of the sun,
 That the Land shall be ours, and that Right shall be done.
 Hear it, ye Tyrants, that hold us in thrall !
 God the great Giver gives freely to All ! "
CHARLES MACKAY.

Yet another of our true poets, but also one who is
comparatively unappreciated, Gerald Massey, is also one
of our friends and supporters ; and he, too, has been so kind
as to embody his thoughts in the following energetic verses
for our benefit :—

The Earth for All.

" *Thus saith the Lord*—You weary me
 With prayers, and waste your own short years :
 Eternal truth you cannot see
 Who weep, and shed your sight in tears.
 In vain you wait and watch the skies,
 No better fortune thus will fall ;
 Up from your knees I bid you rise,
 And claim the Earth for all.

" They eat up Earth, and promise you
 The Heaven of an empty shell.
 'Tis theirs to *say :* 'tis yours to *do*,
 On pain of everlasting hell.

They rob and leave you helplessly,
   For help of Heaven to cry and call :
Heaven did not make your misery,
   When Earth was given for all !

" Behold in bonds your Mother Earth !
   The rich man's prostitute and slave !
Your Mother Earth that gave you birth,
   You only *own* her for a grave.
And will you die like slaves, and see
   Your Mother left a bounden thrall ;
Or live like men to set her free
   As Heritage for all ? "

                        GERALD MASSEY.

Here we have the same idea expressed in different forms by two of our sweetest singers ; and it is to be specially noted that they are both much more than poets ; while Gerald Massey has mastered one of the most difficult of modern studies—Egyptology—on which he has published two very bulky works ; and even the specialists, who reject some of his conclusions, admit that his presentation of the facts shows an enormous amount of research.

Another of our supporters from an early date was Grant Allen, a man of the very highest talents, who, had he not been compelled by circumstances to write novels, reviews, and magazine articles for a living, would probably have become one of our greatest philosophical naturalists and expounders of evolution. But, like myself, he was more than a land nationalizer, and my first knowledge of his political and social views was derived from an article he wrote on the condition of India somewhere about 1880. Through my friend, the late Sir David Wedderburn, I had become aware of the terrible defects of our government of that country owing to the ever-increasing influence of European planters, manufacturers, and capitalists ; and I was also a reader of *The Statesman,* a paper brought out by a gentleman who had been for many years editor of one of the most advanced Calcutta newspapers, and who established it for the purpose of letting Englishman know the real facts as to the government of India. All the statements in this paper were founded upon Government Reports or other official documents, referred to in detail. I knew, therefore, that Grant Allen's views as stated in this paper were correct, and therefore wrote to tell him how pleased I was to find that he was not *only* interested in physical science, as was so often the case with my scientific friends. His reply is so interesting that I will here give the more important parts of it :—

" As to your remarks about the wrong actually perpetrated by us in India, I know only too much about that question. For three years I was employed by W. W. Hunter, Director-General of Statistics for India, in collecting and working up the district accounts and other materials in his possession. Not to put too fine a point upon it, Dr. Hunter is *the literary whitewasher of the Indian Government.* In working up the abundant reports and other documents submitted to me, I had plenty of opportunities for realizing what English rule really meant. In the ruin wrought by our land settlements especially, I collected a large number of facts and statistics ; and I offered John Morley to work them into a paper on ' The Indian Cultivator and his Wrongs ;' but Morley did not care for the subject. The fact is, nobody in England wishes to move in the matter. I sent Knowles a paper two years ago about the same subject, dealing especially with the Ganges Canal—a vast blunder, bolstered up by cunningly contrived balance-sheets, in which deficits are concealed as fresh investments ; but he would not take it. I only got this article into the *Contemporary* by leaving out India, and looking at the question from a purely English point of view. I'm afraid the fact can't be blinked that most Englishmen don't mind oppression as long as the oppressed people are only blacks. A startling outrage, like the Zulu War, wakes them up for a moment ; but chronic and old-standing sores, like India or Barbadoes, do not affect them."

Neither do "chronic and old-standing sores" at home affect them. The slums, slow starvation, murder and suicide from want, one-third of our population living without a sufficiency of the bare necessaries for a healthy life—food clothing, warmth, and rest ; while another third, comprising together those who create the wealth of the nation, have not the amount of relaxation or the certainty of a comfortable old age which, in a country deserving to be called civilized, *every* human being should enjoy. This, however, is a step or two beyond land nationalization, before leaving which I must refer to one application of its principles which any Government declaring itself to be "Liberal" ought at once to make law. I call it—

SECURITY OF THE HOME.

It is an old boast that an Englishman's house is his castle, but never was a boast less justified by facts. In a large number of cases a working man's house might be better described as an instrument of torture, by means of which he can be forced to comply with his landlord's demands, and both in politics and religion submit himself entirely to the landlord's will. So long as the agricultural labourer, the village mechanic, and the village shopkeeper are the yearly or weekly tenants of the great land-owner, the squire, the parson, or the farmer, religious freedom or political independence is impossible. And when those employed in factories or workshops are obliged to live, as they so often are, in houses which are the property of their employers, that employer can force his will upon them by the double threat of loss of employment and loss of a home. Under such conditions a man possesses neither freedom nor safety, nor the possibility of happiness, except so far as his landlord and employer thinks proper. A secure *home* is the very first essential of political security and of social well-being.

Now, all this has been said many times before, and we may go on saying it, and yet be no nearer to a remedy for the evil. But now that every worker, even to the hitherto despised and down-trodden agricultural labourer, has been given the right of some fragment of local self-govern-ment, it is time that, so far as affects the inviolability of the home, the landlord's power should at once be taken away from him. This is the logical sequence of the creation of Parish Councils. For to declare that it is for the public benefit that every inhabitant of a parish shall be free to vote for and to be chosen as a representative of his fellow-parishioners, and at the same time to leave him at the mercy of the individual who owns his house to punish him in a most cruel manner for *using* the privileges thus granted him, is surely the height of unreason and injustice. It is giving a stone in place of bread ; the shadow rather than the substance of political enfranchisement.

Now, it seems to me that there is one very simple and very effectual way of rendering tenants secure, and that is by a short Act of Parliament declaring all evictions, *other than for non-payment of rent*, to be illegal.

And to prevent the landlord from driving away a tenant by raising his rent to an exorbitant amount, all alterations of rent must be approved of as reasonable by a committee of the Parish or District Councils, and be determined on the application of either the tenant or the landlord. Of course, at the first letting of a house or small holding, the landlord could ask what rent he pleased, and if it was exorbitant he would get no tenant. But having once let it, the tenant should be secure as long as he wished to occupy it, and the rent should not be raised, except as allowed by some competent tribunal. No doubt a claim will be made on behalf of the landlords for a compulsory tenancy on the part of the occupier ; that is, that if the tenant has security of occupation, the landlord should have equally security of having a tenant. But the two cases are totally different. Eviction from his home may be, and often is, ruinous loss and misery to the tenant, who is therefore, to avoid such loss, often compelled to submit to the landlord's will. But who ever heard of a tenant, by the threat of giving notice to quit, compelling his landlord to vote against his conscience, or to go to chapel instead of to church ! The tenant needs protection, the landlord does not.[1]

The same results might also be gained (and perhaps more surely) by giving the Parish and District Councils power to take over all houses whose tenants are threatened with eviction, or with an unfair increase of rent ; and that will come some day. But the plan of giving a legal permanent tenure to every tenant is so simple, so obviously reasonable and so free from all interference with the fair money-value of the landlord's property, that, with a little energy and persistent agitation, it might possibly be carried in a few years. Such an Act might be more or less in the following form :—" Whereas the security and inviolability of the HOME is an essential condition of political freedom and social well-being, it is hereby enacted, that no tenant shall hereafter be evicted from his house or homestead for any other cause than non-payment of rent, and every heir or successor of such tenant shall be equally secure so long as the rent is paid." A second clause would provide for a permanently fair rent.

This formed part of my address to the Land Nationalization Society at its annual meeting in 1895, and one would have thought that some Liberal or Radical or Labour Member would have made an effort to get so small yet so far-reaching and beneficial a measure discussed in the House of Commons. But no notice whatever was taken of the suggestion, and we have had for the succeeding ten years, and have to-day, cases of punishment by eviction for political or religious opinions. It is true that it is but a small and isolated portion of the

---

[1] The late Lord Tollemache voluntarily recognized this, and gave his tenant-farmers leases for twenty-one years, determinable at *their* pleasure, but not at *his*.

much greater reform that we advocate, but, unlike most small measures, it goes directly to the root of a shameful oppression, and would do more to elevate the very poor and prepare the way for real reform than many whole sessions of even "liberal" legislation.

For about ten years after I first publicly advocated land nationalization I was inclined to think that no further fundamental reforms were possible or necessary. Although I had, since my earliest youth, looked to some form of socialistic organization of society, especially in the form advocated by Robert Owen as the ideal of the future, I was yet so much influenced by the individualistic teachings of Mill and Spencer, and the loudly proclaimed dogma, that without the constant spur of individual competition men would inevitably become idle and fall back into universal poverty, that I did not bestow much attention upon the subject, having, in fact, as much literary work on hand as I could manage. But at length, in 1889, my views were changed once for all, and I have ever since been absolutely convinced, not only that socialism is thoroughly practicable, but that it is the only form of society worthy of civilized beings, and that it alone can secure for mankind continuous mental and moral advancement, together with that true happiness which arises from the full exercise of all their faculties for the purpose of satisfying all their rational needs, desires, and aspirations.

The book that thus changed my outlook on this question was Bellamy's "Looking Backward," a work that in a few years had gone through seventeen editions in America, but had only just been republished in England. On a first reading I was captivated by the wonderfully realistic style of the work, the extreme ingenuity of the conception, the absorbing interest of the story, and the logical power with which the possibility of such a state of society as that depicted was argued and its desirability enforced. Every sneer, every objection, every argument I had ever read against socialism was here met and shown to be absolutely trivial or altogether baseless, while the inevitable results of such a social state in giving to every human being the necessaries, the comforts, the harmless luxuries, and the highest refinements and social enjoyments of life were made equally clear. As the mere story had engrossed much of my attention, I read the whole book through again to satisfy myself that I had not over-looked any flaw in the reasoning, and that the conclusion was as clearly demonstrated as it at first sight appeared to be. Even as a story I found it bore a second almost immediate perusal, a thing I never felt inclined to give any book before (except, I think, in the case of Herbert Spencer's "Social Statics"), and during the succeeding year I read it a third time, in order to refresh my memory on certain suggestions which seemed to me especially admirable.

From this time I declared myself a socialist, and I made the first scientific application of my conviction in an article on "Human Selection" in the *Fortnightly Review* (September, 1890), in which I showed how such a state as socialism postulates would result in the solution of two great problems, (1) that of gradually reducing the rate of increase of the population through a later period of marriage, and (2) by setting up a process of sexual selection which would steadily eliminate the physically imperfect and the socially and morally unfit. This article called forth several expressions of approval, which I highly value. It forms the last chapter of vol. i. of my "Studies, Scientific and Social."

I now read several other books on socialism, such as Mr. Kirkup's "Enquiry into Socialism," an admirable *résumé*, generally favourable; William Morris's "News from Nowhere," a charming poetical dream, but as a picture of society almost absurd, since nobody seems to work except at odd times when they feel the inclination, and no indication is given of any organization of labour. Gronlund's "Our Destiny" is a beautiful and well-reasoned essay on the influence of socialism on morals and religion, and his "Co-operative Common-wealth," an exposition of constructive socialism, which has given us in its title the shortest and most accurate definition of what socialism really is. "A Cityless and Countryless

World," by Henry Olerich, an American writer, is an excellent exposition of an extreme form of what he calls *co-operative individualism*, which is really voluntary socialism ; and I may here state for the benefit of those ignorant writers who believe that socialism *must* be compulsory, and speak of it as a "form of slavery," that my own definition of socialism is "the voluntary organization of labour for the good of all." All the best and most thoughtful writers on socialism agree in this; and for my own part I cannot conceive it coming about in any other way. Compulsory socialism is to me a contradiction in terms—as much so as would be compulsory friendship. The only modern work I have met with that advocates compulsion in initiating socialism is Mr. F. W. Hayes's "Great Revolution of 1905," a very clever book, but hopelessly out of tune with the socialist ideal by the ruthless compulsion and punishment of the opponents of the supposed social revolution.

Among books which deal rather with the evils of the present system than with constructive socialism, but which nevertheless give eloquent expression to its fundamental ideas and aspirations, I may mention "Darkness and Dawn, the Peaceful Birth of a New Age"—an anonymous work which, in its terrible description of the horrors of the factory system in all its forms and ramifications, is unsurpassed in our language ; and Robert Blatchford's "Merrie England," issued first at a shilling, then at fourpence, then at a *penny*, and of which three quarters of a million copies were sold in about a year.

But the most complete and thoroughly reasoned exposition, both of the philosophy and the constructive methods of socialism, is to be found in Bellamy's later work, "Equality," which comparatively few, even of English socialists, are acquainted with. The book is a sequel to "Looking Back-ward," and contains more than twice the matter. It shows, systematically, how our existing system of competition and individual profit—capitalism and enormous private wealth—directly lead to overwork, poverty, starvation, and crime ; that it is necessarily wasteful in production and cruelly unjust in distribution ; that it fosters every kind of adulteration in manufacture, and almost necessitates lying in trade ; that it involves the virtual slavery of the bulk of the population, and checks or destroys any real progress of the race. It also shows how, even the wealthy few, and also the members of each successive grade of comparative well-being, suffer from it socially, by the extreme restriction in each locality of possible intimate associates and friends ; it shows how we can never attain to the maximum benefits and enjoyment of social intercourse without that absolute equality of economic condition, educational opportunities, and social conventions, which alone put us at ease with our fellow-men ; while the enormous loss to all of us of the infinite varieties of character, ability, and even genius, now forbidden any adequate development by the cruel struggle for existence and the shortened lives, are clearly set forth. And as every one of the wasteful and cruel and debasing influences of our competitive system will cease to exist under a rational socialism, labour will be diminished to an almost inconceivable extent, while every possible enjoyment of nature, of art, and of congenial friendship will be indefinitely increased. Until these two works of Bellamy have been carefully read and thoroughly appreciated, no one can properly realize what such a state of society means ; while to any one who has done so, the stock objections to socialism will be seen to be utterly trivial and absurd.

One of the most striking and convincing chapters in " Equality " is that which describes the means by which, after a majority were in favour of it, and a Socialist Government had been elected, the great change was brought about, and, without any compulsion whatever, was soon welcomed and accepted by the adverse minority. This method is so simple, and so little known, that it may be well to give a brief outline of it here.

It is assumed that, before this period, there had already been a great extension of governmental and municipal industry, all the railways and mines, telegrams and tele-phones being worked by the former ; all water, gas, electric light and power, trams, etc., by the latter. The employees in these, together with all persons connected with the courts, the police, the revenue, and other Government offices, with their families, would comprise a population of several million persons paid by and dependent on general or local governments. The first important step taken is the opening of Government stores to supply all these persons with food, clothing, and other necessaries of life at cost price, and of the best quality, absolutely free from adulteration, just as Robert Owen did for his people at New Lanark. As the numbers to be supplied would be exactly known, as no advertisements would be needed beyond simple price-lists, and as there need be no attractive shops in great thoroughfares at high rents, these necessaries could always be supplied at 25 per cent., and often at 50 per cent. below actual retail prices of the time. Robert Owen at New Lanark, with the comparatively small population of 2500 people, was able to supply goods of similar character at about 30 per cent. below shop prices. As this would be equivalent to an increase of earnings by all these employees, all other socialists, whose votes had brought the Government into power, asked for similar benefits, which were, of course, given them. Then an extension was made to the *manufacture* of the most important articles, such as metal goods of all kinds, china and glass, all the commoner textile fabrics, furniture, house-building, etc., so that in the course of a few years *every* necessary and comfort of life would be obtainable by all socialists at the Government stores, at low prices and of the very best quality. At the same time, the health of all these employees would be safeguarded by every available sanitary appliance and rule ; hours of work would be shortened in proportion to the fatigue or the monotony of the labour, and everything possible would be done to make the worker's life a healthy and enjoyable one. And as all these things would be done at *their own expense*, since all the products of labour would be sold at the price they cost to make and distribute, the non-socialists could not possibly complain, as they would not be called upon to bear any of the expense, but would have to go on purchasing the adulterated and costly products of private competition and capitalism as before.

Is it not a fair supposition which Bellamy makes, that at this stage of progress all the workers, all the wage-earners and employees of the private capitalists would beg to be taken into Government employment so as to share in the well-being of their socialist fellow-workmen? The result would be that, gradually and successively, all industry would become organized under the local authorities in co-operation with the various central stores and manufactories. During this process of extension private capitalists would find it more and more difficult to obtain skilled labour of any kind. They would then find that their former boasted "capital" was *not* the chief factor in the production of wealth ; that though they might have *money*, they would not possess *wealth*. The Government stores would, of course, be used by socialists only, by means of a system of tickets or paper money, as described by Bellamy ; capitalists and their managers would gradually have to join the socialist ranks as organizers or superintendents if they had the capacity, or if they preferred to live idle lives they might go to other countries where the competitive *régime* still prevailed. It may, of course, be *said* that this would not succeed ; that the Government *could not* compete with private capitalists, manufacturers, and shopkeepers. But few people who really *think* of the matter will believe this. The American Trusts *do* succeed in competition with the whole world, because they possess *some* of the advantages a Government would possess in a still greater degree. But *they* result in small traders beggared and workers no longer wanted, and in the production of a hundred or more of multi-millionaires. If a socialist *régime* cannot, in the nature of things, succeed, why are all the great capitalists so dreadfully afraid of allowing any approach to a fair trial of it by municipalities or other local authorities?

After much consideration, however, I have come to the conclusion that this will not (probably) be the way in which socialism will come about in England, and that it would not be the easiest or the best way. I think it more likely that we shall pass through a stage of true "individualism," in which complete "equality of opportunity" will be established. I have sufficiently explained this in my "Studies," vol. ii. chap. xxviii.; and if to this we add the broad scheme of general education outlined by Mr. John Richardson in his admirable little book, ".How it can be done," we shall have prepared the way for the rational society of the future. Equality of opportunity is, as Herbert Spencer has shown in his "Justice," the correlative of natural selection in human society, and has thus a broad foundation in the laws of nature. But Spencer himself did not follow out his principles to their logical conclusion as I have done.

Many good people to-day who are almost horror-struck at hearing that any one they know is a socialist, would be still more amazed if they knew how many of the very salt of the earth belong (or did belong) to this despised and much dreaded body of thinkers. Grant Allen, one of the most intellectual and many-sided men of our time, was one of us; so is Sir Oliver Lodge, one of our foremost students of physical science; and Professor Karl Pearson, a great mathematical evolutionist. Among the clergy we have the Revs. John Clifford, R. C. Fillingham, and many others among the Christian socialists, who are as much socialists as any of us. Among men of university training or of high literary ability we have H. M. Hyndman, Edward Carpenter, J. A. Hobson, Sydney Webb, Hubert Bland, H. S. Salt, J. C. Kenworthy, Morrison-Davidson, and many others. Of poets there are Gerald Massey and Sir Lewis Morris. The labour members of Parliament are almost all socialists; while Margaret Macmillan, the Countess of Warwick, and many less known women are earnest workers for the cause.

I should almost think that Mrs. Humphry Ward was a socialist at heart or as an ideal, or she could not have set forth its principles and the arguments for it so well as she has done in "Marcella." But the weak and illogical conclusion of that and some other books caused me to write to Grant Allen, urging him to write a thorough socialistic story, which I felt sure he could do better than any one I knew. His reply was so interesting from a literary point of view that I give it here. It is the last letter I received from him.

"Hotel Royal, Varenna, Italy, April 24.

"I despair of giving you in writing all the reasons why your suggestion is for me an impossible one. There are eleven thousand; I will content myself with two. The first is practical. I have to write stories which editors will accept and the public will buy. Now, no editor will take a socialistic story—I have tried, and failed; and the public will not buy such stories to a sufficient extent to pay for the trouble. As a general rule, the more in earnest I am about a subject, the less I get for it. The second reason is artistic. A story grows out of a plot or situation, and cannot be forced in the way you describe, as I at least do not know how to force it. Plots come. I could not invent a plot in order to sustain a particular thesis. Thank you so much for the many kind expressions in your letter. Come and see us some day on Hind Head, when you are passing up or down, and we will thrash this matter out more fully.

"With very kind regards,
"Cordially yours,
"GRANT ALLEN."

I do not know to what he alludes when he says he "has tried and failed." "The Woman who Did" was not socialistic, and I can only suppose he refers to a short story of life in a phalanstery, where all children in the least deformed are killed at one year old, for the improvement of the race; and the feelings of the mother for her first-born are vividly described, though, as the law was absolute and known from childhood, it was submitted to uncomplainingly! But neither of these stories had any necessary connection with socialism, and were especially repugnant to our customs and ideals. But there is nothing whatever repugnant in socialism itself, and I cannot believe that a story by a well-known and talented writer would be unsaleable merely because its field of action was a successful socialistic community.

I may conclude this chapter with the answer I recently gave to the question, " Why I am a Socialist ? " I am a socialist because I believe that the highest law for mankind is justice. I therefore take for my motto, " Fiat Justicia Ruat Cœlum ; " and my definition of socialism is, " The use by every one of his faculties for the common good, and the voluntary organization of labour for the equal benefit of all." That is absolute social justice ; that is ideal socialism. It is, therefore, the guiding star for all true social reform.

CHAPTER XXXV

I HAVE already described my first introduction to mesmerism at Leicester, how I found that I had considerable mesmeric power myself, and could produce all the chief phenomena on some of my patients ; while I also satisfied myself that almost universal opposition and misrepresentations of the medical profession were founded upon a combination of ignorance and prejudice. I will here only add that my brother Herbert also possessed the power, and that when we were residing together at Manaos, he used to call up little Indian boys out of the street, give them a copper, and by a little gazing and a few passes send them into the trance state, and then produce all the curious phenomena of catalepsy, loss of sensation, etc., which I have already described. This was interesting because it showed that the effects could be produced without any expectation on the part of the patients, and, further, that similar phenomena followed as in Europe, although these boys had certainly no knowledge of such phenomena. One day, I remember, when we were going out collecting, we entered an Indian's hut, where we had often been before, and my brother quietly began mesmerizing a young man nearly his own age. He did not entrance him, but obtained enough influence to render his arm rigid. This he instantly relaxed, and asked the Indian to lie down on the floor, which he did. My brother then made a pass along his body, and said, " Lie there till we return." The man tried to rise but could not, though several of his relatives were present. We then walked out, he crying and begging to be loosed. Thinking he would certainly overcome the influence we went on, and coming back about two hours later we found the man still on the ground, declaring he could not get up. On a pass from my brother and his saying, "Now get up," he rose easily. We gave him a small present, but he did not seem much surprised or disturbed, evidently thinking we were white medicine-men. Here, again, it seemed to me pretty certain that the induced temporary paralysis was a reality, and by no means due to the imagination of the usually stolid Indian.

During my eight years' travels in the East I heard occasionally, through the newspapers, of the strange doings of the spiritualists in America and England, some of which seemed to me too wild and *outré* to be anything but the ravings of madmen. Others, however, appeared to be so well authenticated that I could not at all understand them, but concluded, as most people do at first, that such things *must* be either imposture or delusion. How I became first acquainted with the phenomena and the effect they produced upon me are fully described in the "Notes of Personal Evidence," in my book on "Miracles and Modern Spiritualism," to which I refer my readers. I will only state here that I was so fortunate as to be able to see the simpler phenomena, such as rapping and tapping sounds and slight movements of a table in a friend's house, with no one present but his family and myself, and that we were able to test the facts so thoroughly as to demonstrate that they were not produced by the physical action of any one of us. Afterwards, in my own house, similar phenomena were obtained scores of times, and I was able to apply tests which showed that they were not caused by any one present. A few years later I formed one of the committee of the Dialectical Society, and again witnessed, under test conditions, similar phenomena in great variety, and in these three cases, it must be remembered, no paid mediums were present, and every means that could be suggested of excluding trickery or the direct actions of any one present were resorted to.

At a later period I paid frequent visits, always with some one or more of my friends as sceptical and as earnest inquirers after fact as myself, to one of the best public mediums for physical phenomena I have ever met with—Mrs. Marshall and her daughter-in-law. We here made whatever investigations we pleased, and tried all kinds of tests. We always sat in full daylight in a well-lighted room, and obtained a variety of phenomena of a very startling kind, as narrated in the book referred to. During the latter part of my residence in London (1865–70) I had numerous opportunities of seeing phenomena with other mediums in various private houses in London. These were sometimes with private, sometimes with paid mediums, but always under such conditions as to render any kind of collusion or imposture altogether out of the question. During this time I was in frequent communication with Sir William Crookes, Mr. Cromwell Varley, Serjeant Cox, Mr. Hensleigh Wedgwood, Mr. E. T. Bennett, Mr. S. C. Hall, Professor and Mrs. de Morgan, Mr. W. Volckman, Rev. C. Maurice Davies, Dr. and Mrs. Edmunds, William Howitt, Mrs. Catherine Berry, and many other friends, who were either interested in or were actively investigating the subject ; and through the kindness of several of them I had many opportunities of witnessing some of the more extraordinary of the phenomena under the most favourable conditions. At a much later period, when I visited America on a lecturing tour, I made the acquaintance of some of the most eminent spiritualists in Boston and Washington, and had many opportunities of seeing phenomena and obtaining tests of a different kind from any that I had seen in England ; and some of these I may refer to later on. What I propose to do now is to give a consecutive outline of my correspondence with some of my scientific and literary friends on this subject, which will, I think, have some historical interest now that investigations into physical phenomena are not treated in the same utterly contemptuous way they were in the early period of my inquiries.

When I had obtained in my own house the phenomena

described in my "Notes of Personal Evidence," I felt sure
that if any of my scientific friends could witness them they
would be satisfied that they were not due to trickery, and
were worthy of a careful examination. I therefore first
invited Dr. W. B. Carpenter to attend some of our sittings,
telling him that I could not guarantee anything without a
series of, say, half a dozen visits. He came one evening, the
only other persons present being the medium—Miss Nichol—
my sister, and myself. After a short time a few raps were
heard on the table, and these were repeated, sometimes in
different tones, and sounding, at request, in any part of the
table. They were not, however, strong, and soon came to
an end. Dr. Carpenter sat quite still, and made hardly any
remark. He knew from my statements that this was a mere
nothing to what often occurred, and though I strongly urged
him to come at least two or three times more, I never could
prevail upon him to come again.

I then tried to get Professor Tyndall to take up the
subject seriously, giving him an account of the results I had
obtained, the tests I had applied, and the general conditions
that seemed favourable or unfavourable. He replied in a
letter which I now have before me, and as it shows how
difficult it then was to get any man of eminence to keep an
open mind on this subject, I think it worth reproducing.

'MY DEAR WALLACE,

"Your sincerity and desire for the pure truth are
perfectly manifest. If I know myself, I am in the same vein.
I would ask one question.

"Supposing I join you, will you undertake to make the
effects evident to my senses? Will you allow me to reject
all testimony, no matter how solemn or respectable? Will
you allow me to touch the effects with my own hands, see
them with my own eyes, and hear them with my own ears?
Will you, in short, permit me to act towards your phenomena
as I act, and successfully act, in other departments of
nature?

"I really wish to see the things able to produce this

conviction in a mind like yours, which I have always con-sidered to be of so superior a quality.

> " I am, very faithfully,
> " JOHN TYNDALL."

I replied to this extraordinary letter by telling him that *I* could "undertake" nothing, but that the phenomena had occurred at various times when many different persons had been present ; that, of course, he could examine and test them as he pleased, but that if he really wished to witness the phenomena in all their variety, I strongly advised him to be a passive spectator on the first two or three visits, and only apply tests and impose conditions at a later period. I asked him to name a day, and he came.

At the very beginning he forgot or purposely acted contrary to my advice. On being asked to sit at the table with my sister, Miss Nichol, and myself, he declined, saying, " I never form part of my experiments. I will sit here and look on "—drawing his chair about a yard away. So we three sat without him, with our hands on the table ; and rather to my surprise the rapping sounds began, and were much stronger and more varied in character than when Dr. Carpenter had heard them. They were, in fact, very varied in tone—some mere ticks, others loud slaps or thumps. But to all this he paid no attention. He joked with Miss Nichol, who was always ready for fun, and, after the raps had gone on some time, he remarked, " We know all about these raps. Show us something else. I thought I should see something remarkable." But nothing else came. Then, after a little talk and more chaff with Miss Nichol, he said " Good night ; " and though I begged him to appoint a day for the next sitting, *he* never came again.

I next tried Mr. G. H. Lewes (whose acquaintance I had made at Huxley's), but he was too much occupied and too incredulous to give any time to the inquiry. During this time I was reading almost everything I could obtain upon the phenomena, and found that there was such a mass of testimony by men of the highest character and ability in every department of human learning, that I thought it would be useful to bring these together in a connected sketch of the whole subject. This I did, and sent it to a secularist magazine, in which it appeared in 1866, and I also had a hundred copies printed separately, which I distributed among my friends. It was called "The Scientific Aspect of the Supernatural," a somewhat misleading title, as in the introductory chapter I argued for all the phenomena, however extraordinary, being really "natural" and involving no alteration whatever in the ordinary laws of nature. Some years later (1874) this was included in my volume on "Miracles and Modern Spiritualism," with an additional chapter, "Notes of Personal Evidence."

The letters I received from those to whom I sent copies of this little pamphlet were interesting though not instructive. Huxley wrote: "I am neither shocked nor disposed to issue a Commission of Lunacy against you. It may be all true, for anything I know to the contrary, but really I cannot get up any interest in the subject. I never cared for gossip in my life, and disembodied gossip, such as these worthy ghosts supply their friends with, is not more interesting to me than any other. As for investigating the matter—I have half a dozen investigations of infinitely greater interest to me—to which any spare time I may have will be devoted. I give it up for the same reason I abstain from chess—it's too amusing to be fair work, and too hard work to be amusing."

To the latter part of this letter no objection can be made, but the objection as to "gossip" was quite irrelevant as regards a book which had not one line of "gossip" in it, but was wholly devoted to a summary of the evidence for facts—physical and mental—of a most extraordinary character, given on the testimony of twenty-two well-known men, mathematicians, astronomers, chemists, physiologists, lawyers, clergymen, and authors, many of world-wide reputation.

Tyndall read the book "with deep disappointment," because it contained no record of my own experiments. *He* knew Baron Reichenbach, and had visited him, and had seen all his apparatus and his methods. It was *he* who had reproached Thackeray for allowing the article about "Home" to appear in the *Cornhill Magazine*, and he added—

"Poor Thackeray was staggered and abashed by the earnestness of my remonstrance regarding the lending the authority of his name to 'Stranger than Fiction,' my great respect for Thackeray rendering my remonstrance earnest."

Then he concludes with a gentle admonition to myself—

"I see the usual keen powers of your mind displayed in the treatment of this question. But mental power may show itself, whether its material be facts or fictions. It is not lack of logic that I see in your book, but a willingness that I deplore to accept data which are unworthy of your attention. This is frank—is it not?

"Yours very faithfully,
"JOHN TYNDALL."

G. H. Lewes, to whom I had sent the little book with an invitation to investigate at my house the phenomena which occurred with my friend Miss Nichol, replied much in the same way as Tyndall—that he was quite ready to examine any serious claim to spiritual power, but that he *had* "thoroughly examined" the phenomena, "had forced Mrs. Hayden to avow herself an impostor," while all other mediums he had tested "were either impostors or dupes." Still, he would come to me if he could have "*all* the conditions of testing the phenomena freely accorded." He "would not permit a medium to determine the conditions or to open the usual loopholes of escape." He would also wish to bring Mr. Herbert Spencer, or some other scientific friend with him; and he concluded, "I pledge you my word that I will publicly state, with all the accuracy I can, whatever phenomena I may witness."

I gladly accepted his offer, only stipulating as before that he should not impose conditions on the *first* occasion, and that he should devote at least six sittings (I think) of an hour each to the investigation before coming to any conclusion. But he never came at all.

Several of my friends about this time urged me strongly to make a personal investigation of the subject. Among these were my old companion, H. W. Bates, and Professor E. B. Tylor. I was doing so at the time, but when I published the results a few years later, and about the same time Sir William Crookes published his much more remarkable investigations, both alike were received with silence, incredulity, or contempt.

Notwithstanding this refusal to accept my offer of a full examination of phenomena which had repeatedly occurred in my presence and had been submitted to varied tests, a year afterwards two of these men of science wrote to the *Pall Mall Gazette* (May 19, 1868), making various accusations against mediums and spiritualists. Mr. Lewes declared that scientific men are never allowed to investigate, but are put off by an evasion of some kind ; and many other things equally untrue. He then suggested that the whole thing could be tested by allowing Professor Tyndall to have one sitting with any medium, and to propose three questions for the spirits to answer correctly. I thereupon wrote to the editor with a full reply, pointing out that Mr. Cromwell Varley, the eminent electrician, had recently published the statement that he *had* been permitted to investigate fully by Mr. Home, with satisfactory results. I then related a series of test experiments in my own house, and asked Mr. Lewes how his statement that others have discovered how the tables are turned (and can turn them), how the raps are produced (and can produce them), how the ropes are untied (and can untie them), can apply to such phenomena as I relate, and to such tests and conditions as I gave, or what bearing Professor Tyndall's proposed "three questions" could have upon them.

This reply was, however, refused publication by the editor, and I wrote to Mr. Lewes suggesting that, for the sake of his own reputation, he should in future, if he wrote publicly on this subject, do so only in such journals as would admit a reply.

As an example of the strange methods of our opponents at this time, I may refer to Mr. Lewes's statement to me that "he had forced Mrs. Hayden to avow herself an impostor." As this was important if true, because this lady was the medium whose phenomena had convinced Professor de Morgan, I inquired further about it, and found from Mr. Lewes's own statement of his experiment that he had asked a series of written questions which were answered through the alphabet by raps in the usual way, most of the answers being either vague or altogether wrong, and the last question was, "Is Mrs. Hayden an impostor?" to which the answer was "Yes." And this ingenious trick he afterwards termed "forcing Mrs. Hayden to avow herself an impostor!"

As it is always of interest to have at first-hand an expression of the frame of mind of eminent men upon this subject, I here give a letter from John Stuart Mill to a gentleman who sent him a tract in which it was stated that he, along with Ruskin, Tennyson, and Longfellow, had become believers in spiritualism, and asking if it were true. This gentleman, Mr. N. Kilburn, of Bishop Auckland, sent me a copy of Mill's reply, which was as follows :—

"It is the first time I ever heard that I was a believer in spiritualism, and I am not sorry to be able to suppose that some of the other names I have seen mentioned as believers in it are no more so than myself.

"For my own part I not only have never seen any evidence that I think of the slightest weight in favour of spiritualism, but I should also find it very difficult to believe any of it on any evidence whatever, and I am in the habit of expressing my opinion to that effect very freely whenever the subject is mentioned in my presence. You are at liberty to make any use you please of this letter."

This was dated "March 18, 1868," but I did not know of it till 1874, or I might have mentioned the subject when I dined with him in 1870. If by "any evidence whatever" Mr. Mill meant testimony of others, I myself, and most spiritualists, were in the same frame of mind when we began our inquiries ; but as he used the word "evidence," he no doubt included personal evidence, and to decide beforehand that he would not believe it is very unphilosophical. Still, he only says *difficult*, not *impossible*, and here, again, I quite agree with him.

At this same period I had letters from other men of various degrees of eminence of a much more satisfactory nature. On receipt of a copy of my pamphlet, Professor de Morgan wrote me as follows:—

"I am much obliged to you for your little work, which is well adapted to excite inquiry. But I doubt whether inquiry by *men of science* would lead to any result. There is much reason to think that the state of mind of the inquirer has something—be it internal or external—to do with the power of the phenomena to manifest themselves. This I take to be one of the phenomena—to be associated with the rest in inquiry into cause. It may be a consequence of action of incredulous feeling on the nervous system of the recipient; or it may be that the volition—say the spirit, if you like—finds difficulty in communicating with a repellent organization; or, maybe, is offended. Be it which it may, there is the fact.

"Now the man of science comes to the subject in utter incredulity of the phenomena, and a wish to justify it. I think it very possible that the phenomena may be withheld. In some cases this has happened, as I have heard from good sources.

"I have had students[1]—a couple of dozen in my life—whose effort always was *not to see it*. As I, their informing spirit, was under contract to make them see it if I could—which the *spirits* we are speaking of are *not*—I generally succeeded in convincing them. In their minds I have studied—with power of experiment arranged by myself—the character of the man of science.

"D'Alembert said, speaking of mathematics—of all things —'*En avant et la foi viendra.*' But I doubt if the man of science of our day can persuade himself of a possibility of his fifth attempt destroying the effect of the failure of the first four.

---

[1] De Morgan was one of the greatest mathematicians of his time, and Professor of Mathematics at University College.

"Your book will set many rational persons suspecting they ought to inquire.

"Yours faithfully,
"A. DE MORGAN."

This seems to me to exhibit the scientific frame of mind, as manifested by Tyndall, Lewes, and W. B. Carpenter, with great perspicuity.

I had some correspondence at this time with William Howitt, and he and Mrs. Howitt came one evening for a *séance* with Miss Nichol, and were much pleased with the curious musical and other phenomena; and I also made the acquaintance of Mr. and Mrs. S. C. Hall, and visited them to attend a *séance* with Home, which, although all present were friends and spiritualists, turned out a failure, owing to the circle being broken by Mr. Hall being called out on urgent business.

But perhaps the most interesting response to a copy of my pamphlet was that from Robert Chambers, which I here give—

"St. Andrews, February 10, 1867.

"DEAR SIR,
"I have received your letter of the 6th inst., and your little volume. It gratifies me much to receive a friendly communication from the Mr. Wallace of my friend Darwin's 'Origin of Species,' and my gratification is greatly heightened on finding that he is one of the few men of science who admit the verity of the phenomena of spiritualism. I have for many years *known* that these phenomena are real, as distinguished from impostures; and it is not of yesterday that I concluded they were calculated to explain much that has been doubtful in the past, and when fully accepted, revolutionize the whole frame of human opinion on many important matters.

*   *   *   *   *   *

"How provoking it has often appeared to me that it

seems so impossible, with such a man, for instance, as Huxley, to obtain a moment's patience for this subject—so infinitely transcending all those of physical science in the potential results!

"My idea is that the term 'supernatural' is a gross mistake. We have only to enlarge our conceptions of the natural, and all will be right.

"I am, dear sir,
"Yours very sincerely,
"ROBERT CHAMBERS."

In the latter part of the year, while attending the meeting of the British Association at Dundee, I visited St. Andrews, and after a geological excursion under the guidance of Sir A. Geikie, and a collation with the university authorities, at which Robert Chambers was present, I had the great pleasure of an hour's conversation with him in his own house. *The Spiritual Magazine,* founded by William Howitt and some friends, was at that time admirably edited by Mr. Thomas Shorter, and my host told me that he always read it through from cover to cover, and that few of the magazines of the day contained so much valuable information and so much good writing as this depised periodical, in which I fully agreed with him.

Two years later (in 1869) I received a letter from him to introduce me to Miss Douglas, a lady much interested in spiritualism, who lived in South Audley Street. Here I attended many *séances*—on one occasion when Home was the medium and Mr. (now Sir William) Crookes was present. As I was the only one of the company who had not witnessed any of the remarkable phenomena that occurred in his presence, I was invited to go under the table while an accordion was playing, held in Home's hand, his other hand being on the table. The room was well lighted, and I distinctly saw Home's hand holding the instrument, which moved up and down and played a tune without any visible cause. On stating this, he said, "Now I will take away my hand"—which he did; but the instrument went on playing, and I saw a detached hand holding it while Home's two hands were seen above the table by all present. This was one of the ordinary phenomena, and thousands of persons have witnessed it ; and when we consider that Home's *séances* almost always took place in private houses at which he was a guest, and with people absolutely above suspicion of collusion with an impostor, and also either in the daytime or in a fully illuminated room, it will be admitted that no form of legerdemain will explain what occurred.

In view of the extraordinary misstatements that were continually made by scientific men, who had influence with the public (and are still made both on this and on other subjects), it will be well to give a short account of one of these, which caused much discussion at the time.

Mr. Home first came to England (since his childhood) early in 1855, and lived for some months with Mr. Cox, of Cox's Hotel in Jermyn Street. Here, among numerous other eminent men, he gave a sitting to Lord Brougham accompanied by Sir David Brewster, "in order to assist in finding out the trick," as Sir David himself stated. About six months afterwards a not quite correct account of this *séance* was given in the *Morning Advertiser*, copied from an American paper, whereupon Sir David wrote to the editor to give his own account, in which he said, "It is quite true that I saw at Cox's Hotel, in company with Lord Brougham, and at Ealing, in company with Mrs. Trollope, several mechanical effects which I was unable to explain. But although I could not account for all these effects, I never thought of ascribing them to spirits stalking beneath the drapery of the table ; and *I saw enough to satisfy myself that they could all be produced by human hands and feet*, and to prove to others that some of them, at least, had such an origin.

" Were Mr. Home to assume the character of the Wizard of the West, I should enjoy his exhibition as much as that of other conjurors ; but when he pretends to possess the power of introducing among the feet of his audience the spirits of the dead, of bringing them into physical communication with their dearest relatives, and of revealing the secrets of the grave, he insults religion and common sense, and tampers with the most sacred feelings of his victims.

"I am, sir,

"Yours, etc.,

"D. BREWSTER."

Here Sir David appeals to religious prejudice, as he had just done in his very weak book in reply to Whewell's "Plurality of Worlds." But his account of the *séance* and the imputations it cast on both Home and his host, Mr. Cox, were at once answered by that gentleman, who declared that, immediately after the *séance*, both Lord Brougham and Sir David had expressed their great astonishment, and that the latter had exclaimed, "Sir, this upsets the philosophy of fifty years." A friend of Mr. Cox and of Home—Mr. Coleman— also wrote, reminding Sir David that very shortly afterwards he and Mr. Cox had called upon him to talk over the subject, and that Sir David declared that what he had seen was "quite unaccountable." Mr. Coleman continues thus:—

"I then asked him, 'Do you, Sir David, think these things were produced by trick?'

"'No, certainly not,' was his reply.

"'Is it a delusion, think you?'

"'No; that is out of the question.'

"'Then what is it?'

"To which he replied, 'I don't know; but spirit is the last thing I give in to.'"

To this Sir David replied by a very long letter, denying some things and explaining others. The most important passages are the following:—

"Mr. Home invited us to examine *if there was any machinery about his person*, an *examination, however, which we declined to make*. When all our hands were upon the table noises were heard—rappings in abundance; and, finally, when we rose up, the table actually rose, *as appeared to me*, from the ground. This result I do not pretend to explain. . . .

"A small hand-bell, to be rung by the spirits, was placed

on the ground near my feet. I placed my feet round it in the form of an angle, to catch any intrusive apparatus. *The bell did not ring;* but when taken across to a place near Mr. Home's feet, it speedily came across, and placed its handle in my hand. This was amusing."

There is also a long account of the phenomena he saw at Ealing in a still more jocular vein, which called forth a very scathing letter from Mr. T. Adolphus Trollope, who had been present. These letters and some others can all be read in full in an appendix to Home's "Incidents in my Life," and as this appendix was drawn up by Dr. Robert Chambers (as I know from private information), the reader may feel satisfied that these letters are given as they were written.

But the chief reason why I have introduced the matter here is, that we possess, fortunately, another account of Sir David Brewster's *séance* at Cox's Hotel, written by himself very shortly afterwards, while the facts were fresh in his memory, in a letter to some member of his own family, and published in the " Home Life of Sir David Brewster " by his daughter, in 1869. At my request my friend Mr. Benjamin Coleman sent me a copy of this contemporary account, dated London, June, 1855. It is as follows :—

"Last of all, I went with Lord Brougham to a *séance* of the new spirit-rapper, Mr. Home, a lad of twenty, the son of a brother of the late Earl of Home. He went to America at the age of seven, and, though a naturalized American, is actually a Scotchman. Mr. Home lives in Cox's Hotel, in Jermyn Street, and Mr. Cox, who knows Lord Brougham, wished him to have a *séance*, and his lordship invited me to accompany him, in order to assist in finding out the trick. We four sat down at a moderately sized table, *the structure of which we were invited to examine.* In a short time *the table shuddered,* and a *tremulous motion ran up all our arms;* at our *bidding these motions ceased and returned.*

"The most unaccountable rappings were produced in various parts of the table, and the table *actually rose from the ground* when no hand was upon it. A larger table was produced, and exhibited similar movements. An accordion was held in Lord Brougham's hand and gave out a single note, but the experiment was a failure; it would not play either in his hand or mine.

"A small hand-bell was then laid down with its mouth on the carpet, and after lying for some time *it actually rang when nothing could have touched it.* The bell was then placed on the other side, still upon the carpet, and *it came over to me and placed itself in my hand. It did the same to Lord Brougham.*

"These were the principal experiments; we could give no explanation of them, and *could not conjecture how they could be produced by any kind of mechanism.* Hands are sometimes seen and felt, the hand often grasps another, and melts away as it were under the grasp.

"The object of asking Lord Brougham and me seems to have been to get our favourable opinion of the exhibition, but though neither of us can explain what we saw, we do not believe that it was the work of idle spirits."

I have italicized certain passages in this early letter to compare with the corresponding parts of the letters Sir David wrote to the *Morning Advertiser* about half a year later, and it will be seen that the discrepancies are very serious. He told the public that he had *satisfied himself* that all could have been done *by human hands and feet;* whereas in his earlier private letter he terms them *unaccountable,* and says that he *could not conjecture* how they were done. Neither did he tell the public of the *tremulous motion up his arms,* while he denied that the bell rang at all, though he had before said that it *actually rang where nothing could have touched it.*

If this case stood alone it would not, perhaps, be worth mentioning, but a similar tendency has prevailed in all the scientific opponents of spiritualism, one example of which I have given in the case of Mr. Lewes's declaration that he had forced Mrs. Hayden to avow herself an impostor, whereas what happened really proved that Mrs. Hayden herself did not consciously give the answers to his questions.

One of the eminent men with whom I became acquainted

through spiritualism was Mr. Cromwell F. Varley, the electrician. Any one who will read his evidence, printed in the Report of the Dialectical Society (1871), will see that he was at first as sceptical as any other scientific man usually is, and ought to be, but, having married a lady who was a medium, phenomena of such a marvellous nature were presented to him in his own home, that he could not help becoming an ardent believer. But he was always a critic and an experimenter, and he assisted Sir William Crookes in applying some of the electrical tests to Mrs. Fay, as described by that gentleman in *The Spiritualist* newspaper of March 12, 1875.

I became acquainted with him in 1868 through a letter from Professor Tyndall referring, I think, to the single test at one *séance* as proposed by G. H. Lewes in the *Pall Mall Gazette* shortly afterwards, and suggesting that Mr. Varley, who had published some of his investigations, might be able to supply such a test. To this letter I replied as follows :—

"May 8, 1868.

"DEAR MR. TYNDALL,

"I do not know Mr. Varley, but I will forward him your note, and he can reply if he thinks proper. I rather doubt if any *single case* would be conclusive to you. Hume's argument is overwhelming against any single case, considered alone, however well authenticated. He himself admits that no facts could possibly be better authenticated than the (so-called) miracles which occurred at the tomb of the Abbé Paris. But when you look at a series of such cases, amounting to thousands in our own day, and a corresponding series extending back through all history, Hume's argument entirely fails, because his major proposition—that such facts are contrary to the *universal experience of mankind*—ceases to be true.

"During the last two years I have witnessed a great variety of phenomena, under such varied conditions that each objection as it arose was answered by other phenomena. The further I inquire, and the more I see, the more impossible

becomes the theory of imposture or delusion. I *know* that
the facts are real natural phenomena, just as certainly as I
know any other curious facts in nature.

"Allow me to narrate *one* of the scores of equally remarkable things I have witnessed, and this one, though it certainly
happened in the dark, is thereby only rendered more difficult
to explain as a trick.

"The *place* was the drawing-room of a friend of mine, a
brother of one of our best artists. The *witnesses* were his
own and his brother's family, one or two of their friends,
myself, and Mr. John Smith, banker, of Malton, Yorkshire,
introduced by me. The medium was Miss Nichol. We sat
round a pillar-table in the middle of the room, exactly under
a glass chandelier. Miss Nichol sat opposite me, and my
friend, Mr. Smith, sat next her. We all held our neighbour's
hands, and Miss Nichol's hands were both held by Mr. Smith,
a stranger to all but myself, and who had never met Miss N.
before. When comfortably arranged in this manner the
lights were put out, one of the party holding a box of
matches ready to strike a light when asked.

"After a few minutes' conversation, during a period of
silence, I heard the following sounds in rapid succession:
a slight *rustle*, as of a lady's dress ; a little *tap*, such as
might be made by setting down a wineglass on the table ;
and a very slight jingling of the drops of the glass chandelier.
An instant after Mr. Smith said, 'Miss Nichol is gone.' The
match-holder struck a light, and on the table (which had
no cloth) was Miss Nichol *seated in her chair*, her head just
touching the chandelier.

"I had witnessed a similar phenomenon before, and was
able to observe coolly ; and the facts were noted down soon
afterwards. Mr. Smith assured me that Miss Nichol simply
glided out of his hands. No one else moved or quitted hold
of their neighbour's hands. There was not more noise than
I have described, and no motion or even tremor of the table,
although our hands were upon it.

"You know Miss N.'s size and probable weight, and can
judge of the force and exertion required to lift her and her

chair on to the exact centre of a large pillar-table, as well as the great surplus of force required to do it almost instantaneously and noiselessly, in the dark, and without pressure on the side of the table which would have tilted it up. Will any of the known laws of nature account for this?

"Yours very faithfully,
"ALFRED R. WALLACE."

Of course I did not expect Professor Tyndall to accept such a fact on my testimony; on the contrary, I described it for the very purpose of arguing that, if he himself had been present, he would probably not have been satisfied that it was not a trick, unless he could have it repeated under varied conditions. Yet he was so illogical as to think that a test phenomenon occurring once only under his or Mr. G. H. Lewes's conditions would settle the whole question—that is, would satisfy the scientific world and the general public that the spiritualistic phenomena were genuine, and that what used to be called "miracles" did happen in our midst to-day. Sir William Crookes's experience, a few years later, proves how totally wrong Tyndall was in this opinion, since his careful experiments, continued for several years, are to this day ignored or rejected by the bulk of scientific and public opinion as if they had never been made!

In order to show Mr. Varley's liberal spirit towards opponents, and also for suggestions of great value, I give here some extracts from a letter I received from him in January, 1869—

"We spiritualists should remember that the way in which science has reached its present brilliant position has been through our philosophers doubting, disbelieving, and testing everything until further disbelief was impossible.

"We privileged ones owe it to the world to present spiritualism to them in a manner so clearly defined and demonstrated, that those who follow us shall be able to make themselves as much masters of the subject as we are.

"What is wanted is to bring together a large number

of harmonious mediums, to form of these several circles of different characters, and to secure the assistance of several clairvoyants.

"Each circle should be under the management of a clever man and each should carry on a continuous and exhaustive examination of the groundwork of the subject. Once establish a clue to the relations existing between the physical forces known to us and those forces by which the spirits are sometimes able to call into play the power by which they produce physical phenomena—once establish this clue there will be no lack of investigators, and the whole subject will assume a rational and intelligible shape to the outside world."

This was written thirty-five years ago, but, though the Society for Psychical Research has done a good deal, the first step has not been taken in the direction here indicated. Now, however, that a research fund is being formed there are better prospects. Much will depend, however, on choosing investigators who will be content for some time to observe the phenomena as they occur under those conditions which have been found most successful by other inquirers. Above all things, it is essential to make friends of the mediums employed, to treat them with the greatest consideration, and strictly to follow the advice of the intelligence that works through them. It was in this way that Sir William Crookes and other successful observers have obtained such striking results, under the most stringent conditions and subject to the most varied tests ; whereas those who begin by treating the mediums as if they were on their trial, and insist upon applying their own conditions at the very outset, usually obtain nothing but the conviction that all spiritualists are fools and all mediums impostors.

In 1872 I reviewed Robert Dale Owen's work, "The Debatable Land between this World and the Next," a sequel to his "Footfalls on the Boundary of Another World," the two forming the best-reasoned and the most logically arranged body of evidence for psychical phenomena in existence. Every example is quoted from the original authority wherever possible, confirmatory testimony has been collected with the greatest care, and the bearing of each upon the general argument is discussed or clearly pointed out. This review brought me a very interesting letter from the author, and later on a communication from Dr. Eugene Crowell, M.D. of New York, with a copy of his exceedingly valuable work, " Primitive Christianity and Modern Spiritualism" (2 vols.), in which almost every miraculous occurrence narrated in the Old or New Testaments is paralleled by well-authenticated phenomena from the records of modern spiritualism, many of them having been witnessed and carefully examined by Dr. Crowell himself.

During the years 1870–80 I had many opportunities of witnessing interesting phenomena in the houses of various friends, some of which I have not made public. Early in 1874 I was invited by John Morley, then editor of the *Fortnightly Review*, to write an article on " Spiritualism " for that periodical. Much public interest had been excited by the publication of the Report of the Committee of the Dialectical Society, and especially by Mr. Crookes's experiments with Mr. Home, and the refusal of the Royal Society to see these experiments repeated. I therefore accepted the task, and my article appeared in May and June under the title " A Defence of Modern Spiritualism." At the end of the same year I included this article, together with my former small book, " The Scientific Aspects of the Supernatural," and a paper I had read before the Dialectical Society in /1871 answering the arguments of Hume, Lecky, and other writers against miracles, in a volume which has had a very considerable sale, and has led many persons to investigate the subject and to become convinced of the reality of the phenomena. In the preface I showed the inaccuracy of Anton Dohrn's supposition that religious prejudices had led me to believe in spiritualism. A third edition of the book, in 1895, contained two new chapters on the nature and purport of apparitions, and also, in a new preface, a brief outline of the remarkable progress of the subject ; so that at the present day a large number of its phenomena, at first denied, and afterwards sneered at or ignored, have now become recognized and included among the undoubted facts of physiological or psychical science.

Among the friends with whom I investigated the subject was Mr. Marshman, at that time Agent-General for New Zealand, and Miss Buckley. Both were friends of Samuel Butler, the author of those remarkable works, "Erewhon" and "Life and Habit." Mr. Marshman invited him to a *séance* at his house, with myself and several other friends; but he thought it all trickery. I sent him a copy of my book, and he wrote me three letters in a week, chiefly to explain that the whole subject bored him. In his first letter he says that Mr. Marshman and Miss Buckley are two of the clearest-headed people he knows, and therefore he cannot help believing there must be something in it. "But," he says, "what I saw at the Marshmans' was impudent humbug." In the second he gives a curious revelation of the state of his mind in a personal anecdote. He writes: "Granted that wonderful spirit-forms have been seen and touched and then disappeared, and that there has been no delusion, no trickery. Well; *I don't care.* I get along quite nicely as I am. I don't want them to meddle with me. I had a very dear friend once, whom I believed to be dying, and so did she. We discussed the question whether she could communicate with me after death. 'Promise,' I said, and very solemnly, 'that if you find there *are* means of visiting me here on earth —that if you *can* send a message to me—*you will never avail yourself of the means, nor let me hear from you when you are once departed.*' Unfortunately she recovered, and never forgave me. If she had died, she would have come back if she could; of that I am certai⟨n⟩ by her subsequent behaviour to me. I believe my instinct was perfectly right; and I will go farther: if ever a spirit-form takes to coming near me, I shall not be content with trying to grasp it, but, in the interest of science, *I will shoot it.*"

The third is a very nice letter, and is a kind of apology for what he thought I might consider rather unreasonable in the others, and I will therefore give it, in order that my readers may not, through me, get a wrong idea of this remarkably gifted though eccentric writer.

"15, Clifford's Inn, E.C., May 27, 1859.

"DEAR SIR,

"Pray forgive me. I am sure I must have said rather more than I ought. A friend was with me when your letter came; I read it to him, and he said, 'If you grant Mr. Wallace's facts—and you do not deny them—he is perfectly right, and your answer does not meet him at all. He tells you that you are engaged on certain investigations in which your opinions must be entirely altered if you accept his facts. You admit this yourself—you do not deny his facts—and say that you do not care,—that is childish.'

"I admitted the truth of what he said; and I feel therefore that an apology is due to you, which pray understand me as making without reserve. I have read the greater part of the book you so kindly gave me, and shall read every word of it. I admire the force and clearness with which it is written, every word of it impressing me that it is written by one who understands his own meaning, and wishes others to understand it; but I cannot pretend that it has kindled in me that inward motion to see and hear more, without which you and I both know no good can come of any investigation.

"If there is that spiritual world independent of matter, which you believe in, a day may come when something will happen to me which will kindle in a moment the right spirit of inquiry; no one will follow it up more promptly or persistently when it is aroused. If that time never comes, it must be taken as a sign that I am not one of those from whom that cause would gain.

"Hoping you will forgive me for any rudeness that I fear I have been guilty of,

"Believe me,

"Yours very truly,

"S. BUTLER."

That seems to me a very pleasant letter, expressing his position very clearly. Of course, he had no rudeness to apologize for, as I told him, and though I do not think we met very often afterwards, we continued very good friends.

While residing at Godalming, I made the acquaintance of William Allingham and his wife—the poet and the artist—who then lived at Witley—I think it was about the years 1886 or 1887. Mr. Allingham told me that Tennyson wished to see me, and would be glad if I would come some day and lunch with him. A day was fixed, and I accompanied Mr. Allingham to the beautifully situated house on Blackdown, near Haslemere, where the poet lived during the summer. Lord Tennyson did not appear till luncheon was on the table, but in the mean time we had seen Lady Tennyson and her son and daughter-in-law, and been shown round the grounds. After luncheon we four men retired to the study, with its three great windows looking south-east over the grand expanse of the finely wooded Weald of Kent. Here Tennyson lit his pipe, and we sat round the fire and soon got on the subject of spiritualism, which was evidently what he had wished to talk to me about. I told him some of my experiences, and replied to some of his difficulties—the usual difficulties of those who, though inclined to believe, have *seen* nothing, and find the phenomena as described so different from what they think they ought to be. He was evidently greatly impressed by the evidence, and wished to see something. I gave him the names of one or two mediums whom I believed to be quite trustworthy, but whether he ever had any sittings with them I did not hear.

Then we talked a little about the tropics and of the scenery of the Eastern islands; and, taking down a volume he read, in his fine, deep, chanting voice, his description of Enoch Arden's island—

> " The mountain wooded to the peak, the lawns
> And winding glades high up like ways to heaven,
> The slender coco's drooping crown of plumes,
> The lightning flash of insect or of bird,

> The lustre of the long convolvuluses
> That coiled around the stately stems, and ran
> Ev'n to the limit of the land, the glows
> And glories of the broad belt of the world,—
> All these he saw; but what he fain had seen
> He could not see, the kindly human face,
> Nor ever hear a kindly voice, but heard
> The myriad shriek of wheeling ocean fowl,
> The league-long roller thundering on the beach,
> The moving whisper of huge trees that branch'd
> And blossom'd to the zenith, or the sweep
> Of some precipitous rivulet to the wave,
> As down the shore he ranged, or all day long
> Sat often in the seaward-gazing gorge,
> A shipwreck'd sailor waiting for a sail :
> No sail from day to day, but every day
> The sunrise broken into scarlet shafts
> Among the palms and ferns and precipices;
> The blaze upon the waters to the east;
> The blaze upon his island overhead ;
> The blaze upon the waters to the west ;
> Then the great stars that globed themselves in heaven,
> The hollower-bellowing ocean, and again
> The scarlet shafts of sunrise—but no sail."

Then he closed the book and asked me if that description was in any way untrue to nature. I told him that so far as I knew from the islands I had seen on the western borders of the Pacific, it gave a strikingly true general description of the vegetation and the aspects of nature among those islands, at which he seemed pleased. Of course, it avoids much detail, but the amount of detail it gives is correct, and it is just about as much as a rather superior sailor would observe and remember.

We then bade him good-bye, went downstairs and had tea with the ladies, and walked back to Haslemere station. I was much pleased to have met and had friendly converse with the most thoughtful, refined, broad-minded, and harmonious of our poets of the nineteenth century.

CHAPTER XXXVI

AMONG my scientific friends there are two with whom I had
some relations in regard to spiritualism of a specially interesting character—St. George Mivart and George J. Romanes—
and to each of these I must devote a few pages.

It was, I think, through my conversation and my first
small book that Mivart became satisfied that the phenomena
were at least partly genuine, and although a Roman Catholic,
he was not afraid to pursue the inquiry. On going to Naples
in the winter of 1870, he wrote me for an introduction to my
friends, Mr. and Mrs. Guppy, who were then staying there.
On the eve of his departure, he wrote telling me what had
happened—

"Nothing could have exceeded the kindness which your
good friends, Mr. and Mrs. Guppy, have shown to me, and I
have felt quite ashamed of the quantity of their time I have
taken up. Besides morning calls and a walk, they have
given me three *séances* (all to myself) and have most kindly
promised to give me a fourth and last this evening, as to-
morrow morning I start on my road northwards.

" At the first *séance* there was nothing but raps—questions
were replied to ; two of which much surprised me, as they
were only asked mentally. A remedy was indicated for an
affection of the teeth, which I have tried and believe will
prove efficacious.

" At the second *séance* (the first dark one I ever attended)
flowers were produced. The door was locked, the room

searched, and all requisite precautions taken. I was not surprised, because of all I had heard from you and others; but the phenomenon was to me convincing. *One* such fact is as good as a hundred.

"At the third *séance* (last night) I preferred to ask questions to having a repetition of the flowers. The value of the answers received time may show. I have received a wrong answer (as to a person being tall), also as to there being a letter awaiting me at my hotel. Altogether the conclusions I have arrived at are as follows :—

"I. I have encountered a power capable of removing sensible objects in a way altogether new to me.

"II. I have encountered an intelligence other than that of the visible assistants.

"III. In my *séances* this intelligence has shown itself capable of reading my thought, but yet either liable to fall into error or else not strictly truthful.

"IV. It has been sometimes capricious, saying *it will not do* what it has afterwards done, and that it will do what, nevertheless, it has failed to perform.

* * * * * *

"I am precluded from saying how much I like your friends, because I think this letter is to be read by them; but I am not precluded from thanking you, my dear Wallace, for the introduction, which I do very heartily, remaining always,

"Yours very truly,

"ST. GEORGE MIVART."

I was somewhat surprised at Mivart's appreciation of the Guppys, because of the great contrast between them : he extremely refined in speech and manners, and somewhat fastidious in his acquaintances ; they both rather brusque and utterly unconventional; yet he evidently recognized in them a straightforwardness of character, kindness of heart, love of truth, and earnestness of purpose, which are vastly more important than any amount of superficial polish. I may here note that he would probably have had more satisfactory results if he had allowed the powers at work to take their own course, instead of attempting to limit the phenomena to answering questions—a form of mediumship which, so far as I remember, was never very prominent or successful with Mrs. Guppy. This is the great fault of all beginners. Instead of being content for a time to *observe* only what happens, they almost always want certain phenomena which alone will satisfy them ; acting on the tacit assumption that *all* mediums and *all* preterhuman intelligences are able to produce at will *all* the various classes of phenomena. Those who follow the more scientific method of beginning with observation only— which, strange to say, the scientific men are hardly ever willing to do—almost always find that their early doubts and suspicions are, one by one, shown to be unfounded, through the occurrence of phenomena which seem specially adapted to answer them.

A few years later my friend visited Lourdes, in order to inquire on the spot as to the marvellous cures said to be effected there, and, if possible, to see some of them himself. While there he wrote me a very interesting letter, giving some account of his inquiries, which, being a Catholic, a well-known writer, and a good French scholar, he had facilities for pursuing which the ordinary English tourist or reporter does not possess. I give here the more important parts of this letter, dated April 5, 1874.

After referring to my *Fortnightly Review* article which I had told him I was writing, he continues, " We are here in a charming country and quiet, pleasant old town, at this season almost empty of visitors. We are here also, as you are, no doubt, fully aware, at the headquarters of a whole series of alleged modern miracles performed, as asserted, through the water which suddenly began to flow while Bernadette Sou-birons was in an ecstatic state in the presence, as she affirms, of an apparition of the B.V.M. [Blessed Virgin Mary].

" I have made such inquiries as I have been able, and find that here, on the spot, the miracles are fully believed in. The clergy were for a long time opposed to the whole thing, and the bishop had to be morally forced to institute an inquiry, he was so little disposed to accept such phenomena as facts. He ended, however, by being fully convinced, as also the curé (a fine, soldier-like man of about sixty-five, somewhat brusque in his manners), who is quite certain as to the marvellous nature of many cures. I have had a long talk with the doctor here (Dr. Dozens) and with two others at Toulouse (Dr. Rogues, No. 8, Rue d'Aussargues, and Dr. Noguès, Rue St. Anne). I will just mention one or two cases, as to the facts of which I have had face-to-face testimony from one or other of these doctors.

"A woman named Blaisette Soupevue of this place, about fifty, had had an affection (blepharite) of the eyes for several years. Both eyelids were partially everted, lashless, and the lower lids had numerous fleshy excrescences. Dr. Dozens attended this case himself, as also a Dr. Vergez. It was pronounced chronic, and all idea of cure abandoned. She washed her eyes with the water on two successive days; on the second her sight was completely restored, her eyelids righted themselves, and the excrescences vanished. Dr. Dozens assures me he examined this carefully himself. From that day her eyelashes began to grow, and she has never been so afflicted since.

"Justin Bontisharts, also of this place, had a rickety child ten years old, which had much atrophied limbs, and had never been able to walk. It got worse, and was thought to be near its death. Dr. Dozens tells me he attended it, and was present when the mother placed it under the stream of the Lourdes water. It was motionless while so held, and the bystanders therefore fancied it was dead already. The mother took it home, placed it in its bed, and noticed that it seemed to be in a tranquil sleep. Next day it woke with a quite different expression of face, craved for food, ate freely, and wanted to get up, but its parents were afraid to let it. The following morning, while they were out to work, it got up, and when they returned was walking about the room, walking quite well, and has done so ever since.

"Louis Bourriettes, a stone quarryman, had his face severely wounded, and his eyes injured by an explosion. One eye got pretty well; the other remained so imperfect that with it alone he could not distinguish a man from any other similar-sized object at a few paces distance, and he was incapable of doing his former work as a stone mason. This continued for twenty years. At the time of his cure he was under Dr. Dozens' care, as his eyes were then getting worse. He washed, and was completely cured in the course of one day Dr. Dozens met him in the street and would not believe he was cured, and tested him by writing with a pencil on a piece of paper that he had an incurable amaurosis of the right eye, and when he read these words to the doctor, the latter was dumfoundered, for Dr. Dozens was a materialist, and disbelieved in all things preternatural at that time. This case is also vouched for by Dr. Vergez, of Barèges.

"M. Lacassogne, now of 6, Rue du Chai des Varine, Bordeaux, formerly of Toulouse, had a son who had for three years been unable to swallow a morsel of solid food. Both the doctors of Toulouse told me of this case, but Dr. Noguès was his principal medical attendant. Dr. Noguès is still an unbeliever, but he told me he felt bound in justice to declare that his patient was a good obedient child of a sanguine temperament, and not at all nervous or hysterical. When wasting to the extreme from imperfect nutrition, he was *instantaneously* cured in the fountain, and has eaten freely solid food ever since. His father was a Voltairean, and was converted by this fact in his family.

"Finally, Dr. Rogues, of Toulouse, told me that his own daughter had recently had a most remarkable cure, and this was also told me by Dr. Dozens, of Lourdes. Dr. Rogues is short-sighted, his sons are short-sighted, and his father is short-sighted. No wonder, then, that his daughter was also short-sighted. It was a case of heredity—congenital short-sightedness. The mother was exceedingly desirous as her daughter grew up that she might be able to see like ordinary people, and took her to Lourdes, when, *in an instant*, she became ordinarily long-sighted. On her return her father would not believe till he had tested her himself by making her read to him at distances which would have been quite impossible at any previous period of her life. The next morning he told me that being very anxious on the subject, he called her as soon as possible to his window, and pointing out a distant inscription told her to read it to him. She said, ' I can't, papa ; it's Latin.' He told her then to read him the letters, which, to his delight, she did. This change had continued permanent up to my visit to him last Tuesday."

To appreciate fully the weight of this evidence, received at first hand from the best of witnesses—the medical men who had attended the patients cured, and who were all more or less strongly prejudiced against the whole thing—the reader should make himself acquainted with some portion of the mass of equally good evidence to be found in various French works, or in the Rev. R. F. Clarke's "Lourdes and its Miracles" (1887). The detailed history of the origin of the spring at Lourdes, and of all the succeeding events, by M. Henri Lasserre, is both interesting and instructive to the spiritualist. His book, "Notre Dame de Lourdes," had gone through one hundred and twenty-six editions in 1892, and had been translated into eleven European languages. It is written from the point of view of an enthusiastic Roman Catholic, and exhibits Bernadotte Soubirons as a modern representative in character and in psychical faculties of Joan of Arc. The second volume, published fourteen years later, under the title " Les Episodes Miraculeux de Lourdes," contains a detailed record with confirmatory documents of five cases of remarkable cures at Lourdes.

In 1862, M. Lasserre himself was cured of an affection of the eyes which rendered him unable to read or write, and which the best specialists in Paris declared to be incurable. Any attempt to read even the largest print, and however shaded from bright light, produced intense pain. He was persuaded to send for some Lourdes water, and received a small bottle. He washed his eyes for a few minutes, drank the remainder, and was instantaneously cured. He declares that he at once read a hundred pages of a book of which, an hour before, he could not have read three lines. This wonderful cure caused him to become the historian of Lourdes,

and he devoted several years to collecting materials direct
from every person on the spot who could give him information, as well as from all contemporary records and official
documents bearing on the question. The book was published
in 1869, and the second volume of "Episodes" in 1883.

The most remarkable feature of these cures is their rapidity,
often amounting to instantaneousness, which broadly marks
them off from all ordinary remedial agencies. One of the
most prominent of these, related by M. Lasserre, is that of
François Macary, a carpenter of Lavaur. He had had vari-
cose veins for thirty years ; they were as thick as one's finger,
with enormous nodosities and frequent bleedings, producing
numerous ulcers, so that it had been for many years impossible for him to walk or stand. Three physicians had declared
him to be absolutely incurable. At sixty years of age he
heard of the cures at Lourdes, and determined to try the
waters. A bottle was sent him. Compresses with this were
applied in the evening to his two legs. He slept well all
night, and early next morning was quite well ; his legs were
smooth, and there was hardly a trace of the swollen veins,
nodosities, and ulcers. The three doctors who had attended
him certify to these facts.

Other cases are of long-continued paralysis, declared
hopeless by the physicians ; one of serious internal injuries
due to an accident, and declared incurable. The lady had
suffered extreme pain for seven years, had been unable to
walk, and every remedy tried had been useless. She had
at various times consulted five doctors, in vain. Two of
these signed statements that she had been cured instantaneously, and could now walk and perform all the ordinary
actions of a healthy person without suffering the slightest
pain. One of them says, " This cure, so sudden, so unpre-
cedented, so unexpected, is for me a fact positively marvellous.
There is in it something of the divine—an intervention beyond
the natural, visible, incontestible, of a nature to baffle the
reason. For nature does not usually proceed thus, and
when she operates she acts always with a wise deliberation.
—A. Maugni" ("Les Episodes Miraculeaux," p. 486).

In an introductory note to "A Lourdes avec Zola," by Felix Lacaze, Dr. Bernheim states that, "We cure at Nancy the same morbid manifestations that Lourdes cures; medical faith acts like religious faith; that is what I know!" And M. Lacaze, throughout his book, imputes all the cures to belief, expectation, faith. But the student of psychical research and of spiritualism, if he examines the records carefully, will see reason to doubt these general statements. He will meet with cases which are so closely parallel with what every experienced inquirer meets with as to indicate a similarity of cause. I allude to the very common occurrence at *séances*, when messages are being given, to so word them as to contradict the expectation of every one present. I have often seen this myself. At other times the inquirer expects a message from a particular person, has gone to the *séance* with the express purpose of obtaining it, but instead gets a message from some one else. All this is clearly for the purpose of answering the common objection—your "expectation" was read by the medium, and produced the wished-for word or message. Now, among the five cases given by M. Lasserre, one of the most striking serves to illustrate this special feature. A paralyzed Abbé of good family, and of the most firm and genuine religious faith, is yet so humble that he does not expect a miracle to be performed in his favour. More to please his family and friends than himself, he goes to Lourdes, and it is so arranged that he shall attend the grand service of the Assumption, when all his clerical friends are convinced he will be cured, and they excite in him the same belief. But though all the ceremonies have been fulfilled, nothing happens, and he resigns himself to the conviction that it is not the will of God that he should be cured. But when attending another service the next day, and not expecting anything, he suddenly feels a conviction that he is well, rises from his couch, kneels down, and prays. From that moment he is perfectly cured. Here we seem to see the *time* of the cure arranged for the very purpose of demonstrating that it is *not* expectation or faith that *causes* the cure, although it may sometimes be a helping *condition*.

It is clear from these accounts by fervent Catholics that they see in all these cures, *not* any special effect of the water—that is only an outward sign—but a real spiritual agency, which they believe to be that of the Virgin Mary. They also clearly recognize that either the power or the will to cure is limited, that only the few are cured, and that those few are not those who are the best, or the most religious, or the most deserving, but are, so far as can be seen, chosen at random. This, again, exactly corresponds with modern spiritualistic phenomena, which evidently depend upon special conditions in the individuals termed mediums, which conditions do not seem to consist in any superiority of mind or character.

There is another point which seems indicated by the detailed narratives of these remarkable cures. This is, that not only are they rare cases, but that they have been, as it were, selected and induced to try the Lourdes water often by a very unusual combination of circumstances. If we look upon these cures as analogous to those of the many "healers" in the modern spiritualistic world—Dr. Newton, the Zouave Jacob, Mr. Spriggs, and many others, and performed probably by a band of spirit-healers of exceptional power, and who wish to produce that effect upon character which such apparently miraculous cures by the supposed direct agency of the Virgin are calculated to produce—it is not improbable that they should be always searching for cases of ordinarily incurable disease, which are yet amenable to their powers. Having found any such, and having satisfied themselves that a cure is possible, and having perhaps already begun to effect such a cure in a way not perceptible to the patient, it then becomes necessary to induce him to make use of the means which will have the desired effect on his own mind and of those who hear of it. Hence the often curious combination of circumstances which first induce the patient to go to Lourdes (or use the water), and then to go at a particular time, even on a particular day. This may be necessary, both because at a particular stage only can the cure be instantaneously or rapidly effected, also because, if delayed, the patient might feel himself getting better, and the moral effects of a cure, supposed to be by the Virgin (or any other saint), be lost. The detailed narratives certainly show that in several cases a moral and religious, as well as a physical, renovation has been effected.

We have here an explanation of these events which is, I submit, much more complete than that which declares them all to be of the same nature as cures occurring through hypnotic suggestion, because in these cases there is no hypnotizer, and often no suggestion or expectation. And when we consider that the cures at the tomb of the Abbé Paris in the early part of the eighteenth century, some of which were even more wonderful than any which have occurred at Lourdes, were equally well attested, and compelled even David Hume to say—referring to one of these— "Had it been a cheat, it would certainly have been detected by such sagacious and powerful antagonists,"[1] we see that we have to do with a phenomenon which is one of the myriad forms of spirit agency.

ROMANES AND DARWIN

I first made the acquaintance of Romanes in a rather curious way. A letter appeared in *Nature* (February 5, 1880) headed "A Speculation regarding the Senses," beginning with this suggestive passage: "On examining the modes of action of the senses, we find a series of advances in refinement. Beginning with *touch*, we find it has primarily to do with *solids* which come into direct contact with the organ. In *taste* a liquid medium is necessary. In *smell* we have minute particles carried by a gas. In *hearing* we have vibrations (longitudinal) in a gas. In *sight*, finally, we find transverse vibrations transmitted by a finer medium, the ether." The writer then goes on to suggest that thought, or brain-vibrations, may also be carried by the ether to other

---

[1] A very full account of these cures is given in Howitt's "History of the Supernatural," and an abstract in my "Miracles and Modern Spiritualism" (pp. 9-12).

brains, and thus produce thought-transference, which, he
suggests, might be termed a kind of "induction of thought,"
and he thinks this is supported by the experiences of most
people, and especially "by the ascertained facts of *clairvoyance*
and *mesmerism*." This letter was signed "M."

In the next issue of *Nature* was a letter signed "F. R. S.,"
objecting to "M." for speaking, in a scientific journal, of the
facts of mesmerism and clairvoyance as being "ascertained,"
adding, however, that they ought to be thoroughly investigated,
that he is prepared to do so if he can find suitable material,
and that he will give wide publicly to his results. He then
says, "If the phenomena should admit of repetition I should
have them witnessed and attested to by a selected number
of the leading scientific men of the day." He therefore
begs for assistance in carrying out his experiments, letters
to be addressed c/o the Editor of *Nature*.

To this request I replied, pointing out to him that many
scientific men, such as Dr. Elliotson, Dr. Gregory, and Dr.
Haddock *had* thoroughly examined and tested mesmerism
and clairvoyance, getting only abuse or ridicule, and that he
was rather sanguine in thinking that any experiments of his
would convince the scientific world, or that they would even
condescend to witness and test them, and referred to my
own experience with Tyndall and Carpenter, and those of
Crookes with the Royal Society. This brought me the following letter from Romanes, and I have now little doubt
that "M." of the first communication to *Nature* was my friend
Mivart, as I do not know any other man likely to have
written on such a subject, and to have spoken in such an
assured way of clairvoyance, which was, of course, a comparatively small matter after his experiences above related.
It is rather curious that these two men should have been
thus brought together without knowing it, and in relation
to a subject as to which neither of them made any public
acknowledgment of what he believed. The majority of
their readers, I have no doubt, look upon them as biologists,
and have no idea that they were also inquirers into
spiritualism.

"18, Cornwall Terrace, Regent's Park,
"February 17, 1880.

"DEAR SIR,

"I am very glad that you have been so kind as to answer my letter in *Nature*, for the fact of your having done so supplies me with an opportunity, which I have long desired to bring about, of obtaining the benefit of your advice upon the methods of conducting an inquiry into the facts of 'spiritualism.' You will not wonder that I should have desired this opportunity when I tell you that one or two facts, which you might consider almost commonplace, have profoundly staggered me, and led me to feel it a moral duty no less than a matter of unequalled interest, to prove the subject further. As a biologist I knew the quality of your scientific work, and the general character of your mind, and knowing also your intellectual attitude towards the subject in which my interest was awakened, I greatly desired to meet you. But by some fate you always seemed to be the only scientific man of the day whom I never did meet, and I felt it would be imprudent to force any questions upon you unsolicited, as I knew Mr. Crookes to be very reticent, and feared you might be the same.

"Now for what you very truly say about the uselessness of any one man, 'however eminent,' trying to prove the truth of the phenomena to the world. This I think is only as it ought to be. The phenomena are of an order so astounding that proof of their reality must rest upon the authority of more than one observer if the proof is to be commensurate with its own requirements. What the precise number of witnesses and what amount of accumulated authority ought to be, or would be, held sufficient to justify a man of the world in accepting the alleged facts as real facts, this is a question I need not consider, for there can be no doubt that *some* such number of witnesses and amount of competent testimony would be sufficient for the purpose. But, looking to the astounding nature of the alleged facts, I do not think that this number and amount have yet been attained. An exceedingly strong case, however, has been made out to

justify full and patient inquiry by at least *several* authoritative persons, and this is what I desire to get done. The
leading men of science have neither time nor inclination to
sift the grain from the chaff of these subjects, but if once
the grain were placed before them we should soon have the
bread. I think you are too despairing on the subject of
prejudice. That prejudice should exist in the matter is only
what common sense would expect, but I am convinced that
it would quickly yield to adequate proof. There is already
more than enough proof were the facts to be proved of any
ordinary kind ; but as they are nothing less than miracles, a
further weight of proof is, I think, required to justify any one
who has not himself witnessed the facts, accepting the latter
on testimony. Therefore it is that in *Nature* I implied that
in my judgment the facts were not yet proven. But pray do
not suppose that I am blind to the importance of the testimony already accumulated. I should rather infer it is you
who are blind to that importance ; I think you underrate the
impression which your own publications and that of your
few scientific co-operators have produced. I know that this
impression is in many minds profound, and has already
prepared the way to a full acceptance by the scientific world
of the facts ; but before this can be, the latter must and ought
to be attested to by some important body of well-known
men.

"You will see, then, that far from imagining that the
world will take my authority on the subject as final, I do not
think that, looking to the nature of the facts, the world *ought*
to do so; and I similarly think that the world is not altogether
wrong in having weighed the amount of proof required to
substantiate a miracle against the weight of authoritative
testimony hitherto forthcoming, and in deciding to await
further testimony.

"I am myself in the position of the world ; I want more
evidence to make me believe. If once I do believe and can
get any repeatable results to show, I shall insist upon the
best men in science and literature coming to see and telling
what they see.

"I am greatly obliged to you for your advice, but some time I should like to have a talk with you to benefit by your large experience of a subject with which I have hitherto had but small acquaintance. Could you fix any date towards the latter end of next month?

"I am, yours truly,
"GEO. J. ROMANES."

After receiving my reply I had another short letter, as follows:—

"I am exceedingly pleased to hear that you are so disposed to assist me with your advice. Time, money, tolerance, and patience I have in abundance, but I lack experience in a subject which, till recently, I rejected as beneath consideration. Therefore, under various circumstances that may arise, I doubt not that your advice may be of much service to me. Thus already you have presented a point of no small importance to me, viz. that I must not count too confidently on being always able to repeat results, even supposing them to be genuine. But, after all, my principal object is to satisfy my own mind upon the subject. If I could obtain any definite evidence of mind unassociated with any observable organization, the fact would be to me nothing less than a revelation—'life and immortality brought to light'—and although I might say to others, 'Come and see,' my chief end would have been attained if I could say, 'I have found that of which the prophets (to wit, Crookes, Wallace, Varley, and the rest) have spoken.'

"I will therefore be most happy to accept your invitation to go to Croydon some day to gain some preliminary ideas on the subject. I shall write again to fix a day."

These two letters express very clearly the writer's position and general ideas, with which I myself was completely in accord. They also are very characteristic of his somewhat wordy and involved style of writing, and of some peculiarities of character. But the first was specially interesting to myself by showing me that my book, which had been published six years before, had really produced *some* effect among men

of science as well as among the general public, many of
whom, I knew, it had induced to investigate and, as a
consequence of their investigation, to become complete
converts. I will here mention a little incident that shows
how people were accustomed to speak on the subject in the
popular tone of contemptuous incredulity, even when they
had reason to accept some of the facts. One evening, while
having tea after a Royal Institution lecture, I heard the late
Professor Ansted and a friend (not knowing I was just
behind them) mention spiritualism, and the professor
remarked, "What a strange thing it is such men as Crookes
and Wallace should both believe in it!" To which the other
replied, with a laugh, "Oh, they are mad on that one subject."
As soon as the friend had turned away I addressed Ansted,
telling him I had heard what he and his friend had been
saying, and asked him if he had any knowledge whatever of
the subject. To which he replied, "Well, not much; but a
neighbour and friend of mine at Great Bealings has had the
most wonderful things happen in his house, which no one has
ever been able to find a cause for. He has often told me
about the bells ringing when no one was in the house. He
was a very clever man, and I am sure what he says is true,
and many people in the neighbourhood were witnesses of it."
This case I had referred to in my book, and it brought it
home to me more vividly to speak with a scientific man who
was a friend of the owner of the house where it occurred, and
had heard it from his own lips. This was shortly before
Professor Ansted's death from an accident, or he might
have become one of the band of " persecuted lunatics "—the
term by which my friend Mr. Guppy used to describe the
despised spiritualists.

To return now to Romanes. He called upon me at
Croydon, and I think I paid him a visit in town, and he then
told me how he had come to take so deep an interest in
spiritualism. Some time previously a member of his own
family—I think either a sister or a cousin—had been found to
have considerable mediumistic power. Through her he had
witnessed a good many of the usual phenomena—movements

and raps by which messages had been spelt out—together with the usual perplexities which beset the beginner ; the messages being sometimes true and sometimes false, sometimes totally unexpected by any one present, at other times seeming to be the reflex of their own thoughts. Yet he was already absolutely convinced that the sounds and motions —the physical part of the phenomena—were not caused in any normal way by any of the persons present, and almost equally convinced that the intelligence manifested was not that of any of the circle. In some cases even his mental questions were replied to. I gave him the best advice I could, and for some years, being fully occupied with my own domestic affairs and literary work, I saw or heard nothing more of the subject he had been so intent upon. At this I was not surprised, as he himself was writing a series of works which gave him his scientific reputation, and I thought it probable that, not getting the evidence he wanted, he had given up the inquiry.

But seven years later, when I was in Canada, I obtained a knowledge of the correspondence between Romanes and Darwin before my interview with the former, as already narrated in Chapter XXX. This was, to me, of extreme interest because it showed how reticent Romanes was, and how little he told me of the evidence he had really obtained some years before, and of the profound impression it had made upon him. The letters then shown me were very long and full of curious details of evidence, the more important of which I took notes of. Darwin's reply was of the usual kind—suggestion of clever trickery ; more investigation required ; had no time to go into it himself, etc. Of course I had no intention of referring to these letters in any way without Romanes' permission, but I thought I might some day ask him why he had not mentioned having written to Darwin when corresponding with me and discussing this very subject. But a year or two later I was surprised by something he wrote as to one of the "thought readers" then exhibiting in London, in a way which implied that all such phenomena were clever trickery by means of muscle-reading, although in his letter to Darwin he had declared that his mental questions had been answered.

But a cause of difference on a scientific question had since arisen between Romanes and myself which led to complica-cation. In 1886 he read a paper to the Linnean Society, which was printed in their *Journal*, entitled "Physiological Selection: an additional suggestion on the Origin of Species." This paper put forth what was really a new theory of the origin of infertile races, which was supposed to account for the infertility that so generally occurs between allied species. It was very complex, and led to much discussion, and before leaving for America I had criticized it in the September issue of the *Fortnightly Review*. Later, I gave what I considered a proof of its entire fallacy in my "Dar-winism" (published in 1889), and many other writers had also given reasons for rejecting it. This rejection of a theory which he evidently thought very highly of seems to have been very unexpected and to have somewhat ruffled his temper, as was very natural, or he would not, I think, have written of me as he did, especially if we consider the letters he had sent me four years previously. In an article in the *Nineteenth Century*, of May, 1890, he repeats a statement which he had made before in other periodicals in the follow-ing words:—"He presents an alternative theory to explain the same class of facts. Yet this theory is purely and simply, without any modification whatsoever, a restatement of the first principles of physiological selection, as these were origi-nally stated by myself." To this and to a repetition of it in the American magazine, *The Monist*, of October, I replied in *Nature*, and I need only say here that the essential parts of my theory were founded partly on facts established by Darwin, and partly on a mathematical demonstration that sterility could be increased by natural selection. This last argument was stated by me in nearly the same form in letters to Darwin in 1868, eighteen years before Romanes set forth his theory of physiological selection (see "More Letters of Charles Darwin," vol. i. pp. 288–297). Further,

while this last theory has now, I believe, no supporters
my own view, so far as I know, has not been shown to be
unsound; and I do not think that the accusation of direct
and barefaced plagiarism is now aecepted by any naturalist
who has taken the trouble to follow the whole discussion.

But much worse than this was the following passage
referring to my "Darwinism," where he says it is in the concluding chapter of my book "that we encounter the Wallace
of spiritualism and astrology, the Wallace of vaccination and
the land question, the Wallace of incapacity and absurdity"
(*Nineteenth Century*, May, 1890, p. 831).

To this I made no public reply, since I was sure that all
whose opinion I valued would condemn this mode of discussing the problems of science. But I thought it afforded
an excellent opportunity to let my critic know what I
thought of his behaviour, and perhaps puzzle and frighten
him a little by exhibiting an acquaintance with facts which
he evidently wished to conceal. I accordingly wrote him the
following letter:—

"Parkstone, July 18, 1890.

"DEAR MR. ROMANES,

"Some time back I read your article in the *Nineteenth Century* for May, but I have been so much occupied
that I have, till now, had no time to write about it. Whether
or no it was good taste for you to appeal to the political
and medical prejudices of your readers in a matter purely
scientific—by referring to my advocacy of land nationalization and opposition to vaccination—I leave others to judge.
I am quite satisfied myself that, in a not distant future, I
shall have ample credit given me on both these points. But
as to your appeal to popular *scientific* prejudice by referring
to my belief in spiritualism and astrology (which latter I
have never professed my belief in), I have something to say.

"In the year 1876 you wrote two letters to Darwin, detail-
ing *your* experiences of spiritual phenomena. You told him
that you had had *mental questions* answered with no paid
medium present. You told him you had had a message

from Mr. J. Bellew, which message was worded in a manner so unexpected that it was, till completed, thought to be erroneous. And you declared your *belief* that some non-human intelligence was then communicating with you. You also described many physical phenomena occurring in your own house with the medium Williams. You saw 'hands,' apparently human, yet not those of any one present. You saw hand-bells, etc., carried about; you saw a human head and face above the table, the face with mobile features and eyes. Williams was held all the time, and your brother walked round the table to prove that there was no wire or other machinery (in your own room !), yet a bell, placed on a piano some distance away, was taken up by a luminous hand and rung, and carried about the room !

" Can you have forgotten all this ?

" In your second letter to Darwin you expressed your conviction of the *truth* of these facts, and of the existence of spiritual intelligences, of *mind without brain*. You said that these phenomena had altered your whole conceptions. Formerly you had thought there were *two* mental natures in Crookes and Wallace—one sane, the other lunatic ! Now (you said) you belonged to the same class as they did.

" Tell it not in Gath ! There are, then, two Romanes as well as two Wallaces. There is a Romanes ' *of incapacity and absurdity !!* ' But he keeps it secret. He thinks no one knows it. He is ashamed to confess it to his fellow-naturalists ; but he is *not* ashamed to make use of the ignorant prejudice against belief in such phenomena, in a scientific discussion with one who has the courage of his opinions, which he himself has not.

"Yours truly,
" ALFRED R. WALLACE."

His answer, written from Scotland on July 21, was as follows :—

" DEAR MR. WALLACE,

          " I am truly sorry to observe the tone of injury which pervades your letter of the 18th inst., just received. It certainly did not occur to me that I was hitting below the belt in alluding to matters so notorious; but after receiving this expression of your own opinion upon the matter, I shall assuredly never do so again. Unfortunately what has been done cannot be undone; but perhaps you will allow me to say that, rather than have offended you in this way, I would have suppressed the article altogether. Perhaps, also, I may add that in giving public expression to my opinion on the relative nature of your different lines of publication it seemed to me that I was only making 'fair comment.' If you were to say that you thought my writings on Darwinism betokened 'incapacity and absurdity,' but my experiments in physiology the reverse, I do not think I should at all object. This, however, is a matter of feeling about which it would be fruitless to argue. So all that I can now do is to express my sorrow, and promise never to allude to this subject again.

"'Astrology' I alluded to, because you once told me that you were investigating it. You refused to hear argument against it, and left me with the impression that you believed in it.

"Touching my correspondence with Mr. Darwin, fourteen years is a long time to remember details, and I kept no copies. But I do clearly remember two points. The first is that the letters were to be strictly *private*, and the next is that they were to be regarded as *provisional*. Now, after these letters were written, further work with Williams showed him to be an impostor. I spent an immense deal of time and trouble over the matter, and in the end withdrew the opinions expressed in these letters.

"If you have gained your knowledge of their contents by any occult process, I hope you will publish them as evidence, which in that case I would not be wanting in courage to back. But otherwise, in the event of your publishing them, I should require to know the source from which they were obtained. That it was not from Mr. Darwin himself, I am already satisfied; if it was from any member of his family, the conditions under which they were written, and some time afterwards, with my permission, submitted to their perusal, must have

been forgotten. In any case, I do not know that you ought
to have read them—but am not sorry that you did, if only to
show you that, although too credulous in the first instance, I
was at any rate not unopen to an honest conviction.

"Yours truly,

"GEO. J. ROMANES."

"Parkstone, July 27, 1890.

"DEAR MR. ROMANES,

"You are mistaken in thinking I wrote under a
sense of injury, and I do not think my letter showed it. I
merely pointed out that to *assume*, without any attempt at
proof, that my writings on vaccination and land nationalization showed incompetence and absurdity was appealing to
ignorant prejudice, and was therefore both unscientific and in
bad taste. My writings on these subjects are public property.
Pray, therefore, refer to them as much as you like, when you
have read them and can refer to them and criticize them
with knowledge of their facts and arguments. But this is a
comparatively small matter. The important part of my letter
and your reply refers to the spiritualistic phenomena.

"You now say you have found Williams to be an impostor.
But I presume you did not write to Darwin, trusting to
Williams' *honesty*, or to any statements that *he* made. You
set forth your own *observations* and *precautions* in proof of
the facts. Have you found out *how* the things you saw in
your own room and in the presence of your own friends were
done ? Can you tell me how the bust and face, 'with move-
able features and eyes,' appeared above your own table while
Williams was sitting beside you and firmly held ? Can you
tell me how the 'luminous hand' was formed and worked,
which lifted a bell from a distant piano, rang it, and carried
it about, your brother walking round the table to see that no
wires had mysteriously fixed themselves in the room ? You
knew then, as well as you know now, that almost all mediums
are accused of imposture ; but you gave your experience as
evidence which did not admit of being explained by imposture.
How is this altered now, if you can no more explain this to

me than you could to Darwin? And were your own relatives impostors when you obtained answers to your mental
questions? Do not those experiments prove a non-human
intelligence now as they did when you wrote to Darwin? If
you cannot now explain these things, your change of opinion
has no logical justification. If you can explain them, I call
upon you to do so—if not to the scientific world, yet to
me, whom you have publicly accused of incompetence and
absurdity because I believe that phenomena of exactly the
same character as yours are *realities* and cannot be proved
to be impostures. As to your letters, copies of them were
handed to me to read by a person to whom (I was told) they
had been given without restrictions, and who was thus quite
justified in showing them to me. Of course I have treated
them, and shall treat them, as private letters; but they
interested me so much that I made full notes immediately
after carefully reading them, so that I possess their substance
and many of their very expressions. After the way you have
referred in print to my belief in such phenomena most persons
would think I was quite justified in making known the fact
of the existence of these letters and their general tenor. I
hope, however, you will not render any such course necessary.
I think your proper course would be to publish the letters
together with the full details of your discovery of the imposture, a discovery so complete as to induce you to change
those convictions you so earnestly and solemnly expressed to
Darwin.

" Hoping you will do so,

" I remain, yours faithfully,

" ALFRED R. WALLACE."

The next letter I will give the substance of. He stated
that soon after having written to Darwin he detected Williams
cheating. He then had a cage made of perforated zinc, and
when Williams sat in it nothing happened. This fact, he says,
logically justified his change of opinion. It could not be a
supernatural power, or why should the interposition of a perforated zinc cage have suspended the power? There was

therefore nothing to publish. As to the answers to mental questions, he only got them when his own hands were on the table. He therefore concluded and still believes that he himself gave the initiatory impulse to move the table, which the other sitters involuntarily intensified and carried out.

I will give my reply because it points out some of the common fallacies of beginners in coming to hasty conclusions from a few isolated facts.

"DEAR MR. ROMANES,

"As I do not wish to continue this correspondence, I will confine myself to pointing out why I consider your present position to be logically untenable and unscientific. You admit you cannot explain what took place in your own house, but you say, 'not being able to explain' is very far from admitting it to have been done by supernatural means (I would say 'supernormal' or 'preterhuman' rather than supernatural; but that is a detail). You then describe the 'cage' you had made, with the result that nothing happened when the medium sat within it; and you imply that if phenomena *had* occurred when Williams was within it you *would* have admitted something 'supernatural.' But why? Simply because, in your own words, you could not explain 'how the trick was done.' To me, and I think to most persons, what *did* occur—the 'luminous hand,' lifting a bell at a distance, etc., etc.—was just as inexplicable, and just as much a proof of something beyond 'trick,' as would have been some physical effect produced *outside* the cage while Williams was in it.

"Again, it is not 'scientific' to treat your own limited experience as if it stood alone, and to refuse to admit all evidence from other inquirers in corroboration. Although *your* cage-test did not succeed, it did succeed with others. Mr. Adshead, a gentleman of Belper, had a wire cage made, and Miss Wood sat in it in his own house, many times, and under these conditions many forms of men, women, and children, appeared in the room. A similar cage was afterwards used by the Newcastle Spiritual Evidence Society, for a year or more, and Miss Wood sat in it weekly. It was *screwed* up from the outside, yet all the usual phenomena of materialization occurred just the same as when no cage was used. At other times Miss Wood sat in the circle visible to all, yet other figures of various apparent ages came out of the cabinet. Then again Mr. Varley, the electrician, applied the electrical test to Miss Cook, she forming part of the circuit, yet all the usual phenomena occurred. Crookes again used the same test, with the same result ; and he also saw Miss Cook and the materialized form 'Katie' at the same time, in his own house, and he photographed the latter. All these facts and many others of like nature have been published, and are known to all inquirers, and every investigator knows that your failure to obtain phenomena under the test, was no proof of any dishonesty in the medium, or of impossibility of obtaining the phenomena under such conditions. Such tests often require to be tried many times before success is attained. To me, and I believe to most inquirers, it will appear in the highest degree *unscientific* to reject phenomena that could not possibly be due to imposture, and to ignore the hundreds of corroborative tests by other equally competent observers, and then, after this, to call all such observers (by implication) fools or lunatics !

"Yet, again, your attempted explanation of the 'mental question' test does *not* apply to the Bellew case, where you expressly state that some of the words while being spelled out were challenged by all present as being wrong, and were yet insisted on by the unknown intelligence, and resulted, contrary to the expectation of all, in—'I, John Bellew, *fear* no *being*.'

"Yours truly,

"ALFRED R. WALLACE."

In reply to this, I received another long and very argumentative letter, admitting that from my point of view and greater experience, my arguments were very strong, but that from his point of view, with his "bias against the preter-human," his refusal to accept any evidence, unless it could be repeated under "several reasonable alterations of conditions, designed to exclude merely human powers of trickery," his objections and his incredulity were quite logical and scientific. He also urged that the mental tests and that of the unexpected answer about Bellew did not require any other intelligence, because equally unexpected things and sayings occurred in dreams, in which we ourselves supply the whole of the matter dreamt of. He therefore thought "that a man may, unconsciously, or subconsciously, supply the other side of a dialogue when he is wideawake, just as well as he can when he is fast asleep." This shows how ingenious was my correspondent as a dialectician, and rendered me disinclined to carry on a further correspondence which seemed likely to be a long one. He quite overlooked, however, the circumstance that our correspondence began, not on account of his being unconvinced by what he witnessed, but by using the fact that I, after much longer experience and a much wider acquaintance with the subject, had been convinced, as a weapon against me in a scientific argument.

However, on the whole, he took my criticism, and even my ridicule, in very good part—better, in fact, than I expected —and he was completely mystified when I told him that my knowledge of his letters did not come directly or indirectly through any of Darwin's family. In order to relieve *their* minds of such a supposition, I told them how I got to see copies of the letters.

In this letter, however, he gave me an account of a "sack trick" he had seen, which he thought as wonderful as anything he saw with Williams, but which he persuaded the performer to show him the secret of. As I think this may interest my readers, I will give it in his own words.

"But for the fact that he is now dead, I could have introduced you to an American medium who would have gone to your own house, and allowed you to furnish your own cabinet, handcuffs, canvass sack, twine, sealing-wax, and seal. Having fastened his hands together behind his back by means of handcuffs as tightly as possible, you might have taken him to the cabinet, placed him inside the sack, tied the mouth of the sack as tightly with the string as you could, and sealed the knots and likewise the two ends of the string to the outside of the sack. Lastly, you might have shut and locked the cabinet door. Then after a period varying from one to two minutes, you would have heard the medium knock, and on opening the door would have found him outside the sack with his hands handcuffed behind his back as before—the mouth of the sack being wide open, and all the knots and seals intact. This performance the medium would repeat any number of times. Having seen him do this I was completely baffled (as I was with Williams), and so would you have been unless you can suggest 'how it was done,' and unless I add, what I do now, that I persuaded *this man* to *explain* the trick."

In reply to this I pointed out that the "sack" and handcuff trick involved only *one* essential operation, that of quickly slipping his hands in and out of the handcuffs, and that this was probably done partly by a natural mobility of the bones of the wrists and hand, partly by induced suppleness by long practice. That being done while being put into the sack inside the cabinet when the movement of his arms would not be noticed, he had only to insert one or two fingers in the neck of the sack while it was being tied, and all the rest was easy. Nothing was needed or done but to slip out of the handcuffs and slip off the string tied round the neck of the sack.

In the case of Williams, solid objects were moved which were a long way from the medium, and two self-moving objects — a luminous hand, and a head and face with movable features, were produced and seen by all while the medium was held and one of the party looked on outside the circle. And I asked him what became of these solid objects afterwards?

In his reply he said I was substantially right about the way the "sack" trick was done. Also, that several years afterwards Darwin wrote to him that Williams had been detected by some one striking a light! He therefore felt quite justified in disbelieving all he had once thought so convincing.

Thus ended our correspondence on this subject; and, I suppose, as a kind of *amende honorable*, my correspondent asked me, the next year, to allow him to have a photograph of myself for a forthcoming book of his on the Darwinian theory! This I declined with thanks.

CHAPTER XXXVII

SPIRITUALISTIC EXPERIENCES IN ENGLAND AND
AMERICA

THE publication of my book in 1874, not only brought me an extensive correspondence on the subject, but led to my being invited to take part in many interesting *séances*, and making the acquaintance of spiritualists both at home and abroad. As what I witnessed was often very remarkable, and forms a sort of supplement to the " Notes of Personal Evidence " given in my book, and also because these phenomena have had a very important influence both on my character and my opinions, it will be necessary here to give a brief outline of them.

I attended a series of sittings with Miss Kate Cook, the sister of the Miss Florence Cook, with whom Sir William Crookes obtained such very striking results. The general features of these *séances* were very similar, though there was great variety in details. They took place in the rooms of Signor Randi, a miniature painter, living in Montague Place, W., in a large reception-room, across one corner of which a curtain was hung and a chair placed inside for the medium. There were generally six or seven persons present. Miss Cook and her mother came from North London. Miss C. was always dressed in black, with lace collar, she wore laced-up boots, and had earrings in her ears. In a few minutes after she had entered the cabinet, the curtains would be drawn apart and a white-robed female figure would appear, and sometimes come out and stand close in front of the curtain. One after another she would beckon to us to come

up. We then talked together, the form in whispers ; I could
look closely into her face, examine the features and hair,
touch her hands, and might even touch and examine her ears
closely, which were *not* bored for earrings. The figure had
bare feet, was somewhat taller than Miss Cook, and, though
there was a general resemblance, was quite distinct in features,
figure, and hair. After half an hour or more this figure would
retire, close the curtains, and sometimes within a few seconds
would say, "Come and look." We then opened the curtains,
turned up the lamp, and Miss Cook was found in a trance in
the chair, her black dress, laced-boots, etc., in the most perfect
order as when she arrived, while the full-grown white-robed
figure had totally disappeared.

Mr. Robert Chambers introduced me to a wealthy Scotch
lady, Miss Douglas, living in South Audley Street, and at
her house I attended many *séances*, and met there Mr.
Hensleigh Wedgwood, and several other London spiritualists.
Perhaps the most interesting of these were a series with Mr.
Haxby, a young man engaged in the post-office and a remarkable medium for materializations. He was a small man,
and sat in a small drawing-room on the first floor separated
by curtains from a larger one, where the visitors sat in a
subdued light. After a few minutes, from between the
curtains would appear a tall and stately East Indian figure
in white robes, a rich waistband, sandals, and large turban,
snowy white, and disposed with perfect elegance. Sometimes
this figure would walk round the room outside the circle,
would lift up a large and very heavy musical box, which he
would wind up and then swing round his head with one
hand. He would often come to each of us in succession, bow,
and allow us to feel his hands and examine his robes. We
asked him to stand against the door-post and marked his
height, and on one occasion Mr. Hensleigh Wedgwood
brought with him a shoe-maker's measuring-rule, and at our
request, Abdullah, as he gave his name, took off a sandal,
placed his foot on a chair, and allowed it to be accurately
measured with the sliding-rule. After the *séance* Mr. Haxby

removed his boot and had *his* foot measured by the same rule, when that of the figure was found to be full one inch and a quarter the longer, while in height it was about half a foot taller. A minute or two after Abdullah had retired into the small room, Haxby was found in a trance in his chair, while no trace of the white-robed stranger was to be seen. The door and window of the back room were securely fastened, and often secured with gummed paper, which was found intact.

On another occasion I was present in a private house when a very similar figure appeared with the medium Eglinton before a large party of spiritualists and inquirers. In this case the conditions were even more stringent and the result absolutely conclusive. A corner of the room had a curtain hung across it, enclosing a space just large enough to hold a chair for the medium. I and others examined this corner and found the walls solid and the carpet nailed down. The medium on arrival came at once into the room, and after a short period of introductions seated himself in the corner. There was a lighted gas-chandelier in the room, which was turned down so as just to permit us to see each other. The figure, beautifully robed, passed round the room, allowed himself to be touched, his robes, hands, and feet examined closely by all present—I think sixteen or eighteen persons. Every one was delighted, but to make the *séance* a test one, several of the medium's friends begged him to allow himself to be searched so that the result might be published. After some difficulty he was persuaded, and four persons were appointed to make the examination. Immediately two of these led him into a bedroom, while I and a friend who had come with me closely examined the chair, floor, and walls, and were able to declare that nothing so large as a glove had been left. We then joined the other two in the bedroom, and as Eglinton took off his clothes each article was passed through our hands, down to underclothing and socks, so that we could positively declare that not a single article besides his own clothes were found upon him. The result was published in the *Spiritualist* newspaper, certified by the names of all present.

Yet one more case of materialization may be given, because it was even more remarkable in some respects than any which have been here recorded. A Mr. Monk, a non-conformist clergyman, was a remarkable medium, and in order to be able to examine the phenomena carefully, and to preserve the medium from the injury often caused by repeated miscellaneous *séances*, four gentlemen secured his exclusive services for a year, hiring apartments for him on a first floor in Bloomsbury, and paying him a moderate salary. Mr. Hensleigh Wedgwood and Mr. Stainton Moses were two of these, and they invited me to see the phenomena that occurred. It was a bright summer afternoon, and everything happened in the full light of day. After a little conversation, Monk, who was dressed in the usual clerical black, appeared to go into a trance; then stood up a few feet in front of us, and after a little while pointed to his side, saying, " Look." We saw there a faint white patch on his coat on the left side. This grew brighter, then seemed to flicker, and extend both upwards and downwards, till very gradually it formed a cloudy pillar extending from his shoulder to his feet and close to his body. Then he shifted himself a little sideways, the cloudy figure standing still, but appearing joined to him by a cloudy band at the height at which it had first begun to form. Then, after a few minutes more, Monk again said " Look," and passed his hand through the connecting band, severing it. He and the figure then moved away from each other till they were about five or six feet apart. The figure had now assumed the appearance of a thickly draped female form, with arms and hands just visible. Monk looked to-wards it and again said to us " Look," and then clapped his hands. On which the figure put out her hands, clapped them as he had done, and we all distinctly heard her clap following his, but fainter. The figure then moved slowly back to him, grew fainter and shorter, and was apparently absorbed into his body as it had grown out of it.

Of course, such a narration as this, to those who know nothing of the phenomena that gradually lead up to it, seems mere midsummer madness. But to those who have for years obtained positive knowledge of a great variety of facts equally strange, this is only the culminating point of a long series of phenomena, all antecedently incredible to the people who talk so confidently of the laws of nature. I will here just remark that in the four cases of materialization now recorded, with four different mediums, four different kinds of tests were obtained without any interference with the conditions needed for the production of the phenomena. In the first, with Miss Cook, the figure was positively distinguished by unpierced ears, while the circumstances were such that the medium could not possibly have resumed her dress and concealed the robes of the figure in the few seconds only that sometimes elapsed between its disappearance and the examination of the medium. With Mr. Haxby, the measurements both of body and foot were so different as to prevent any possibility of personation by the medium. With Mr. Eglinton, the impromptu and thorough search after the *séance* rendered personation equally impossible ; while the last case, in which the whole process of the formation of a shrouded figure was seen in full daylight, absolutely precluded any normal mode of production of what we saw. I may mention that Mr. Wedgwood assured me that in the course of their long investigation they had had far more wonderful results. In some cases, instead of a shrouded and somewhat shadowy female figure, a tall robed male figure was produced, while Mr. Monk was in a deep trance, and in full view.

This figure would remain with them for half an hour or more, would touch them, and allow of close examination of his body and clothing, and was so thoroughly, though temporarily material, that it could exert considerable force, sometimes even lifting a chair on which one of them was seated, and thus carrying him around the room.

Now, however, that the whole series of similar phenomena have been co-ordinated, and to some extent rendered intelligible, by Myer's great work on " Human Personality," it is to be hoped that even students of physical science will no longer class all those who have either witnessed such phenomena or express their belief in them, as insane or idiotically credulous, without even attempting to show how, under the same conditions, such effects can be produced.

Before leaving the subject of my experiences of spiritualism and spiritualists in England, I will give a case of the strange phenomenon called the "double," so well authenticated and so instructive as to deserve to be here recorded. About the year 1874, Mr. Pengelly, of Torquay, had sent me his very interesting critical article, "Is it a Fact?" in which, to my great surprise, I found an anecdote describing the strange appearance of the doubles of several persons to a friend of his (apparently), as he says he can vouch for it. When, as narrated in Chapter XXVI., we dined together at Glasgow, I took the opportunity of asking him privately whether I was right in my conjecture that the person to whom the event happened was himself, thinly disguised under the pseudonym, Mr. Hazelwood (Pengelly meaning in Cornish the head of the hazel-grove). He replied, "You are right;" which led me to read it again with still greater interest.

In 1883, thinking the case would be one suited for the Psychical Research Society, I sent the paper to my friend, F. W. H. Myers, telling him what Mr. Pengelly had told me; and Miss Pengelly has allowed me to copy a letter from Mr. Myers to her father, thanking him for the additional information he had received about the case, and saying, that as the distance at which the figures were seen was so small, "It is almost inconceivable that you could have mistaken other persons for your own family at that distance, especially as the train must have been almost or quite at a standstill." But he did not publish the case, and it was probably, among the mass of other matter, forgotten. I now give the story as it occurs in Mr. Pengelly's paper, "Is it a Fact?" (p. 32).

"The following case, for which I can vouch, may serve to illustrate this.[1] It will be found to be supported by both personal and circumstantial evidence:—*Mr. Hazelwood*, of Torquay, having a few years since to go to Dawlish, informed

[1] The disbelief in witchcraft, notwithstanding the mass of good testimony supporting it.

his wife that he should return by the train due at Torre Station at a certain hour, and suggested that she might, with their two children, walk up to meet him, which she agreed to do. On the return journey there was in the same carriage with him a gentleman who had known *Mrs. Hazelwood* for several years, and who knew her children also. It should be remarked that the family were in mourning, and that the children were a boy and a girl, the former being the older. At Newton Station Mr. Hazelwood bought a newspaper, and was reading it during the remainder of the journey. He had, for a time, forgotten the arrangement made with his wife, and he states that he certainly had not spoken of it to any one. As the train drew near Torre Station his companion said, 'There's Mrs. Hazelwood and your two children standing on the bank.' He at once looked in the direction indicated, and distinctly saw a party, which he had not the least doubt were his wife and boy and girl, standing on the hedge or bank, which, under Chapel Hill, overlooks the railway. On leaving the station, instead of walking towards Torquay, he went in the opposite direction, on the Newton Road, to join them. On his way he met a man who had known Mrs. Hazelwood from her childhood, and who volunteered the remark, 'You are going to join your wife and family, I suppose. They are just above here, standing on the hedge.' He proceeded to the spot, and to his surprise found the party had left, and were nowhere to be seen. After some fruitless search he proceeded to his own house, and found his family just starting to meet him at the station, they having forgotten the hour at which the train was due. Notwithstanding the fact that three persons, who knew them well, were prepared to swear that they had seen Mrs. Hazelwood and her children at a particular spot, notwithstanding the further fact that this was just the spot where they had previously, and without the knowledge of two of the witnesses, agreed to be at the time, *it was not a fact* that Mrs. Hazelwood and her children had on that day been standing on the hedge overlooking the railway near the station at Torre."

This is one of a large class of appearances termed

"doubles," some of the most curious of which I have made use of in my chapters on "Phantasms" in my book on "Spiritualism." This one is especially valuable, as being recorded by a gentleman who was remarkable for the great care he gave to attain accuracy in all his work; and it was published under a well-understood pseudonym, in a place where he had lived nearly all his life. But what is especially remarkable is that the two independent witnesses had no expectation of seeing the parties where the phantasms appeared, while they themselves having mistaken the hour the train was due could have had no special anxiety as to being in time. And two of the persons seen being children, the theory of the phantasms being caused by the *three* "second selves" or "subliminal personalities," is very difficult to conceive.

Among the eminent men whose first acquaintance and valued friendship I owe to our common interest in spiritualism was F. W. H. Myers, whose great work on "Human Personality and its Survival of Bodily Death" has so recently appeared. I think I must have met him first at some *séances* in London, and he asked me to call on him at his rooms in Bolton Row, Mayfair. I think this was in 1878. I spent several hours with him, discussing various aspects of spiritualistic phenomena. He told me a great deal about the long series of experiments with the celebrated Newcastle mediums, Miss Wood and Miss Fairlamb, both under twenty, and whose powers had been discovered only two years previously, who were engaged for twelve months by Professor Sidgwick, Mr. Gurney, and himself, for a long series of *séances* in Newcastle, in London (at Mr. Balfour's house in Carlton Gardens), and in Cambridge at Professor Sidgwick's rooms. He showed me several MSS. books full of notes of these *séances* of which he was the reporter, and drew my attention to some which I read through. In addition, he described to me the complete tests which were applied in order to render it certain that the phenomena were not produced by the mediums themselves. For example, a curtain across the corner of a room formed the cabinet. In this was placed a mattress and pillow on the bare floor. The medium's wrists were tied securely with tapes, leaving two ends a foot or more long. These ends were tacked down to the floor, then covered with sealing-wax and sealed. Under these conditions one or more forms came out from the curtains, sometimes to a considerable distance and touched each one present. The light was just sufficient to see the figures, which were sometimes those of children, at other times of adults. Other phenomena also occurred in the room. Afterwards the medium was found either awake or still entranced, with the tapes, knots and seals all apparently untouched.

But this was not thought sufficient to exclude imposture. The medium might have provided herself with tape, tacks, wax, and a copy of the seal, and by practice and ingenuity be able to restore things to their original state after coming out and personating the figures. To render this impossible (or rather much less credible), at each *séance* the width, quality, or colour of the tape was different, the sealing-wax was of another colour, or a different seal was used, so that on no two occasions were the conditions alike. Yet still the phenomena went on occurring. Then a hammock was procured for the medium to lie in, and this hammock, by means of pulleys, was connected with a weighing machine, so that the medium could not possibly leave it without instant detection. Yet still the phenomena were produced, and the medium was found afterwards comfortably lying in her hammock.

Of course such phenomena as these, however well established, were entirely out of place at so early a period of the inquiries of the persons, who soon afterwards founded the Society for Psychical Research. They wanted to create a science—to make sure of the first steps before they went on to the second ; and, above all things, not to go on too fast, so that the educated but sceptical public might be able to follow them. They have now worked in this way for nearly a quarter of a century ; they have published a wonderful collection of well-attested evidence, and yet they are only now beginning to approach very carefully and sceptically even the simpler physical phenomena which hundreds of spiritualists, including Sir William Crookes and Professor Zollner, demonstrated more than thirty years ago. As to the more advanced phenomena, such as the disintegration and reproduction of matter, and the various forms of materialization of the human form or its parts, Mr. Myers himself, in his great work, only alludes to them and indicates their possibility, without laying special stress on the fact of their occurrence. The equally well-attested phenomena of psychic photography are entirely unnoticed, though they would easily be fitted into the great structure he has erected based upon phenomena which he considered to be demonstrated facts.

This method of very slow advance was, no doubt, necessary for the purpose of establishing what is really a new science, and in the establishment of this science a foremost place will always be given to Frederic Myers. He was the first English writer to attempt to educe order out of the vast chaos of psychic phenomena, to connect them with admitted physical and physiological laws, and to formulate certain hypotheses which would serve to connect and explain a considerable portion of them.[1] Yet there are indications that even his careful examination of evidence and tentative suggestions are still in advance of most of his fellow researchers; as shown by the fact that since his greatly lamented death the publications of the society have become retrograde rather than progressive, by devoting space to the publication of mere inconclusive or suspicious phenomena which are absolutely worthless, and by needlessly pointing out that certain facts may possibly be explained by imposture or delusion. Nevertheless, for the advanced "Researcher," Myers's great work will long serve as a vast reservoir of classified information and a guide to further scientific research; while the spiritualist will equally value it, and by its light will be able to interpret the more advanced and more marvellous phenomena with which he is acquainted.

When talking to me about the remarkable *séances* with the

---

[1] Robert Dale Owen's works, at a much earlier period, attempted the same thing with more limited materials, but with remarkable success.

two Newcastle mediums, the entire series of which he attended and recorded very carefully in the notebooks he shewed me, he laid great stress upon the extremely rigid precautions that were taken against the possibility of imposture, and conveyed to me the impression that he himself was quite convinced of the genuineness of the whole series. He also told me that Professor and Mrs. Sidgwick only attended a portion of the series, and that, unfortunately, several of the most astounding and conclusive of the phenomena occurred in their absence.

This is important, because Mrs. Sidgwick has published an account of those she attended (in the *Proceedings of the Society for Psychical Research*, vol. iv.), and has laid great stress upon the inconclusive nature of the tests applied, though she admits that it was exceedingly difficult, "but not perhaps impossible," to impute the results to imposture. Under these circumstances, it seems to me, that if these records of the whole series of *séances* by Mr. Myers are still in existence and can be obtained, it is the duty of the Society, in justice to the mediums (one of whom is still living), and in the interests of science, to make the entire series public. Under the light of our more advanced knowledge to-day, such a record by so careful and so critical an observer, of so long-continued an inquiry, must contain a mine of invaluable facts.

MY EXPERIENCES IN AMERICA.

During a lecturing tour in the United States in 1886–87, I stayed some time in three of the centres of American spiritualism—Boston, Washington, and San Francisco, and made the acquaintance of many American spiritualists and inquirers, with whom I attended many remarkable *séances*. At Boston I met the Rev. Minot J. Savage, whose latest work, "Can Telepathy Explain?" contains such a collection of personal experiences as have fallen to the lot of few inquirers; Mr. F. J. Garrison, a son of the great abolitionist; Mr. E. A. Brackett, a sculptor, and author of a remarkable book on "Materialized Apparitions"; Dr. Nichols, author of

"Whence, Where, and Whither"; Professor James, of Harvard, and several others.

I attended several *séances* at the house of Mrs. Ross, a very good medium for materializations, in the company of one or more of my friends. I will state what occurred on one of these occasions. The *séance* took place in a front downstairs room of a small private house, opening by sliding doors into a back room, and by an ordinary door into the passage. The cabinet was formed by cloth curtains across the corner of the room from the fireplace to the sliding door. One side of this was an outer wall, the other the wall of the back room, where there was a cupboard containing a quantity of china. I was invited to examine, and did so thoroughly— front room, floor, back room, rooms below in basement, occupied by a heating apparatus ; and I am positive there were no means of communication other than the doors for even the smallest child. Then the sliding doors were closed, fastened with sticking-plaster, and privately marked with pencil. The ten visitors formed a circle opposite the cabinet, and I sat with my back close to the passage door and opposite the curtain at a distance of about ten feet. A red-shaded lamp was in the furthest corner behind the visitors, which enabled me to see the time by my watch and the out-lines of every one in the room ; and as it was behind me the space between myself and the cabinet was very fairly lighted. Under these circumstances the appearances were as follows :—

(1) A female figure in white came out between the curtains with Mrs. Ross in black, and also a male figure, all to some distance in front of the cabinet. This was apparently to demonstrate, once for all, that, whatever they were, the figures were not Mrs. Ross in disguise.

(2) After these had retired three female figures appeared together, in white robes and of different heights. These came two or three feet in front of the curtain.

(3) A male figure came out, recognized by a gentleman present as his son.

(4) A tall Indian figure came out in white mocassins ; he

danced and spoke ; he also shook hands with me and others, a large, strong, rough hand.

(5) A female figure with a baby stood close to the entrance of the cabinet. I went up (on invitation), felt the baby's face, nose, and hair, and kissed it—apparently a real, soft-skinned, living baby. Other ladies and gentlemen agreed.

Directly the *séance* was over the gas was lighted, and I again examined the bare walls of the cabinet, the curtains, and the door, all being just as before, and affording no room or place for disposing of the baby alone, far less of the other figures.

At another special *séance* for friends of Dr. Nichols and Mr. Brackett, with Professor James and myself—nine in all, under the same conditions as before, eight or nine different figures came, including a tall Indian chief in war-paint and feathers, a little girl who talked and played with Miss Brackett, and a very pretty and perfectly developed girl, " Bertha," Mr. Brackett's niece, who has appeared to him with various mediums for two years, and is as well known to him as any near relative in earth-life. She speaks distinctly, which these figures rarely do, and Mr. Brackett has often seen her develop gradually from a cloudy mass, and almost instantly vanish away. But what specially interested me was, that two of the figures beckoned to me to come up to the cabinet. One was a beautifully draped female figure, who took my hand, looked at me smilingly, and on my appearing doubtful, said in a whisper that she had often met me at Miss Kate Cook's *séances* in London. She then let me feel her ears, as I had done before to prove she was not the medium. I then saw that she closely resembled the figure with whom I had often talked and joked at Signor Randi's, a fact known to no one in America.

The other figure was an old gentleman with white hair and beard, and in evening-dress. He took my hand, bowed, and looked pleased, as one meeting an old friend. Considering who was likely to come, I thought of my father and of Darwin, but there was not enough likeness to either. Then

at length I recognized the likeness to a photograph I had of
my cousin Algernon Wilson, whom I had not seen since we
were children, but had long corresponded with, as he was an
enthusiastic entomologist, living in Adelaide, where he had
died not long before. Then I looked pleased and said, "Is
it Algernon?" at which he nodded *earnestly*, seemed *very*
much pleased, shook my hand vigorously, and patted my
face and head with his other hand.

These two recognitions were to me very striking, because
they were both so private and personal to myself, and could
not possibly have been known to the medium or even to any
of my friends present. I may state here that a few months
afterwards, a party of twelve gentlemen went to a *séance* at
Mrs. Ross's, determined to seize hold of the alleged spirit
forms, which they believed to be all confederates, and thus
expose the supposed imposture. It was agreed that some were
to seize the Indian, others to hold Mr. and Mrs. Ross, others
the women and children performers, while the remainder
were to assist when called upon and secure any "properties"
they could find in the cabinet. They carried out the first
part of their programme successfully, but notwithstanding
they were twelve men against two men, one woman, two
boys and a little girl (according to their own account), they
appear to have been entirely overmatched in the struggle, for
they did not succeed either in securing or identifying any one
of them, or in carrying away any of the alleged paraphernalia of imposture. They further declared, as if it were an
observed fact, that the assistants, young and old, entered the
cabinet by a sliding portion of the mop-board (or skirting, as
we call it). Immediately this was published in the Boston
papers, Mr. Brackett and some other friends of Mrs. Ross
called on the landlord of the house and asked him to go with
them, taking a carpenter with them, to see if the tenants had
made any such alteration as described by the would-be
exposers. The examination was made, and it was declared
that there was no such opening as alleged, nor had any been
made and closed up again. I wrote a letter to the *Banner of
Light*, pointing out these facts, and I urged, that the utter

failure of twelve men, who went for the express purpose of detecting and identifying confederates, utterly failing to do so or to secure any tangible evidence of their existence, is really a very strong proof that there were no confederates to detect.

To any one who has carefully studied Mr. Myers's monumental volumes, and given due weight to the whole of the evidence he adduces for the reality of such phenomena as are here narrated and what is known of the various stages that lead up to them ; and considering the proof that even detached hands are capable of moving material objects, it will, I think, appear probable that some such result as here occurred was to be anticipated. I cannot remember a single instance in which a confederate has been secured by such a seizure, though cases have occurred in which the seizure of the spirit form has resulted in the seizure of the medium— which is not remarkable if we remember the amount of evidence showing that these forms originate from the body of the medium, and either visibly or invisibly return to it. Also, considering the demonstrated fact that clothing, flowers, hair, and other objects pertaining to or brought by these psychic forms have sometimes a permanent, sometimes a temporary existence, the fact of any such objects being found on or near a medium is of itself no proof whatever that they were brought by the medium for purposes of imposture, except on the assumption that no such phenomena were *possible*, in which case no evidence one way or the other is required, since the question has been already decided against the medium.

In Washington, where I resided several months, I made the acquaintance of Professor Elliott Coues, General Lippitt, Mr. D. Lyman, Senator and Mrs. Stanfield, Mr. T. A. Bland the Indians' friend, and Mrs. Beecher Hooker, all thorough spiritualists, as well as many others unknown to fame. With the three former gentlemen I attended the *séances* of a very remarkable public medium, Mr. P. L. O. A. Keeler, and both witnessed phenomena and obtained tests of a very interesting kind. The medium was a young man of the clerk or tradesman class, with only the common school education, and with no appearance of American smartness. The arrangement of his *séances* was peculiar. The corner of a good-sized room had a black curtain across it on a stretched cord about five feet from the ground. Inside was a small table on which was a tambourine and hand-bell. Any one, before the *séances* began or afterwards, could examine this enclosed space, the curtain, the floor, and the walls, as I did myself, the room being fully lighted, and was quite satisfied that there was absolutely nothing but what appeared at first sight, and no arrangements whatever for ingress or egress but under the curtain into the room. The curtain, too, was entire from end to end, a matter of importance in regard to certain phenomena that occurred. Three chairs were placed close in front of this curtain on which sat the medium and two persons from the audience. Another black curtain was passed in front of them across their chests so as to enclose their bodies in a dark chamber, while their heads and the arms of the outer sitter were free. The mediums two hands were placed on the hands and wrist of the sitter next him.

The *séance* began with purely physical phenomena. The tambourine was rattled and played on, then a hand appeared above the curtain, and a stick was given to it which it seized. Then the tambourine was lifted high on this stick and whirled round with great rapidity, the bell being rung at the same time. All the time the medium sat quiet and impassive, and the person next him certified to his two hands being on his or hers. On one occasion a lady, a friend of Professor Elliott Coues and a woman of unusual ability and character, was the sitter, and certified at all critical times during the whole *séance* that the medium's hands were felt by her. After these and many other things were performed, the hand would appear above the curtain, the fingers moving excitedly. This was the signal for a pencil and a pad of note-paper (as commonly used in America); then rapid writing was heard, a slip of paper torn off and thrown over the curtain, sometimes two or three in rapid succession, and in the direction of certain sitters. The director of the *séance* picked them up, read the name signed, and asked if any one knew it, and when claimed it was handed to him. In this way a dozen or more of the chance visitors received messages which were always intelligible to them and often strikingly appropriate. I will give some of the messages I thus received myself.

On my second visit a very sceptical friend went with us, and seeing the writing-pad on the piano marked several of the sheets with his initials. The medium was very angry and said it would spoil the *séance*. However, he was calmed by his friends. When it came to the writing the pad was given to me over the top of the curtain to hold. I held it just above the medium's shoulder, when a hand and pencil came *through the curtain*, and wrote on the pad as I held it. It is a bold scrawl and hard to read, but the first words seem to be, " Friends were here to write, but only this one could. . . . A. W." Another evening, with the same medium, I received a paper with this message, " I am William Martin, and I come for Mr. William Wallace, who could not write this time after all. He wishes to say to you that you shall be sustained by coming results in the position you have taken in the Ross case. It was a most foul misrepresentation."

This, and other writing I had afterwards, are to me striking tests in the name William Martin. I never knew him, but he was an early friend of my brother who was for some time with Martin's father to learn practical building, the latter being then engaged in erecting King's College. When I was with my brother learning surveying, etc., he used often to speak of his friend Martin, but for the last forty-five years I had never thought of the name and was greatly surprised when it appeared. About a month later I had the following message from the elder Martin, written in a different hand :—

" MR. WALLACE,

            "Your father was an esteemed friend, and I like to come to you for his sake. We are often together. How strange it seems to us here that the masses can so long exist in ignorance. Console yourself with the thought that though ignorance, superstition and bigotry have withheld from you the just rewards to which your keen enlightenment and noble sacrifices so fully entitle you, the end is not yet, and a mighty change is about to take place to put you where you belong.

"WILLIAM MARTIN."

I have no evidence that this Mr. Martin was a friend of my father, but the fact that my brother William was with him as stated (which must have been a favour), renders it probable. On the same evening there were a number of messages to about a dozen people all in different hand-writings, several of which were recognized. My friend General Lippitt had a most beautiful message which he allowed me to copy, as it was a wonderful test and greatly surprised and delighted him. His first wife had died twenty-seven years before in California. She was an English lady and he was greatly attached to her. This is the message :—

"DARLING FRANCIS,
        "I come now to greet you from the high spheres to which I have ascended. Do you recall the past? Do you remember this day? This day I used to look forward to and mention with such pride? This, my darling, is my birthday anniversary. Do you not remember? Oh how happy shall we be when reunited in a world where we shall see as we are seen and know as we are known.

"ELIZABETH LIPPITT."

General Lippitt told me it *was* his first wife's birthday, that he had not recollected it that day, and that no one in Washington knew the fact but himself.

A German gentleman who was present had a message given him, which was not only written, as he declared, in excellent German, but was very characteristic of the friend from whom it purported to come.

On this evening most wonderful physical manifestations occurred. A stick was pushed out *through* the curtain. Two watches were handed to me *through* the curtain, and were claimed by the two persons who sat by the medium. The small tambourine, about ten inches diameter, was pushed *through* the curtain and fell on the floor. These objects came through different parts of the curtain, but left no holes as could be seen at the time, and was proved by a close examination afterwards. More marvellous still (if that be possible), a waistcoat was handed to me over the curtain, which proved to be the medium's, though his coat was left on and his hands had been held by his companion all the time; also about a score of people were looking on all the time in a well-lighted room. These things *seem* impossible, but they are, nevertheless, facts.

Before passing on from my Washington friends, I wish to give one curious test which occurred to General Lippitt recently, and an account of which he sent to me in February, 1894. In his early life he had lived in Paris, and had become acquainted with several members of the Bonaparte family, and had rendered some services to them. This was only known to himself, but it accounted (to him) for the fact that he had, through different mediums, received messages from some of them, and from Napoleon III. In August, 1893, he had *séances* with a medium previously unknown to him, and received on a slate under test conditions a long message in French, purporting to come from Napoleon III., and to give his last dying thoughts. A facsimile of this is given in a Chicago paper, and is written as if it were an ordinary prose message; but on copying it out I found that it was in rhyme, and, so far as I could judge, very forcible, and even pathetic verse. I therefore sent a copy of it to Mr. F. Myers, asking him what he thought of it, and whether it was correctly written. In reply he told me that he had paid special attention to the rules of French poetry, and that this was correct verse such as no one but a Frenchman could have written. General Lippitt, who was a good French scholar, observes that there is only one error in it—the omission of the final "e" in the word *profonde* near the end, which is doubtless an oversight, when all the other refinements of the language, as well as the numerous accents, are correct. General Lippitt also prints a certificate that the medium knew no French ; but that is quite unnecessary in view of the test conditions. Esprit C., who signs it, is one of the medium's guides who knows French.

> "L'Heure sonne ! on la compte ; elle n'est déjà plus :
> L'airain n'annonce, hélas ! que des moments perdus.
> Son redoutable son m'épouvante, m'éveille ;
> Et c'est la voix du temps qui frappe à mon oreille.
> S'il ne m'abuse point, le lugubre métal
> De mon heure dernière a donné le signal :
> C'est elle ! . . . où retrouver tant d'heures écoulées ?
> Vers leur source lointaine elles sont refoulées ;
> Le seul effroi me reste et l'espoir est banni.
> Il faut mourir, finir, quand je n'ai rien finí,
> Où vais-je ? et quelle scène a mes yeux se déploie
> Des bords du lit funébre, où palpite sa proie
> Aux lugubres clartés de son pâle flambeau,
> L'impitoyable mort me montre le tombeau.
> Eternité profonde : Océan sans rivage :
> De ce terme fatal c'est toi que j'envisage ;
> Sur le fleuve du temps, quoi ? c'est là que je cours?
> L'éternité pour l'homme? il vit si peu de jours."
>
> Esprit C.

At San Francisco my time was short, and my experiences were limited to a slate-writing *séance* of a striking and very satisfactory nature. I went with my brother John who had lived in California nearly forty years, and who, the day before, had bought a folding-slate bound with list to shut noiselessly. The *séance* was in the morning of a bright sunny day, and we sat at a small table close to a window. Mr. Owen, the editor of the *Golden Gate*, with a friend (a physician), accompanied us ; but they sat a little way from the table, looking on. The medium, Mr. Fred Evans, was quite a young man, whose remarkable gift had been developed under Mr. Owen's guidance.

From a pile of small slates on a side-table four were taken at a time, cleaned with a damp sponge, and handed to us to examine, then laid in pairs on the table. All our hands were then placed over them till the signal was given, and on ourselves opening them writing was found on both slates. Two other pairs were then similarly placed on the table, on one of which the medium drew two diagonal pencil lines, and on that slate writing was produced in five different colours— deep blue, red, light green-blue, pale red-lilac, deep lilac, and these could be seen all superposed upon the pencil cross-lines. My brother's folding-slate was then placed upon the floor a foot or two away from the table, and after we had conversed for a few minutes, keeping it in sight, it was found to be written on both the inner sides. It then occurred to me to ask the medium whether writing could be produced on paper placed between slates. After a moment's pause, as if asking the question of his guides, he told me to take a paper pad, tear off six pieces, and place them all between a pair of slates. This I did, and we placed our hands over them as before, and in a few minutes, on opening them, we found six portraits in a peculiar kind of crayon drawing.

I will now describe what were the writings and drawings we obtained, which are all now before me. The first was a letter filling the slate in small, clear, and delicate writing, of which I will quote the concluding portion : "I wish I could describe to you my spirit home. But I cannot find words suitable in your earthly language to give it the expression it deserves. But you will know all when you join me in the spirit world. . . . Your loving sister, Elizabeth Wallace. Herbert is here."

Here are two family names given, the first being one which no one present could have known, as she died when we were both schoolboys. The opening and concluding parts of the letter show that it was addressed specially to myself. The next was addressed to my brother, referring to me as "brother Alf," and is signed "P. Wallace." This we cannot understand, as we have no relative with that initial, except a cousin, Percy Wilson. It is, I think, not improbable that in transferring the message through the medium, and perhaps through a spirit-scribe (as is often said to be the case), the surname was misunderstood owing to the latter supposing that the communicant was a brother.

The next slate contains a message signed "Judge Edmonds," addressed to myself and Mr. Owen, on the general subject of spirit manifestations. It is written very distinctly in a flowing hand.

The next is the slate written in five colours, and signed "John Gray," one of the well-known early advocates of spiritualism in America. It is also on the general subject of spirit-return. Then comes a slate containing a portrait and signature of "Jno. Pierpont," one of the pioneers of spiritualism, and around the margin three messages in different handwritings. One is from Stanley St. Clair, the spirit-artist, who says he has produced the portrait for me, at the request of the medium. The others are short messages from Elizabeth Wallace and R. Wallace, the latter perhaps one of the unknown Scotch uncles of my father, the other beginning, "God bless you, my boys," is probably from our paternal grandmother, who is buried at Laleham. The last is my brother's folding-slate, containing on one side a short farewell from "John Gray," the signature being written three times in different styles and tints; the other side is a message signed, "Your father, T. V. Wallace." This, again, was a test, as no one present would have been able to give my father's unusual initials correctly, and as he was accustomed to sign his name.

The six portraits on paper with the lips tinted are those of Jno. Pierpont (signed) ; Benjamin Rush (an early spiritualist, signed) ; Robt. Hare, M.D., whose works I had quoted (signed) ; D. D. Home, the celebrated medium who had died the year before—a likeness easily recognized ; a girl (signed "The Spirit of Mary Wallace"), probably my sister who had died the year before I was born, when eight years old ; and a lady, who was recognized as Mrs. Breed, a medium of San Francisco. These are all rather rude outlines, in somewhat irregular and interrupted dashes, but they are all lifelike, and considering that they must have been precipitated on the six surfaces while in contact with each other between the slates, as placed by myself, are exceedingly curious. The whole of these seven slates and six papers were produced so rapidly that the *seance* occupied less than an hour, and with such simple and complete openness, under the eyes of four observers, as to constitute absolutely test conditions, although without any of the usual paraphernalia of tests which were here quite unnecessary. A statement to this effect was published, with an account of the *séance*, signed by all present.

During the last fifteen years I have not seen much of spiritualistic phenomena ; but those who have read the account of my early investigations in my book on the subject, and add to them all that I have indicated here, will see that I have reached my present standpoint by a long series of experiences under such varied and peculiar conditions as to render unbelief impossible. As Dr. W. B. Carpenter well remarked many years ago, people can only believe new and extraordinary facts if there is a place for them in their existing " fabric of thought." The majority of people to-day have been brought up in the belief that miracles, ghosts, and the whole series of strange phenomena here described cannot exist ; that they are contrary to the laws of nature ; that they are the superstitions of a bygone age ; and that therefore they are necessarily either impostures or delusions. There is no place in the fabric of their thought into which such facts can be fitted. When I first began this inquiry it was the same with myself. The facts did not fit into my then existing fabric of thought. All my preconceptions, all my knowledge, all my belief in the supremacy of science and of natural law were against the possibility of such phenomena. And even when, one by one, the facts were forced upon me without possibility of escape from them, still, as Sir David Brewster declared after being at first astounded by the phenomena he saw with Mr. Home, " spirit was the last thing I could give in to." Every other possible solution was tried and rejected. Unknown laws of nature were found to be of no avail when there was always an unknown intelligence behind the phenomena—an intelligence that showed a human character and individuality, and an individuality which almost invariably *claimed* to be that of some person who had lived on earth, and who, in many cases, was able to prove his or her identity. Thus, little by little, a place was made in my fabric of thought, first for all such well-attested facts, and then, but more slowly, for the spiritualistic interpretation of them.

Unfortunately, at the present day most inquirers begin at the wrong end. They want to see, and sometimes do see the most wonderful phenomena first, and being utterly unable to accept them as facts denounce them as impostures, as did Tyndall and G. H. Lewes, or declare, as did Huxley, that such phenomena do not interest them. Many people think that when I and others publish accounts of such phenomena, we wish or require our readers to believe them on *our* testimony. But that is not the case. Neither I nor any other well-instructed spiritualist expects anything of the kind. We write not to convince, but to excite to inquiry. We ask our readers not for *belief*, but for doubt of their own infallibility on this question ; we ask for inquiry and patient experiment before hastily concluding that we are, all of us, mere dupes and idiots as regards a subject to which we have devoted our best mental faculties and powers of observation for many years.

CHAPTER XXXVIII

THE ANTI-VACCINATION CRUSADE

I WAS brought up to believe that vaccination was a scientific procedure, and that Jenner was one of the great benefactors of mankind. I was vaccinated in infancy, and before going to the Amazon I was persuaded to be vaccinated again. My children were duly vaccinated, and I never had the slightest doubt of the value of the operation—taking everything on trust without any inquiry whatever—till about 1875–80, when I first heard that there were anti-vaccinators, and read some articles on the subject. These did not much impress me, as I could not believe so many eminent men could be mistaken on such an important matter. But a little later I met Mr. William Tebb, and through him was introduced to some of the more important statistical facts bearing upon the subject. Some of these I was able to test by reference to the original authorities, and also to the various Reports of the Registrar-General, Dr. Farr's evidence as to the diminution of small-pox *before* Jenner's time, and the extraordinary misstatements of the supporters of vaccination. Mr. Tebb supplied me with a good deal of anti-vaccination literature, especially with "Pierce's Vital Statistics," the tables in which satisfied me that the claims for vaccination were enormously exaggerated, if not altogether fallacious. I also now learnt for the first time that vaccination itself produced a disease, which was often injurious to health and sometimes fatal to life, and I also found to my astonishment that even Herbert Spencer had long ago pointed out that the first compulsory Vaccination Act had led to an increase of sm all-pox. I then began to study the Reports of the Registrar-General myself, and to draw out curves of small-pox mortality, and of other zymotic diseases (the only way of showing the general course of a disease as well as its annual inequalities), and then found that the course of the former disease ran so generally parallel to that of the latter as to disprove altogether any *special protective effect* of vaccination.

As I could find no short and clear statement of the main statistical facts adverse to vaccination, I wrote a short pamphlet of thirty-eight pages, entitled "Forty-five Years of Registration Statistics, proving Vaccination to be both Useless and Dangerous." This was published in 1885 at Mr. W. Tebb's expense, and it had the effect of convincing many persons, among whom were some of my personal friends.

A few years later, when the Royal Commission on Vaccination was appointed, I was invited to become a member of it, but declined, as I could not give up the necessary time, but chiefly because I thought I could do more good as a witness. I accordingly prepared a number of large diagrams, and stated the arguments drawn from them, and in the year 1890 gave my evidence during part of three days. As about half the Commissioners were doctors, most of the others gave way to them. I told them, at the beginning of my evidence, that I knew nothing of medicine, but that, following the principle laid down by Sir John Simon and Dr. Guy, that "the evidence for the benefits of vaccination must now be statistical," I was prepared to show the bearing of the best statistics only. Yet they insisted on putting medical arguments and alleged medical facts to me, asking me how I explained this, how I accounted for that ; and though I stated again and again that there were plenty of medical witnesses who would deal with those points, they continually recurred to them ; and when I said I had no answer to give, not having inquired into those alleged facts, they seemed to think they had got the best of it. Yet they were so ignorant of statistics and statistical methods that one great doctor held out a diagram, showing the same facts as one of mine, and asked me almost triumphantly how it was that mine was so different. After comparing the two diagrams for a few moments I replied that they were drawn on different scales, but that with that exception I could see no substantial difference between them. The other diagram was on a greatly exaggerated vertical scale, so that the line showing each year's death-rate went up and down with tremendous peaks and chasms, while mine approximated more to a very irregular curve. But my questioner could not see this simple point ; and later he recurred to it a second time, and asked me if I really meant to tell them that those two diagrams were both accurate, and when I said again that though on different scales both represented the same facts, he looked up at the ceiling with an air which plainly said, " If you will say that you will say anything ! "

The Commission lingered on for six years, and did not issue its final report till 1896, while the evidence, statistics, and diagrams occupied numerous bulky blue-books. The most valuable parts of it were the appendices, containing the tables and diagrams presented by the chief witnesses, together with a large number of official tables and statistics, both of our own and foreign countries, affording a mass of material never before brought together. This enabled me to present the general statistical argument more completely and forcibly than I had done before, and I devoted several months of very hard work to doing this, and brought it out in pamphlet form in January, 1898, in order that a copy might be sent to every member of the House of Commons before the new Vaccination Act came up for discussion. This was done by the National Anti-Vaccination League, and I wrote to the half-dozen members I knew personally, begging them to give one evening to its careful perusal. But so far as any of their speeches showed, not one of the six hundred and seventy members gave even that amount of their time to obtain information on a subject involving the health, life, and personal freedom of their constituents. Yet I *know* that in no work I have written have I presented so clear and so conclusive a demonstration of the fallacy of a popular belief as is given in this work, which was entitled " Vaccination a Delusion : its Penal Enforcement a

Crime, proved by the Official Evidence in the Reports of the Royal Commission." This was included in the second part of my "Wonderful Century," published in June, 1898, and was also published separately in the pamphlet form, as it continues to be ; and I feel sure that the time is not far distant when this will be held to be one of the most important and most truly scientific of my works.

The great difficulty is to get it read. The subject is extremely unpopular ; yet as presented by Mr. William White in his " Story of a Great Delusion," it is seen to be at once a comedy and a tragedy. The historian of epidemic diseases, Dr. C. Creighton, the man who best knows the whole subject, and should be held to be the greatest living authority upon it, terms vaccination "a grotesque delusion." To inoculate a healthy child (or adult) with an animal disease, under the pretence of protecting it from another disease, the risk of having which is not one in a thousand, would, if now proposed for the first time, be so repugnant to every principle of sane medicine, as well as to common sense, that its proposer would be held to be a madman. The publication of this essay in the "Wonderful Century" (as one of the "failures") did lead to its being read by a considerable number of persons, and, as I know, of making many converts. With the hope of getting it read by Sir John Gorst, I sent a copy of my pamphlet to Mr. F. W. H. Myers, asking him to be so good as to read it carefully. In reply he wrote, " I will read your pamphlet most carefully ; will write and tell you how it affects me ; and will, in any case, send it on with your letter and a letter of my own to Sir John Gorst, whom I know well, and whom I agree with you in regarding as the most accessible member of the Government. If I am converted, it will be wholly *your* doing. I have read much on the subject—Creighton, etc., and am at present strongly pro-vaccination ; at the same time, there is no one by whom I would more willingly be converted than yourself."

The letter then goes on to quite another matter, and I may give the remainder further on.

Two days later he wrote me again :—

"I can see no answer to your statistics and arguments. Of course I should like to see what the doctors can *say in reply,* as it is difficult to believe in such a widespread blunder. But so far as the statistics with which you deal go—and that is very far—I cannot imagine a convincing answer. I am much obliged to you for letting me see the pamphlet; and I shall hand it on to Sir John Gorst, with your letter and a letter of my own.

"The unveracity of W. B. Carpenter, and especially of Ernest Hart, ought not to surprise me after what I already knew of their standards in controversy ; but it is staggering all the same."

Such a letter from so clear a reasoner and so thoroughly honest and impartial a writer, was very satisfactory to me ; but some months later, in September 1898, I received the following quite unsolicited testimonial from a perfect stranger to me—Lord Grimthorpe—an opponent in politics, but being a King's Counsel and a mathematician, as well as an able writer, was well fitted to form an opinion upon a rather complex statistical problem.

The letter is as follows :—

"Batch Wood, St. Albans, September 14, 1898.
"*To* DR. ALFRED R. WALLACE.

"SIR,

"I dare say you will excuse my troubling you with this letter on a subject on which I do not profess to be an expert, but on which it may again be my duty to form a legislative judgment. Last session I was not able to go up and sit through two probably late debates to vote ; and, indeed, I had not then made up my mind as I have now, though I had written a short letter to the *Times* on the vacillation of the Government about the Vaccination Bill.

"Since then I have been reading the chapter about it in your recent book, the "Wonderful Century," and the subsequent letters in the *Times;* and those of yesterday, especially Dr. Bond's, move me to tell you that, absurd as his

356 MY LIFE [CHAP.

statement about your 'only three converts' is, he and his
associates may add me to their number. I do not profess to
have wandered through the thickets of the Commissioners'
contradictory Reports, but I have long learnt in controversies
involving facts, to take more account of the style of the con-
troversialists, and their apparent regard for truth, than of their
assertions and references to other people, and the final
balance of voting. Specially I had to do so in the somewhat
similar controversy in the *Times* which lasted several months
in 1887–8 in which, from the accident of being put in the
chair of a hospital meeting that had been called to turn out
some doctors for homœopathic heresy, I had gradually to
take a leading part, being helped by information from the
experts on both sides as the dispute went on. Finally the
Times pronounced that I had completely proved the charges
of medical conspiracy and tyranny, which the 'orthodox'
party had been called upon at the meeting to answer, and
declined to attempt, except by their own dicta.

"Such letters as that of Dr. Bond, even without the
answers to it, always go a long way to persuade me that the
author has no solid case ; and I regard them as mere con-
troversial fireworks, throwing no real light on the subject of
discussion. In most controversies involving facts, it soon
becomes apparent to competent judges, after hearing the
professed experts, on which side is the balance of truth and
honesty, as it is very clearly in one of a very different kind
which has been going on in the *Times* for two months, on
what is called clerical and episcopal lawlessness, in which the
writers on one side think themselves at liberty to assert
anything that is 'necessary for their position' (as their great
founder avowed fifty years ago), and take their chance of being
refuted.

"In your dispute, as in that, the really decisive facts are
becoming more and more extant from the intolerable mass of
assertions and references to other people's writings which are
worth very little in the face of current genuine evidence, such
as you and other writers on your side have produced in
manageable form, and which the other side have now had

plenty of time to refute if they can, but certainly have not. In such a case neither past nor present majorities go for much. Indeed, a heavy discount may generally be taken off as due to laziness, and the desire of most people to take the apparently strongest side. I can only say that the more the vaccinationists go on writing and talking as they have done for a long time, the more they are likely to be wrong and conscious that they are so.

"Lest I should be thought to include your 'appendix' of a socialistic nostrum or 'Remedy for Want' in my general approval of your book, I think it prudent to add that I consider it more demonstrably wrong and ruinous to any country that should adopt it than any disease that has ever been propagated ; but I am not going to discuss that. I only add that you may either publish this if you like, or announce me as a 'fourth convert' to anti-vaccination under your treatment—and such as Dr. Bond's.

"Yours obediently,
"GRIMTHORPE."

"Batch Wood, St. Albans, October 4, 1898.

"*To* DR. F. T. BOND, Gloucester.

"SIR,

"I am much obliged to you for the copies of your and Mr. Wallace's letters to the *Echo*. I have read them carefully, and compared them with the chapter on Vaccination in his "Wonderful Century," and I have no hesitation in giving my verdict as a 'juryman,' and not as an 'expert,' in his favour.

"I take it for granted that you have made as good a case as anybody can on your side, and I have less doubt than before that we (I mean Parliament) have done right in putting an end, as the late Act practically has, to punishing parents for refusing to have their children vaccinated at the risk, as they believe, of doing them more harm than good. The few magistrates who are taking upon themselves to judge of the rightness of the belief will have to be taught that they are breaking and not obeying or executing the law. Nobody will pay a shilling for a certificate that he conscientiously (which only means really) believes what he does not. If he did not he would let them be vaccinated.

" Yours obediently,

" GRIMTHORPE."

" F. T. Bond, Esq., M.D., Gloucester."

This letter is of the more importance, because Dr. Bond was the only medical advocate of vaccination who attempted any extended criticism of my work. He wrote long letters in scores of newspapers all over the kingdom, some of which I replied to, showing in every case of any importance that he had either misrepresented or misunderstood my argument, and had sometimes been guilty of misquotation. The great features of my statistical argument were never dealt with by him or any other critic. The medical journals were content with pointing out minute and quite unimportant slips in medical nomenclature or classification, and though my work has now been before the public seven years, and has been widely circulated, no attempt at rebutting the main statistical argument has been made, no disproof has been adduced of the long series of misstatements of fact or fallacies of reasoning which I have charged against the whole body of official and medical advocates of vaccination. To Myers's very natural remark, " I should like to see what the doctors can say in reply," I can now answer, " They have made no reply ; and their single representative who attempted to do so, only succeeded in convincing an able and independent inquirer that they had ' no case.' "

In 1904, in view of a possible general election, I carefully condensed my pamphlet into a twenty-four page tract, which treats the subject under the following seven headings :—

(1) Why Doctors are not the best judges of the results of Vaccination.

(2) What is proved by the best Statistical Evidence available.

(3) London Death-rates during Registration.

(4) Death-rates in England and Wales during the period of Registration.

(5) Thirty years of decreasing Vaccination in Leicester, and its Teachings.

(6) The Army and Navy: a Demonstration of the Uselessness of Vaccination.

(7) How to deal with Medical Pro-vaccinators.

This has been written specially to instruct voters, and to enable them to catechise their parliamentary candidates and any medical or other upholders of vaccination. I will here give the last three paragraphs of what will probably be my last word on this subject.

"Doctors and Members of Parliament are alike grossly ignorant of the true history of the effects of vaccination. They require to be taught; and nothing is so likely to teach them as to show them the diagrams I have referred to in this short exposition of the subject—those of *London* for thirty years before and after vaccination—of *England and Wales* during the period of official registration—of *Leicester* which has almost abolished small-pox by refusing to be vaccinated for thirty years—and for the *Army and Navy*—which, though thoroughly *re-vaccinated* and therefore (according to the doctors) as well protected as they possibly can be, yet die of small-pox at least as much as badly vaccinated Ireland, and many times more than unvaccinated Leicester.

"A doctor who *has not* studied these most vital statistics has no right to an opinion on this subject.

"A candidate for Parliament who *will not* give the necessary time and attention to study them, but is yet ready to vote for penal laws against those who know infinitely more of the question than he does, is utterly unworthy to receive a single vote from any self-respecting constituency."

CHAPTER XXXIX

A CHAPTER ON MONEY MATTERS—EARNINGS AND LOSSES
—SPECULATIONS AND LAW-SUITS

Up to the age of twenty-one I do not think I ever had a sovereign of my own. I then received a small sum, perhaps about £50, the remnant of a legacy from my grandfather, John Greenell. This enabled me to get a fair outfit of clothes, and to keep myself till I got the appointment at the Leicester school. While living at Neath as a surveyor I did little more than earn my living, except during the six months of the railway mania, when I was able to save about £100. This enabled me to go to Para with Bates, and during the four years on the Amazon my collections just paid all expenses, but those I was bringing home with me would probably have sold for £200. My agent, Mr. Stevens, had fortunately insured them for £150, which enabled me to live a year in London, and get a good outfit and a sufficient cash balance for my Malayan journey.

My eight years in the Malay Archipelago were successful, financially, beyond my expectations. Celebes, the Moluccas, the Aru Islands, and New Guinea were, for English museums and private collections, an almost unknown territory. A large proportion of my insects and birds were either wholly new or of extreme rarity in England; and as many of them were of large size and of great beauty, they brought very high prices. My agent had invested the proceeds from time to time in Indian guaranteed railway stock, and a year after my return I found myself in possession of about £300 a year. Besides this, I still possessed the whole series of private collections, including large numbers of new or very rare species, which, after I had made what use of them was needed for my work, produced an amount which in the same securities would have produced about £200 a year more.

But I never reached that comfortable position. Owing to my never before having had more than enough to supply my immediate wants, I was wholly ignorant of the numerous snares and pitfalls that beset the ignorant investor, and I unfortunately came under the influence of two or three men who, quite unintentionally, led me into trouble. Soon after I came home I made the acquaintance of Mr. R., who held a good appointment under Government, and had, besides, the expectation of a moderate fortune on the death of an uncle. I soon became intimate with him, and we were for some years joint investigators of spiritualistic phenomena. He was, like myself at that time, an agnostic, well educated, and of a more positive character than myself. He had for some years saved part of his income, and invested it in various foreign securities at low prices, selling out when they rose in value, and in this way he assured me he had in a few years doubled the amount he had saved. He studied price-lists and foreign news, and assured me that it was quite easy, with a little care and judgment, to increase your capital in this way. He quite laughed at the idea of allowing several thousand pounds to lay idle, as he termed it, in Indian securities, and so imbued me with an idea of his great knowledge of the money-market, that I was persuaded to sell out some of my bonds and debentures and buy others that he recommended, which brought in a higher interest, and which he believed would soon rise considerably in value. This change went on slowly with various success for several years, till at last I had investments in various English, American, and foreign railways, whose fluctuations in value I was quite unable to comprehend, and I began to find, when too late, that almost all my changes of investment brought me loss instead of profit, and later on, when the great depression of trade of 1875–85 occurred, the loss was so great as to be almost ruin.

In 1866 one of my oldest friends became secretary to a

small body of speculators, who had offices in Pall Mall, and who, among other things, were buying slate quarry properties, and forming companies to work them. Two of these properties were in the neighbourhood of Dolgelly and Machynlleth, and a party of us went down to see them. We were shown the outcrop of the slate rock, followed it across the country, were told it was of fine quality, and that there was a fortune in it if properly developed. My friend's employer seemed to know all about it, and as many large fortunes had been made out of slate quarries, it seemed quite a safe thing. The slate was undoubtedly there, as small portions had been worked and split up, and we saw the piles of slates, and were assured that the quality would be still better as it was worked deeper. One of the veins had been found to be excellent for billiard tables, and for all kinds of slate tanks, as it could be got out in slabs of almost any size, and only wanted sawing and planing machines worked by a small mountain brook close by to become very profitable. Of course we were shown reports by specialists, who declared that the slate rock was abundant ·and good, and if properly developed would be profitable.

I was persuaded to take shares, and to be a director of these companies, without any knowledge of the business, or any idea how much capital would be required. The quarries were started, machinery purchased, call after call made, with the result in both cases that, after four or five years of struggle, the capital required and the working expenses were so great that the companies had to be wound up, and I was the loser of about a thousand pounds.

While this was going on a still more unfortunate influence became active. My old friend in Timor and Singapore, Mr. Frederick Geach, the mining engineer, came home from the East, and we became very intimate, and saw a good deal of each other. He was a Cornishman, and familiar with tin, lead, and copper mining all his life, and he had the most unbounded confidence in good English mines as an investment. He had shares in some of the lead-mines of Shropshire and Montgomeryshire, and we went for a walking tour in that beautiful country, visited the mines, went down the shafts by endless perpendicular ladders, and examined the veins and workings with the manager, who had great confidence in its value, and was a large shareholder. "Here," said Geach, "you can see the vein of lead ore. It is very valuable, and extends to an unknown depth. This is not a probability, it is a certainty." And so I was persuaded to buy shares in lead-mines, and gradually had a large portion of my capital invested in them.

But here again, neither I nor Geach, nor hardly one in England, knew of the insidious foe that in a remote part of the world was preparing the way for the ruin of English lead-mining. This was twofold : the great development of mining in Spain by English capitalists ; and, more important still, the enormous amount of silver-mining in Nevada, United States, where the ore contained lead and silver combined, so that as the works extended large quantities of lead accumulated as a kind of waste product, and much of this was exported to Europe, and so lowered the price of lead that most of the British lead-mines became unprofitable. About 1870 the price of lead began to fall, and has continued to fall, as has silver, ever since. The result of all this was that by 1880 a large part of the money I had earned at the risk of health and life was irrecoverably lost.

While these continued misfortunes were in progress I was involved in two other annoyances, causing anxiety and worry for years, as well as a very large money loss. The first was with a dishonest builder, who contracted to build my house at Grays, and who was paid every month according to the proportion of the work done. One day, when the house was little more than half finished, he did not appear to pay his men, and as they would not continue to work without their money I paid them. He did not appear the next week, and sent no excuse, so the architect gave him notice that I should complete the building myself, and that, according to the agreement, he would be responsible for any cost beyond the contract price. After a few weeks he appeared, and wanted to go on, but that we declined. The house cost me somewhat more than the contract price, and when it was finished I sent him word he could have his ladders, scaffold-poles, boards, etc., though, according to the agreement, they were to be my property on his failure to finish the building.

I soon found, however, that he had not paid for a large portion of the materials, and bills kept coming in for months afterwards for bricks, timber, stone, iron-work, etc., etc. The merchants who had trusted him found that he had no effects whatever, as he lived as a lodger with his father ; and from all I heard, was accustomed to take contracts in different places round London, and by not paying for any materials that he could get on credit, make a handsome profit. But the height of his impudence was to come. About five years after the house was finished, I received a demand through a lawyer for (I think) between £800 and £900 damages for not allowing this man to finish the house! I wrote, refusing to pay a penny. Then came a notice of an action at law ; and I was obliged to put it in a lawyer's hands. All the usual preliminaries of interrogatories, affidavits, statements of claim, replies, objections, etc., etc., were gone through, and on every point argued we were successful, with costs, which we never got. The case was lengthened out for two or three years, and then ceased, the result being that I had to pay about £100 law costs for what was merely an attempt to extort money. That was my experience of English *law*, which leaves the honest man in the power of the dishonest one, mulcts the former in heavy expenses, and is thus the very antithesis of *justice*.

The next matter was a much more serious one, and cost me fifteen years of continued worry, litigation, and persecution, with the final loss of several hundred pounds. And it was all brought upon me by my own ignorance and my own fault— ignorance of the fact so well shown by the late Professor de Morgan—that "paradoxers," as he termed them, can never be convinced, and my fault in wishing to get money by any kind of wager. It constitutes, therefore, the most regrettable incident in my life. As many inaccurate accounts have been published, I will now state the facts, as briefly as possible, from documents still in my possession.

In *Scientific Opinion* of January 12, 1870, Mr. John Hampden (a relative of Bishop Hampden) challenged scientific men to prove the convexity of the surface of any inland water, offering to stake £500 on the result. It contained the following words : " He will acknowledge that he has forfeited his deposit if his opponent can exhibit, to the satisfaction of any intelligent referee, a convex railway, river, canal, or lake." Before accepting this challenge I showed it to Sir Charles Lyell, and asked him whether he thought I might accept it. He replied, " Certainly. It may stop these foolish people to have it plainly shown them." I therefore wrote accepting the offer, proposing Bala lake, in North Wales, for the experiment, and Mr. J. H. Walsh, editor of the *Field*, or any other suitable person, as referee. Mr. Hampden proposed the Old Bedford canal in Norfolk, which, near Downham Market, has a stretch of six miles quite straight between two bridges. He also proposed a Mr. William Carpenter (a journeyman printer, who had written a book upholding the " flat earth " theory) as his referee ; and as Mr. Walsh could not stay away from London more than one day, which was foggy, I chose Mr. Coulcher, a surgeon and amateur astronomer, of Downham Market, to act on my behalf, Mr. Walsh being the umpire and referee.

The experiment finally agreed upon was as follows : The iron parapet of Welney bridge was thirteen feet three inches above the water of the canal. The Old Bedford bridge, about six miles off, was of brick and somewhat higher. On this bridge I fixed a large sheet of white calico, six feet long and three feet deep, with a thick black band along the centre, the lower edge of which was the same height from the water as the parapet of Welney bridge ; so that the centre of it would be as high as the line of sight of the large six-inch telescope I had brought with me. At the centre point, about three miles from each bridge, I fixed up a long pole with two red discs on it, the upper one having its centre the same height above the water as the centre of the black band and of the telescope, while the second disc was four feet lower down. It is evident that if the surface of the water is a perfectly straight line for the six miles, then the three objects—the telescope, the top disc, and the black band—being all exactly the same height above the water, the disc would be seen in the telescope projected upon the black band; whereas, if the six-mile surface of the water is convexly curved, then the top disc would appear to be decidedly higher than the black band, the amount due to the known size of the earth being five feet eight inches, which amount will be reduced a little by refraction to perhaps about five feet.

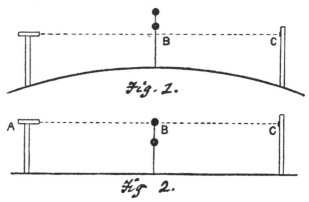

The above diagrams illustrate the experiment made. The curved line in Fig. 1, and the straight line in Fig. 2, show the surface of the canal on the two theories of a round or a flat earth. A and C are the two bridges six miles apart, while B is the pole midway with two discs on it, the upper disc, the telescope at A, and the black line on the bridge at C, being all exactly the same height above the water. If the surface of the water is truly flat, then on looking at the mark C with the telescope A, the top disc B will cover that mark. But if the surface of the water is curved, then the upper disc will appear above the black mark, and if the disc is more than four feet above the line joining the telescope and the black mark, then the lower disc will also appear above the black mark. Before the experiment was made a diagram similar to this was submitted to Mr. Hampden, his referee Mr. Carpenter, and Mr. Walsh, and all three agreed that it showed clearly what should be seen in the two cases, while the former declared their firm belief that Fig. 2 showed what *would* be seen.

When the pole was set up and the mark put upon the bridge, Mr. Carpenter accompanied me, and saw that their heights above the water were the same as that of the telescope resting on the parapet of the bridge. What was seen in the large telescope was sketched by Mr. Coulcher and signed by Mr. Carpenter as correct, and is shown in the following diagram which was reproduced in the *Field* newspaper (March 26, 1870), and also in a pamphlet by Carpenter himself. But

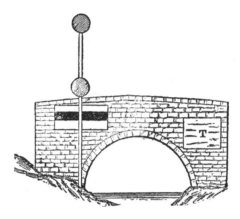

"Signed by Mr. Carpenter."—*Dr. Coulcher's Report.* "*Signed!*"

he declared that this proved nothing, because the telescope was not levelled, and because it had no cross-hair!

At his request to have a spirit-level in order to show if there was any "fall" of the surface of water, I had been to King's Lynn and borrowed a good Troughton's level from a surveyor there. This I now set up on the bridge at exactly the same height above the water as the other telescope, and having levelled it very accurately and called Mr. Carpenter to see that the bubble was truly central and that the least movement of the screws elevating or depressing it would cause the bubble to move away, I adjusted the focus on to the distant bridge, and showing also the central staff and its two discs.

Mr. Coulcher looked at it, and then Mr. Carpenter, and the moment the latter did he said "Beautiful! Beautiful!" And on Mr. Hampden asking him if it was all right, he replied that it was perfect, and that it showed the three points in "a perfect straight line;" "as level as possible!" And he actually jumped for joy. Then I asked Mr. Coulcher and Mr. Carpenter both to make sketches, which they did. We then fixed a calico flag on the parapet to make it more

THE "BEDFORD LEVEL" SURVEY.—SKETCHES BY THE
TWO REFEREES.

Copied from the *Field* for March 26, 1870.

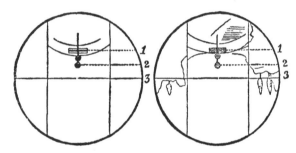

These two views, as seen by means of the *inverting* telescope, are exact representa- tions of the sketches taken by Mr. Hampden's Referee, and attested by Dr. Coulcher as being correct in both cases: first, from Welney Bridge; and secondly, from the Old Bedford Bridge.

visible, and drove back with the instruments to Old Bedford bridge, where I set up the level again at the proper height above the water, and again asked both the referees to make sketches of what was seen in the level-telescope. This they did. Mr. Carpenter's was rather more accurately drawn, and Mr. Coulcher signed them as being correct, and both are reproduced here.

For those who do not understand the use of a level, it may be necessary to explain that the cross-hair in the optical axis of the telescope marks the true level of any object at a distance with regard to the telescope. Any point that is seen above the cross-hair is above the level, any point seen below the cross-hair is below the level, and in the latter case the line from the telescope to it slopes downwards. To show this "true level" is the whole purpose of the instrument called a surveyor's level, and it does show it with wonderful accuracy. The mere fact, therefore, that the top disc on the pole was apparently more below the cross-hair than the two discs were apart, proved that the surface of the water was not flat, or continuously extended in a straight line. And again the fact that the distant signal was again about the same distance, apparently below the middle one, as that was below the telescope of the level, shows that the surface of the water did not merely slope down in a straight line, but was curved downwards with regard to its surface at the starting-point.

The following diagram will illustrate this :—

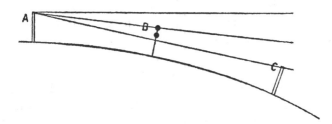

The lower curved line represents the supposed curved surface of the water. The points A B C are three points equi-distant above that surface. The top line from A is the level line shown by the cross-hair in the level-telescope. If the water surface had been truly level, the two points B and C must have been cut by the cross-hair. But even if the cross-hair did not show the true level, but pointed upwards, and the water was truly level, then the distant mark, being the same height above the water as the top disc at half the distance and the telescope, these two objects must have appeared in a straight line, the nearer one covering the more distant. It should appear on the straight line drawn from the eye at A through B, whereas it appears a long way below it, thus proving curvature, the essential point to be shown.

Thus the view in the large telescope and in the level-telescope both told exactly the same thing, and, moreover, proved that the curvature was very nearly of the amount calculated from the known dimensions of the earth. Mr. Hampden declined to look through either telescope, saying he trusted to Mr. Carpenter; while the latter declared positively that they had won, and that we knew it; that the fact that the distant signal *appeared* below the middle one as far as the middle one did below the cross-hair, proved that the three were in a straight line, and that the earth was flat, and he rejected the view in the large telescope as proving nothing for the reasons already stated.

At first Mr. Hampden refused to appoint an umpire, because my referee, Mr. Coulcher, refused to discuss the question with Mr. Carpenter; but after a few days he agreed that Mr. Walsh should be the umpire, after receiving the reports of the two referees. He had, in fact, unbounded confidence in what Mr. Carpenter told him, and firmly believed that the experiments had demonstrated the flat earth, and that no honest man could think otherwise.

But Mr. Walsh decided without any hesitation that I had proved what I undertook to prove. He published the whole of the particulars with the reports of the referees and their sketches in the *Field* of March 18 and 26, while a considerable correspondence and discussion went on for some weeks later. At Mr. Hampden's request he allowed Mr. Carpenter to send in a long argument to show that the experiments were all in Mr. Hampden's favour, and having considered them, he wrote to Mr. Hampden that he should hand me the stakes on a certain day if he had no other reason to adduce why he should not do so. Thereupon Mr. Hampden wrote to him *demanding his money back* on the ground that the decision was unjust, and ought to have been given in his favour. In thus writing to Hampden and receiving his demand for his deposit to be returned, Mr. Walsh made a great mistake, which had serious consequences for me. The law declares that all wagers are null and void, and that money lost by betting is not recoverable at law. But the judges have decided that when a wager is given against him by the umpire, the loser can claim his money back from the stakeholder if the latter has not already paid it away to the winner.

Hence, if a loser immediately claims his money from the stake-holder, the law will enforce the former's claim on the ground that it is *his* money, and the fact that he has lost it in a quite fair wager is beyond the cognizance of the law. Neither I nor Mr. Walsh knew of this, although he had decided and paid many wagers; but this resulted in my having to pay the money back five years later, as will be presently described.

I will now briefly state what were Hampden's proceedings for the next fifteen or sixteen years. He first began abusing Mr. Walsh in letters, post-cards, leaflets, and pamphlets, as a liar, thief, and swindler. Then he began upon me with even more virulence, writing to the presidents and secretaries of all the societies to which I belonged, and to any of my friends whose addresses he could obtain. One of his favourite statements in these letters was, "Do you know that Mr. A. R. Wallace is allowing himself to be posted all over England as a cheat and a swindler?" But he soon took more violent measures, and sent the following letter to my wife :—

"MRS. WALLACE,

"Madam—If your infernal thief of a husband is brought home some day on a hurdle, with every bone in his head smashed to pulp, you will know the reason. Do you tell him from me he is a lying infernal thief, and as sure as his name is Wallace he never dies in his bed.

"You must be a miserable wretch to be obliged to live with a convicted felon. Do not think or let him think I have done with him.

"JOHN HAMPDEN."

For this I brought him up before a police magistrate, and he was bound over to keep the peace for three months, suffering a week's imprisonment before he could find the necessary sureties. But as soon as the three months were up, he began again with more abuse than ever, distributing tracts and writing to small local papers all over England. I now began to receive letters from friends, and also from perfect strangers, asking me if I knew what was said about me everywhere. I will now give a summary of the steps I was obliged to take with the results, or rather absence of results, that followed.

In 1871, Mr. Walsh prosecuted Hampden for libel. He was convicted at the Old Bailey, and bound over to keep the peace for one year.

In January, 1871, I brought an action for libel in order to give Hampden the opportunity of justifying, if he could, his language towards me. He did not defend the action, but suffered judgment to go by default, and the jury gave me a verdict with £600 damages. But whatever property he had had been transferred to his son-in-law (a solicitor), so I could not get a penny, and had to pay the costs of the suit which, though undefended, were heavy.

In October, 1872, I prosecuted him at the Old Bailey for further libels. He was respited on publicly apologizing in several newspapers.

On January 13, 1873, he was brought up again for fresh libels, and was again respited on publishing a fuller apology and complete recantation of all his charges, as follows :—

" PUBLIC APOLOGY.—I, the undersigned John Hampden, do hereby absolutely *withdraw* all libellous statements published by me, which have reflected on the character of Mr. Alfred Russel Wallace, and apologize for having published them ; and I promise that I will not repeat the offence.— JOHN HAMPDEN."

This was published in several of the London daily papers and in various country papers in which any of his letters had appeared, and the judge gave him a serious warning that if brought up again he would be imprisoned.

Some months afterwards, however, he began again with equally foul libels, and I had him brought up under his recognizances, when he was sentenced to two months' imprisonment in Newgate.

But within a year he began again as violently as ever, and on March 6, 1875, he was indicted at Chelmsford Assizes for fresh libels, and on proof of his previous convictions and apologies, he was sentenced to one year's imprisonment and to keep the peace, under heavy recognizances and sureties, for two years more. (A full report is given in the *Chelmsford Chronicle*, March 12, 1875.)

Through the interest of his friends, however, he was liberated in about six months; and thereupon, in January, 1876, he brought an action against Mr. Walsh to recover his deposit of £500, and this action he won, on the grounds already stated; and as I had signed an indemnity to Mr. Walsh, I had to pay back the money, and also pay all the costs of the action, about £200 more. But as I had a judgment for £687 damages and costs in my libel suit against Hampden, I transferred this claim to Mr. Walsh as a set-off against the amount due by him. Hampden, however, had already made himself a bankrupt to prevent this claim being enforced, and had assigned all his actual or future assets to his son-in-law.

There were now legal difficulties on both sides. I was advised that the bankruptcy was *fraudulent*, and could be annulled; but to attempt this would be costly, and the result uncertain. On the other hand, it was doubtful whether my claim against Hampden would not be treated as an ordinary creditor's claim in the bankruptcy. There was, therefore, a consultation of the solicitors, and a voluntary arrangement was arrived at. I was to pay all the costs of the suit and £120, amounting to £277; while £410 still remained nominally due to me from Hampden.

These terms were formally agreed to by Hampden and his son-in-law, and were duly carried out. Of course I had also to pay Mr. Walsh's costs in the action and my own lawyer's bill for the settlement, as well as those of the action for libel, and the various criminal prosecutions of Hampden I had been compelled to undertake.

Notwithstanding this settlement, however, Hampden was by no means silenced. The very day after his recognizances expired, in 1878, he began again with his abusive post-cards, circulars, and other forms of libel. In 1885 he wrote and printed a long letter to Huxley, as President of the Royal Society, chiefly on his biblical discussion with Mr. Gladstone, in a postscript to which he writes as follows :—

"I have thoroughly exposed that degraded blackleg, Alfred Russel Wallace, as I would every one who publicly identifies himself with such grossly false science, which he had the audacity to claim to be true! If this man's experiment on the Bedford canal was founded on fact, then the whole of the Scriptures are false, from the first verse to the last. But your whole system is based upon falsehood and fraud, and refusal of all discussion ; and such characters as Wallace seem to be your only champions." And he has an appendix on "Modern Education conducted on Wrong Principles," in which we find such gems as this :—

"When Mr. Mundella and Mr. Gladstone were schoolboys, the educational professors were all newly indoctrinated with the pretentious learning of the 'Principia' of Newton. The Bible was not regarded as of any authority upon such subjects, and a flood of writers were all extolling the immortal genius of the 'incomparable mathematician.' Newton and his apple-tree were spoken of as the foundation of all true philosophy. The plausibly sounding phrases 'Attraction' and 'Gravitation' were in every pedagogue's mouth, and the poor children were birched into repeating them every hour of their lives." And so on for three closely printed pages.

About this time he printed one thousand copies of a two-page leaflet, and sent them to almost every one in my neighbourhood whose address he could obtain, including most of the masters of Charterhouse School, and the residents as well as the tradesmen of Godalming. It was full of—" scientific villainy and roguery,"—"cheat, swindler, and impostor."—"My specific charge against Mr. A. R. Wallace is that he obtained possession of a cheque for £1,000 by fraud and falsehood of a party who had no authority to dispose of it."—"As Mr. Wallace seems wholly devoid of any sense of honour of his own, I shall most readily submit the whole matter to any two or more disinterested parties,

and adhere most absolutely and finally to their decision."—
"I will compel him to acknowledge that the curvature of
water which he and his dupes pretend was proved on the
Bedford Level, *does not exist!* And this Mr. Wallace saw
with his own eyes." And so on in various forms of repetition
and abuse. To save trouble, I drew up a short circular
stating the main facts already given here for the information
of those who had received Hampden's absurdly false libels,
and thereafter took no further notice of him.

One day about this time we happened to have several
friends with us, and as we were at luncheon, I was called to
see a gentleman at the door. I went, and there was Hampden! I was so taken aback that my only idea was to get rid
of him as soon as possible, but I afterwards much regretted
that I did not ask him in, give him luncheon, and introduce
him as the man who devoted his life to converting the world
into the belief that the earth was flat. We should at least
have had some amusement; and to let him say what he had
to say to a lot of intelligent people might have done him
good. But such "happy thoughts" come too late. He had
come really to see where I lived, and as our cottage and
garden at Godalming, though quite small, were very pretty,
he was able to say afterwards that I (the thief, etc.) was
living in luxury, while he, the martyr to true science, was in
poverty.

He continued to circulate his postcards and tracts, and
to write to all manner of people, challenging them to prove
that the earth was not flat, for several years after. The last
of his efforts which I have preserved is an eight-page tract,
which he distributed at the Royal Geographical Society's
Exhibition of Geographical Appliances, in December, 1885,
in which he attacks all geographical teaching in his usual
style, and declares that "at the present moment they are
cowering beneath the inquiring gaze of one single truth-
seeker, JOHN HAMPDEN, the well-known champion of the
Mosaic cosmogony, as against the infidel theories and superstitions of the pagan mystics, who is, at the end of fifteen
years' conflict, still holding his ground against all the

professional authorities of England and America; and the single fact that during the whole of that time, no one but a degraded swindler has dared to make a fraudulent attempt to support the globular theory, is ample and overwhelming proofs of the worthless character of modern elementary geography." And again : " Surveyors and civil and military engineers are offered £100 for the discovery of any portion of the earth's curvature, on land or water, railway or canal, of not less than five or ten miles, within one hundred miles of the metropolis. Why does not Mr. A. R. Wallace do again what he says he has done before ? " And in a list of advertisements of books, etc., supporting his views he has this one : " *Scientific Information wanted.* A gentleman of ample means and inquisitive disposition offers £100 for particulars setting forth conclusively the grounds on which Sir Isaac Newton's Globular Theory was presumably established or asserted to be the fact."

And this man was educated at Oxford University ! Seldom has so much boldness of assertion and force of invective been combined with such gross ignorance. And to this day a society exists to uphold the views of Hampden, Carpenter, and their teacher, " Parallax ! "

The two law suits, the four prosecutions for libel, the payments and costs of the settlement, amounted to considerably more than the £500 I received from Hampden, besides which I bore all the costs of the week's experiments, and between fifteen and twenty years of continued persecution—a tolerably severe punishment for what I did not at the time recognize as an ethical lapse.

There is one other small money matter which I wish to put on record here, because, though it involves only the small sum of sixpence, it affords an example of official meanness, and what really amounts to petty larceny, which can hardly be surpassed. In 1865 the British Museum purchased from me some specimen (I think a skeleton) for which they agreed to pay £5. Two years later I received the following printed form :—

"Principal Librarian and Secretary's Office.
"British Museum, W.C., June 24, 1867.

"Sir,

"If you will send your own stamped receipt to this
Office, you will be paid the amount due to you by the
Trustees of the British Museum, £5 0s. 0d.

"I am, sir,

"Your very obedient Servant,

"THOMAS BUTLER,

"Assist. Secretary."

"Mr. A. R. Wallace."

I, of course, complied with the request and sent the
stamped receipt, and by return of post had the following
written communication :—

"Mr. Butler begs to transmit the enclosed P.O. order for
£4 19s. 6d. to Mr. Wallace, and the amount of it, with the
cost of the order (6d.), makes up the sum due by the Trustees
to Mr. Wallace.

"British Museum, June 25, 1867."

This amazing little dodge (for I can call it nothing else)
completely staggered me. I was at first inclined to return
the P.O. order, or to write asking for the 6d., and if necessary
summon Mr. Butler (or the Trustees) to a County Court for
the 6d. due. But I was busy, and did not want to enter upon
what I felt sure would be a long correspondence and endless
trouble and expense. I therefore determined to keep the
two incriminating documents, and some day print them.
That day has now come ; and it may be interesting to learn
whether this preposterous and utterly dishonest method of
paying part of an admitted debt, after obtaining a receipt for
the whole, continues to be practised in this or any other
public institution.

It was while these troubles in the Hampden affair were at
their thickest that my earnings invested in railways and mines

continued depreciating so constantly as to be a source of great anxiety to me, and every effort to extricate myself by seeking better investments only made matters worse. It was at this time that the endeavour to get the Epping Forest appointment failed, and had it not been for the kindness of a relative, Miss Roberts, of Epsom, a cousin of my mother's, with whose family I had been intimate from my boyhood, I should have been in absolute want. She had intended to leave me £1000 in her will, but instead of doing so transferred it to me at once, and as it was in an excellent security, and brought me in from £50 to £65 a year, it was most welcome. I had sold my house at Grays fairly well, and in 1880 bought a piece of land and built a cottage at Godalming, so that I had a home of my own ; but I had now to depend almost entirely on the little my books brought me in, together with a few lectures, reviews, and other articles. I had just finished writing my "Island Life," and had no idea that I should ever write another important book, and I therefore saw no way of increasing my income, which was then barely sufficient to support my family and educate my two children in the most economical way. From this ever-increasing anxiety I was relieved through the grant of a Civil Service pension of £200, which came upon me as a very joyful surprise. My most intimate and confidential friend at this time was Mrs. Fisher (then Miss Buckley), and to her alone I mentioned my great losses, and my anxiety as to any sure source of income. Shortly afterwards she was visiting Darwin, and mentioned it to him, and he thought that a pension might be granted me in recognition of my scientific work. Huxley most kindly assisted in drawing up the necessary memorial to the Prime Minister, Mr. Gladstone, to whom Darwin wrote personally. He promptly assented, and the next year, 1881, the first payment was made. Other of my scientific friends, I believe, signed the memorial, but it is especially to the three named that I owe this very great relief from anxiety for the remainder of my life.

I have already stated that what at the time appeared to

XXXIX] MONEY MATTERS 379

be the great misfortune of the loss of about half of my whole
Amazonian collections by the burning of the ship in which I
was coming home, was in all probability a blessing in disguise, since it led me to visit the comparatively unknown
Malay Archipelago, and, perhaps, also supplied the conditions
which led me to think out independently the theory of natural
selection. In like manner I am now inclined to see in the
almost total loss of the money value of those rich collections,
another of those curious indications that our misfortunes are
often useful, or even necessary for bringing out our latent
powers. I am, and have always been, constitutionally lazy,
without any of that fiery energy and intense power of work
possessed by such men as Huxley and Charles Kingsley.
When I once begin any work in which I am interested, I
can go steadily on with it till it is finished, but I need some
definite impulse to set me going, and require a good deal of
time for reflection while the work is being done. Every important book I have undertaken has been due to an impulse
or a suggestion from without. I spent five years in quiet
enjoyment of my collections, in attending scientific meetings,
and in working out a few problems, before I began to write
my "Malay Archipelago," and it was due to the repeated
suggestions of my friends that I wrote my "Geographical
Distribution of Animals."

But if the entire proceeds of my Malayan collections had
been well invested, and I had obtained a secure income of
£400 or £500 a year, I think it probable that I should not
have written another book, but should have gone to live
further in the country, enjoyed my garden and greenhouse
(as I always have done), and limited my work to a few lectures
and review articles, but to a much less extent than I actually
have done. It was the necessity of earning money, owing to
my diminishing income, that caused me to accept invitations
to lecture, which I always disliked ; and the same reason
caused me to seek out subjects for scientific or social articles
which, without that necessity, would never have been written.
Under such conditions as here supposed, my dislike to lecturing
would probably have increased, and I should never have

ventured on my lecturing tour in America, in which case I
should not have written "Darwinism," and, I firmly believe,
should not have enjoyed such good health as I am now doing.
Then, too, I should probably not have accepted Dr. Lunn's
invitation to lecture at Davos, and my two later books would
never have come into existence.

Of course this is all conjecture, but it seems to myself
highly probable. At all events, I feel perfectly sure that
without the spur of necessity I should not have done much
of the work I have done. I have always had a great desire
to see many of the beauty-spots of the world. Some of them
I have seen, but usually under strict limitation of time and
means. I have longed to visit the old volcanoes of Mont
Doré or the Eifel, both for their geology and their rich
flora ; the Dolomites and the Italian lakes ; Pompei, and
Rome, and the lovely Riviera ; Sicily and Greece ; while
the little I have seen of Switzerland has made me wish to
see more. If I had had the means I should probably have
spent a good part of each winter, spring, or summer, in these
countries, and should have found such constant delight in
them, and in my garden at home, to which I should have
brought home every year new floral treasures, that I should
not have felt the want of any other occupation, and should
probably have written nothing but an occasional review or
magazine article. If, therefore, my books and essays have
been of any use to the world—and though I cannot quite
understand it, scores of people have written to me telling me
so—then the losses and the struggles I have had to go
through have been a necessary discipline calculated to bring
into action whatever faculties I possess. I may be allowed
here to give an extract from one of these letters on my
literary work, nearly the last I received from my lamented
friend F. W. H. Myers. He writes (April 12, 1898) :—

"I am glad to take this opportunity of telling you something about my relation to one of your books. I write now
from bed, having had severe influenzic pneumonia, now going
off. For some days my temperature was 105°, and I was very
restless at night—anxious to read, but in too sensitive and

fastidious a state to tolerate almost any book. I found that almost the only book which I could read was your 'Malay Archipelago.' Of course I had read it before. In spite of my complete ignorance of natural history there was a certain uniqueness of charm about the book, both moral and literary, which made it deeply congenial in those trying hours. You have had few less instructed readers ; but very few can have dwelt on that simple, manly record with a more profound sympathy."

Other people, quite strangers, have also told me that they have read it over and over again, and always take it with them on a journey. This is the kind of thing I cannot understand. It is true, if I open it myself I can read a chapter with pleasure ; but, then, to me it recalls incidents and feelings almost forgotten, and renews the delights of my wanderings in the wilderness and of my intense interest in the wonderful and beautiful forms of plant, bird, and insect life I was continually meeting with. Others have written in almost equally laudatory terms of my books on " Land Nationalization " and on " Spiritualism," which have introduced them to new spheres of thought ; while others, again, have been equally pleased with parts of my " Wonderful Century " and " Man's Place in the Universe." I am thus forced to the conclusion that my books have served to instruct and to give pleasure to a good many readers, and that it is therefore just possible that my life may have been prolonged, and conditions modified so as to afford the required impulse and the amount of time for me to write them.

Of course, such a suggestion as this will seem foolishness or something worse to most of my readers ; but for those who are imbued with the teachings of modern spiritualism, and to others who vaguely believe in spiritual guidance on general religious grounds, such forms of what used to be termed special providence will not be wholly rejected.

CHAPTER XL

MY CHARACTER—NEW IDEAS—PREDICTIONS
FULFILLED

I HAVE already (in Chapter XV.) given an estimate of my character when I came of age. I will now make a few further remarks upon it as modified by my changed views of life, owing to my becoming convinced of the reality of a spirit world and a future state of existence.

Up to middle age, and especially during the first decade after my return from the East, I was so much disinclined to the society of uncongenial and commonplace people that my natural reserve and coldness of manner often amounted, I am afraid, to rudeness. I found it impossible, as I have done all my life, to make conversation with such people, or even to reply politely to their trivial remarks. I therefore often appeared gloomy when I was merely bored. I found it impossible, as some one had said, to tolerate fools gladly ; while, owing to my deficient language-faculty, talking without having anything to say, and merely for politeness or to pass the time, was most difficult and disagreeable. Hence I was thought to be proud, conceited, or stuck-up. But later on, as I came to see the baneful influence of our wrong system of education and of society, I began to realize that people who could talk of nothing but the trivial amusements of an empty mind were the victims of these social errors, and were often in themselves quite estimable characters.

Later on, when the teachings of spiritualism combined with those of phrenology led me to the conclusion that there were no absolutely bad men or women, that is, none who, by a rational and sympathetic training and a social system which gave to all absolute equality of opportunity, might not become useful, contented, and happy members of society, I became much more tolerant. I learnt also to distrust all first impressions; for I repeatedly came to enjoy the society of people whose appearance or manner had at first repelled me, and even in the most apparently trivial-minded was able to find some common ground of interest or occupation. I feel myself that my character has continuously improved, and that this is owing chiefly to the teaching of spiritualism, that we are in every act and thought of our lives here building up a character which will largely determine our happiness or misery hereafter; and also, that we obtain the greatest happiness ourselves by doing all we can to make those around us happy.

As I have referred in various parts of this volume to ideas, or suggestions, or solutions of biological problems, which I have been the first to put forth, it may be convenient if I here give a brief account of the more important of them, some of which have, I think, been almost entirely overlooked.

1. The first and perhaps the most important of these is my independent discovery of the theory of natural selection in 1858, in my paper on "The Tendency of Varieties to depart indefinitely from the Original Type." This is reprinted in my "Natural Selection and Tropical Nature;" and it has been so fully recognized by Darwin himself and by naturalists generally that I need say no more about it here. I have given a rather full account of how it first occurred to me in Chapter XXII. of this work.

2. In 1864 I published an article on "The Development of Human Races under the Law of Natural Selection," the most original and important part of which was that in which I showed that so soon as man's intellect and physical structure led him to use fire, to make tools, to grow food, to domesticate animals, to use clothing, and build houses, the

action of natural selection was diverted from his body to his mind, and thenceforth his physical form remained stable while his mental faculties improved. This paper was greatly admired by Mr. Darwin and several other men of science, who declared it to be entirely new to them ; but owing to its having been published in one of my less known works, "Contributions to the Theory of Natural Selection," it seems to be comparatively little known. Consequently, it still continues to be asserted or suggested that because we have been developed physically from some lower form, so in the future we shall be further developed into a being as different from our present form as we are different from the orang or the gorilla. My paper shows *why* this will not be ; *why* the form and structure of our body is permanent, and that it is really the highest type now possible on the earth. The fact that we have not improved physically over the ancient Greeks, and that most savage races—even some of the lowest in material civilization—possess the human form in its fullest symmetry and perfection, affords evidence that my theory is the true one.

3. In 1867 I gave a provisional solution of the cause of the gay, and even gaudy colours of many caterpillars, which was asked for by Darwin, and which experiment soon proved to be correct. This is referred to in Chapter XXI. of the present volume, and is fully described in my "Natural Selection and Tropical Nature," pp. 82–86. The principle established in this case has been since found to be widely applicable throughout the animal kingdom.

4. In 1868 I wrote a paper on "A Theory of Birds' Nests," the chief purport of which was to point out and establish a connection between the colours of female birds and the mode of nidification which had not been before noticed. This led to the formulation of the following law, which has been very widely accepted by ornithologists: *When both sexes of birds are conspicuously coloured, the nest conceals the sitting bird ; but when the male is conspicuously*

coloured and the nest is open to view, the female is plainly coloured and inconspicuous. No less than fifteen whole families of birds and a number of the genera of other families belong to the first class, of brightly coloured birds with sexes alike, and they all build in holes or make domed nests. Most of these are tropical, but the woodpeckers and kingfishers are European. In the second class, however brilliant the male may be, if the nest is open to view, the female is always plainly coloured, sometimes so much so as to be hardly recognizable as the same species. This is especially the case in such birds as the brilliant South American chatterers and the Eastern pheasants and paradise birds. This law is of especial value, as showing the exceptional need of protection of female birds as well as butterflies, and the remarkable way in which the colours of both classes of animals have become modified in accordance with this necessity. This paper forms chapter vi. of my " Natural Selection and Tropical Nature."

5. In the great subject of the origin, use, and purport of the colours of animals, there are several branches which, I believe, I was the first to call special attention to. The most important of these was the establishment of the class of what I termed " Recognition colours," which are of importance in affording means for the young to find their parents, the sexes each other, and strayed individuals of returning to the group or flock to which they belong. But perhaps even more important is the use of these special markings or colours during the process of the development of new species adapted to slightly different conditions, by checking intercrossing between them while in process of development. It thus affords an explanation of the almost universal rule, that closely allied species differ in colour or marking even when the external structural differences are exceedingly slight or quite undiscoverable. The same principle also explains the general symmetry in the markings of animals in a state of nature, while under domestication it often disappears : difference of colour or marking on the two sides would render recognition difficult. This principle was first stated in my article on

" The Colours of Animals and Sexual Selection " (in " Natural Selection and Tropical Nature," 1878) and more fully developed in "Darwinism." I am now inclined to think that it accounts for more of the variety and beauty in the animal world than any other purpose yet discovered.[1]

I may here add that I believe I was first to give adequate reasons for the rejection of Darwin's theory of brilliant male coloration or marking being due to female choice.

6. The general permanence of oceanic and continental areas was first taught by Professor J. D. Dana, the eminent American geologist, and again by Darwin in his " Origin of Species;" but I am, I believe, the only writer who has brought forward a number of other considerations, geographical and physical, which, with those of previous writers, establish the proposition on almost incontrovertible grounds. My exposition of the subject is given in " Island Life " (chap. vi.), while some additional arguments are given in my " Studies " (vol. i. chap. ii.). The doctrine may be considered as the only solid basis for any general study of the geographical distribution of animals, and it is for this reason that I have made it the subject of my careful consideration.

7. In discussing the causes of glacial epochs I have adopted the general views of Mr. James Croll as to the astronomical causes, but have combined them with geographical changes, and have shown how the latter, even though small in amount, might produce very important results. In particular I have laid stress on the properties of air and water in *equalizing* temperature over the earth, while snow

---

[1] A correspondent, Mr. G. Norman Douglass, writing from the British Embassy, St. Petersburg, in 1894, sent me the following translation of a passage in Schopenhauer's " Die Welt als Wille und Vorstellung (Zur Teleologie) " which curiously anticipates my views :—

" One accounts for the wonderfully varied and vividly glowing coloration of the plumage of tropical birds, although only in a very general way, by the stronger influence of light between the tropics—as its *causa efficiens*. As its *causa finalis*, I should say that these brilliant plumages are the full-dress uniforms by means of which the individuals of the numberless species, often belonging to one and the same genus, recognize each other, so that every male finds its female."

and ice, by their immobility, produce *cumulative* effects ; and thus a lowering of temperature of a few degrees may lead to a country being ice-clad which before was ice-free. This is a vital point which is of the very essence of the problem of glaciation ; yet it has been altogether neglected in the various mathematical or physical theories which have recently been put forward. My own discussion of the problem in chapter viii. of "Island Life" has never, so far as I know, been controverted, and I still think it constitutes the most complete explanation of the phenomenon yet given.

During a discussion in *Nature*, so late as 1896, Professor G. H. Darwin and Mr. E. P. Culverwell adduced some new calculations as to the amount of diminished sun-heat due to eccentricity, as invalidating Croll's arguments ; whereupon I pointed out that their facts had not the importance they supposed, because they took no account of the cumulative effects of snow and ice above referred to (*Nature*, vol. liii. p. 220). Sir Robert Ball also, quite independently, made the same objection as myself.

8. In 1880 I published my "Island Life," and the last chapter but one is "On the Arctic Element in South Temperate Floras," in which I gave a solution of the very remarkable phenomena stated by Sir Joseph Hooker in his "Introductory Essay on the Flora of Australia." My explanation is founded on known facts as to the dispersal and distribution of plants, and does not require those enormous changes in the climate of tropical lowlands during the glacial period on which Darwin founded his explanation, and which, I believe, no biologist well acquainted either with the fauna or the flora of the equatorial zone has found it possible to accept. I am informed by my friend Mr. Francis Darwin that this chapter was especially noticed in Germany at the time of its first appearance, but he can hear of no detailed criticism of it, except one by H. von Jhering in *Engler's Botan-Jahrbücher* (vol. xvii., 1893), of which he has kindly sent me a translation of the more important passages. This is not the place to reply to the criticism, which would require

a chapter. I can only say here that the writer has not a sufficient grasp of the elementary laws of distribution to enable him to grapple with the subject. One example of this will suffice. He says, " Plants are not, as a fact, carried far by wind, Corsican, Sardinian, and Sicilian plants not occurring in Italy." No one who understands the first principles of evolution by natural selection could have made such a statement. And as to his alleged " fact," I have given overwhelming evidence against it in my book.

Mr. Darwin informs me, however, that he thinks the great German botanist, Engler, is favourable to my views ; but what is very much more important is that Sir Joseph Hooker himself accepts them, and I have his permission (February, 1905) to quote the following passages referring to the whole book, from a letter written in 1880, and to say that he has not changed his opinion :—

" I think you have made an immense advance to our knowledge of the ways and means of distribution, and bridged many great gaps. Your reasoning seems to me to be sound throughout, though I am not prepared to receive it in all its details."

And again : " I very much like your whole working of the problem of the isolation and connection of New Zealand and Australia *inter se*, and with the countries north of them ; and the whole treatment of that respecting north and south migration over the Globe is admirable."

For those who have not my "Island Life," there is a compact statement of the whole argument in my " Darwinism," pp. 361–373.

9. In 1881 I put forth the first idea of mouth-gesture as a factor in the origin of language, in a review of E. P. Tylor's " Anthropology," and in 1895 I extended it into an article in the *Fortnightly Review*, and reprinted it with a few further corrections in my " Studies," under the title, " The Expressiveness of Speech or Mouth-Gesture as a Factor in the Origin of Language." In it I have developed a completely new principle in the theory of the origin of language by showing

that every motion of the jaws, lips, and tongue, together with inward or outward breathing, and especially the mute or liquid consonants ending words which serve to indicate abrupt or continuous motion, have corresponding meanings in so many cases as to show a fundamental connection. I thus enormously extend the principle of onomatopœiæ in the origin of vocal language. As I have been unable to find any reference to this important factor in the origin of language, and as no competent writer has pointed out any fallacy in it, I think I am justified in supposing it to be new and important. Mr. Gladstone informed me that there were many thousands of illustrations of my ideas in Homer.

10. In 1890 I published in the *Fortnightly Review* an article on "Human Selection," and in 1892 (in the Boston *Arena*) one on "Human Progress, Past and Future." These deal with different aspects of the same great problem—the gradual improvement of the race by natural process ; and they were also written partly for the purpose of opposing the various artificial processes of selection advocated by several English and American writers. I showed that the only method of advance for us, as for the lower animals, is in some form of natural selection, and that the only mode of natural selection that can act alike on physical, mental, and moral qualities will come into play under a social system which gives equal opportunities of culture, training, leisure, and happiness to every individual. This extension of the principle of natural selection as it acts in the animal world generally is, I believe, quite new, and is by far the most important of the new ideas I have given to the world.

A short summary of these papers appears in my thirty-third chapter ; but every one interested in the deepest social problems should read the articles themselves (in my "Studies "), which give a very condensed statement of the whole argument.

11. In an article on "The Glacial Erosion of Lake Basins" (in the *Fortnightly Review*, December, 1893), I brought

together the whole of the evidence bearing upon the question, and adduced a completely new argument for this mode of origin of the valley lakes of glaciated countries. This is founded on their surface and bottom contours, both of which are shown to be such as would necessarily arise from ice-action, while they would not arise from the other alleged mode of origin—unequal elevation or subsidence.

12. In a new edition of "Stanford's Compendium, Australasia," vol. i., when describing the physical and mental characteristics of the Australian aborigines, I stated my belief that they were really a low and perhaps primitive type of the Caucasian race. I further developed the subject in my "Studies," and illustrated it by photographs of Australians and Ainos, of the Veddahs of Ceylon, and of the Khmers of Cambodia—all outlying members of the same great human race. This, I think, is an important simplification in the classification of the races of man.

Bees' cells.—But besides these more important scientific principles or ideas, there are a few minor ones which are of sufficient interest to be briefly mentioned. In the article on the "Bees' Cell" (referred to in Chapter XXVIII.), I called attention to a circumstance that had been, I think, unnoticed by all previous writers. An immense deal of ingenuity and of mathematical skill had been expended in showing that the two layers of hexagonal cells, with basal dividing-plates inclined at a particular angle, gave the greatest economy of space and of material possible ; and the instinct of the bees in building such a comb to contain their store of honey was held to show that it was a divinely bestowed special faculty. But all these writers omitted to take into account one fact, which shows their whole argument to be fallacious. This is, that the combs are suspended *vertically*, and that when full of honey the upper rows of cells have to support at least ten times as much weight as the lowest rows. But there is no corresponding difference in the thickness of the walls of the cells ; so that, as the upper rows are strong enough, the lower must be quite unnecessarily strong, and there is thus a great waste of wax. The whole conception of a supernatural faculty for the purpose of economizing wax is thus shown to be fallacious. Darwin's explanation entirely obviates this difficulty, since it depends on the bees possessing intelligence enough to reduce all the cellwalls to a nearly uniform thickness, being that which is sufficient under all circumstances to support the weight of the whole mass of comb and honey.

The supposed " homing " instinct of dogs, etc.—In the year 1873 one of the many discussions on this subject took place in *Nature.* I had suggested the immense importance of the sense of smell in enabling dogs to find their way back along a route they had been carried in a basket or covered cart ; but, of course, there are cases which this will not explain. I gave a summing up of the whole subject, and added a new and very remarkable case which happened to my friend Dr. Purland, whose amusing letters I have given in Chapter XXVIII. This case is as follows :—

" My friend lost a favourite little dog when he was living in Long Acre. Three months afterwards he removed to a house in another street about half a mile distant—a place he had not contemplated going to, or even seen, before the loss of the dog. Two months later (five months after the loss of the dog) a scratching was heard at the front door, and on opening it the dog rushed in, having found out its master in the new house. My friend was so astonished that he went next day to Long Acre to an acquaintance who lived nearly opposite the old house (then empty), and told him his little dog had come back. ' Oh,' said this person, ' I saw the dog myself yesterday. He scratched at your door, barked a good deal, then went to the middle of the street, turned round several times, and started off towards where you now live.' My friend cannot tell how much time elapsed between the dog's leaving the old and arriving at the new house. If every movement of this dog could have been watched from one door to the other, much might have been learnt. Could it have obtained information from other dogs? Could the odour of persons and furniture linger two months in the streets?"

It is evident that at least twelve hours were occupied in finding the new place, leaving time for a good deal of trial and error. One suggestion now occurs to me. There was a rather circuitous omnibus route leading from Long Acre to very near the new house. The dog may have often seen its master travelling in a 'bus, and may even have gone with some of the family. He may, therefore, have followed the 'bus route, seeking all the way for indications, till at last he crossed the recent track of his master or of some other member of the family, and by scent followed it up to the door. The following passage concludes my letter to *Nature*:—

"I venture to hope that some persons having means and leisure will experiment on this subject in the same careful and thorough way that Mr. Spalding experimented with his fowls. The animal's previous history must be known and recorded ; a sufficient number of experiments, at various distances and under different conditions, must be made ; and a person of intelligence and activity must keep the animal in sight, and note down its every action till it arrives home. If this is done, I feel sure that a satisfactory theory will soon be arrived at, and much of, if not all, the mystery that now attaches to this class of facts be removed." This suggestion I have made several times during the last thirty years, but I cannot learn that any one has yet carried it out. It is strange that while thousands of dogs' lives are sacrificed annually to establish some minute point in physiology, no one can be found to carry out a few pleasurable and interesting experiments to ascertain in what manner and by the use of what faculties lost animals habitually find their way home.

An analogous problem to this is that of the migration of birds, which also has been almost always imputed to some special *instinct* or peculiar faculty other than that of the ordinary senses. On this question I wrote to *Nature* as follows (October 8, 1874) : "It appears to me probable that here, as in so many other cases, 'survival of the fittest' will be found to have had a powerful influence. Let us suppose that with any species of migratory bird breeding can, as a rule, be only safely accomplished in a certain area; and, further, that during a large part of the rest of the year sufficient food cannot be obtained in that area. It will follow that those birds which do not leave the breeding area at the proper season will suffer, and ultimately become extinct; which will also be the fate of those which do not leave the subsistence area at the proper time. Now, if we suppose that the two areas were (for some remote ancestor of the existing species) coincident, but through geological and climated changes gradually diverged from each other, we can easily understand how the habit of incipient and partial migration at the proper season would become hereditary (through the action of natural selection), and so fixed as to appear to be what we term an instinct. It will probably be found that every gradation still exists in various parts of the world, from a complete coincidence to a complete separation of the breeding and subsistence areas; and when the life-histories of a sufficient number of species are thoroughly worked out, we shall find every link between species which never leave a restricted area in which they breed and live the whole year round, to those other cases in which the two areas are very widely separated. The actual causes that determine the exact time, year by year, at which certain species migrate, will, of course, be difficult to ascertain. I would suggest, however, that they will be found to depend upon the climatal changes which most affect each species. The change of colour, or the fall of certain leaves; the change to the pupa state of certain larvæ; prevalent winds or rains, or even the decreased temperature of earth or water, may all have their influence. Ample materials must now exist, in the case of European birds, for an instructive work on this subject. The two areas should be carefully determined for a number of species; the times of their movements should be compared with those of the natural phenomena likely to influence them; the past changes of surface, of climate, and of vegetation should be taken account of; and there seems no reason to doubt that such a mode of research would throw much light on the problem."

In an article on "The Problem of Instinct" in my "Studies" (vol. i. chap. xxii.), I have supplemented the above theory as to *why* birds migrate, by another as to *how* they migrate, and trace it wholly to experience, the young birds following the old ones ; but an enormous proportion of the young fail to make the outward or the homeward journey safely.

I have given a summary of these three papers here, because the views I set forth explain some of the most remarkable cases of what have been termed instincts among the higher animals, as being really due to instruction and imitation, together with the exercise of specially acute faculties of smell or sight, of memory and a moderate amount of intelligence. It is because I go farther in this direction than any other writer I am acquainted with that I put this subject among my "new ideas."

In 1894 I wrote an article for the *Nineteenth Century* on the question of the proper observance of Sunday, which I have reprinted in my "Studies" under the title, "A Counsel of Perfection for Sabbatarians." In this short article I define clearly, I think for the first time, what the "work" so strictly and impressively forbidden really is, and then show how utterly inconsistent are the great majority of sabbatarians, who themselves break the commandment both in letter and spirit, while they loudly condemn others for acts which are not forbidden by it. I also show how the commandment can be and should be strictly kept by all who believe it to be a Divine command, and point out the good results which would follow such a mode of obeying it. That the idea was new and its reasoning unanswerable may be perhaps inferred from the fact that no reply, so far as I know, was made to it ; while a well-known writer was so impressed by it that he made his own bed the following Sunday in accordance with its suggestions.

One other new idea of quite a different nature I will refer to here, because I think that publicity may yet lead to its adoption and to the consequent annual saving of life and property. I was led to it by having seen the effects of the explosion of a powder barge on the Regent's Canal when I was living in the neighbourhood (some time in the sixties) ; and again while living at Grays and often passing the great magazine at Purfleet, where there had been an explosion some years before. On reading of the elaborate and costly precautions at all such magazines, and of explosions occurring somewhere almost every year notwithstanding all precautions, it occurred to me that there was a simple way of rendering such explosions impossible, and at the same time reducing largely the cost of storing all explosives.

The plan was to store all gunpowder, cartridges, and other explosives in metal drums, either hexagonal or circular in form and of uniform size and height, fitted at top with an air-tight cap of a size suited to the kind of explosive it contained. These drums would be arranged in rows in shallow, open tanks, filled with water so as to cover the lids, the water being kept at a uniform level by an inflow and overflow. Such tanks would need no protection whatever, except against thieves, and no precautions whatever would be required. For the conveyance of powder, etc., trucks and barges with water-tanks could be used, and in factories all explosive materials should be kept under water, so that if an explosion occurred during the actual processes of manufacture it would be strictly limited, and could not extend either to the stores of material or of the finished product, since if the water were all blown away by the concussion the contents would remain uninjured.

I drew up a careful statement of the advantages of this plan, with a drawing of the proposed drum, and sent it through a friend to Sir Thomas Brassey, then a Lord of the Admiralty, requesting him to lay it before the proper authorities. In reply I received a memorandum from the Director of Naval Ordnance, referring me to the " Treatise of Ammunition, 1881 " (a copy of which was sent), as to " the present service powder-cases." He added that the plan would be difficult, and perhaps impossible on board ship, on account of the extra space required. The last paragraph was—

"For permanent depôts of powder like Upnor the idea seems worthy of attention, and Mr. Wallace might address the War Office on the subject after informing himself as to the present service powder-cases.

"F. A. HERBERT.

"20. 6. 82."

As the Treatise sent merely showed that copper drums were in use something similar to those I suggested, but the interminable pages of instructions and precautions made no reference whatever to water-storage, I did not trouble myself to send my plan to the War Office. I, however, sent it to a few newspapers, where it appeared, and I received in consequence a letter from the editor of the *Ironmonger* approving of the plan for large stores of powder, but fearing it could not be applied to retail dealers, where explosions, often fatal, were continually occurring, almost always through "gross negligence."

It thus appears that good authorities could see no practical objections to the plan in most cases, neither did they deny the *absolute* security that would be obtained by it ; yet the crop of explosions, with loss of life, goes on every few years, and till some one in authority takes it up, will, I presume, continue.

PREDICTIONS FULFILLED.

Having devoted three chapters to an account of my various experiences in connection with modern spiritualism, which have, however, been far less extraordinary than those of many of my friends, I may not improperly conclude this record of my life and experience with a statement of a few of the predictions which I have received at different times, and which have been to some extent fulfilled.

In 1870 and the following years several communications in automatic writing were received through a member of my family purporting to be from my brother William, with whom I had lived so many years. In some of these he referred to

my disappointments in obtaining employment and to my
money losses, always urging me not to trouble myself about
my affairs, which would certainly improve ; but I was not to
be in a hurry.   These messages never contained any proofs
of identity, and I did not therefore feel much interest in them,
and their ultimate fulfilment, though in quite unexpected
ways cannot be considered to be of any great importance.

Some years later, when we were living at Dorking, my
little boy, then five years old, became very delicate, and
seemed pining away without any perceptible ailment.   At
that time I was being treated myself for a chronic complaint
by an American medium, in whom I had much confidence ;
and one day, when in his usual trance, he told me, without
any inquiry on my part, that the boy was in danger, and that
if we wished to save him we must leave Dorking, go to a
more bracing place, and let him be out-of-doors as much as
possible and "have the smell of the earth."   I then noticed
that we were all rather languid without knowing why, and
therefore removed in the spring to Croydon, where we all felt
stronger, and the boy at once began to get better, and has had
fair health ever since.

Some time afterwards I accompanied a lady friend of
mine to have a *séance* with the same medium, she being quite
unknown to him.   Among many other interesting things, he
told us that something would happen before very long which
would cause us to see less of each other, but would not affect
our friendship.   We neither of us could guess what that could
be, but a year or two later the lady married a very old friend,
a widower, whose wife at the time of the prediction was, I
think, alive, while he was living in a distant colony without
any expressed intention of coming home.   After the marriage
they went to live in Devonshire, and for some years we only
met at very long intervals.   These two cases seem to me to
be genuine clairvoyance or prediction.

But much more important than the preceding are certain
predictions which were made to me in April, 1896, and which
have been fulfilled during the succeeding eight years.   At

that time I was living at Parkstone in rather poor health and subject to chronic asthma, with palpitations and frequent bronchitis, from which I never expected to recover. I had given up lecturing, and had no expectation of ever writing another book, neither had I the least idea of leaving the house I was living in, which I had purchased and enlarged a few years before. It was under these circumstances that a medium I had visited once in London, Madame G——, was staying with friends at Wimborne, and came to see me, and offered to give me a *séance*. One of her controls, an old Scotch physician, advised me about my health, told me to eat fish, and assured me that I was not coming to their side for some years yet, as I had a good deal of work to do here. The other control, named "Sunshine," an Indian girl, who seemed to be able to get information from many sources, was very positive in her statements. She said, "You won't live here always. You will come out of this hole. You will come more into the world, and do something public for spiritualism." I replied, "You are quite wrong. I shall never leave this house now, and I shall not appear in public again." But she insisted that she was right, and said, "You will see ; and when it comes to pass, remember what I told you." She then said, "Fanny [my sister] sends her love. She loved you more than any one in the world." This I knew to be true, though during her life I did not so fully realize it. Then Sunshine gave me her parting words, speaking slowly and distinctly : "The third chapter of your life, and your book, is to come. It can be expressed as Satisfaction, Retrospection, and Work." These three words were spoken very impressively, and I wrote them at once in a small note-book with capital letters, though I had no notion whatever of what they could refer to, and no belief that they would be in any way fulfilled.

Yet two months later the first step in the fulfilment was taken through Dr. Lunn's invitation to give a lecture at Davos, and my acceptance of it, due mainly to the temptation of a week in Switzerland free of cost and with a pleasant party. As already described (Chapter XXXII.), this lecture was the starting-point of all my subsequent work. The very next year brought me renewed health and strength to do the work, as already described. Another year passed, and I received a pressing invitation to take the chair and give a short address to the International Congress of Spiritualists, which I felt myself unable to refuse, and thus, as I had been told I should, I "did something public for spiritualism." Yet another year, and a great desire for life more in the country than at Parkstone (where we were being surrounded by new building operations) led me to join some friends in trying to find a locality for a kind of home-colony of congenial persons ; and though the plan was never carried out, it led ultimately to my finding the site on which to build my present house, and thus "get out of that hole," as I had been told by Sunshine that I should do. And now, looking back upon the eight years of renewed health I have enjoyed, and with constant interesting work, how can this be better described than as "the third chapter of my life ; " while " Man's Place in the Universe "—a totally new subject for me—may well be termed the "third chapter of my book," that is, of my literary work. Again, this wholesome activity of body and mind, the obtaining a beautiful site where I am surrounded by grass and woodland, and have a splendid view over moor and water to distant hills and the open sea, with abundance of pure air and sunshine, the building of a comfortable house in one of the choicest spots in the whole district—surely all this was well foretold in the one word "Satisfaction." What has chiefly occupied me in this house—an Autobiography extending over three-quarters of a century—is admirably described by the word "Retrospection." And the whole of this process has involved, or been the result of, continuous and pleasureable "Work."

I will only add here that during the whole of this "third chapter of my life" I had entirely forgotten the particular words of the prediction which I had noted down at the time, and was greatly surprised, on referring to them again for the purpose of this chapter, to find how curiously they fitted the subsequent events. Of course it may be said that every one who reaches my age enjoys "retrospection," but that kind of general looking back to the past is very different from the detailed Retrospection I have had to make in searching out the many long-forgotten incidents and details of my very varied life as here recorded ; and the Work this has involved, and the Satisfaction I have had in writing, fully justify the solemn emphasis with which the prediction was made.

I now bid my readers, who have travelled with me so far a hearty Farewell.

ADDENDUM

CHAPTER XXXIIIa

EXCURSIONS AND EXAMINATIONS

WHILE endeavouring to give an account of all matters which occupied or interested me during the latter half of my life, I somewhat hastily concluded my MSS. and sent it to press without any reference to two matters which were of some importance to myself, and one of them of some general interest. These are, the various holiday excursions I took with my father-in-law, Mr. William Mitten, whose deep enthusiasm for nature and extensive knowledge of plants in general, and mosses in particular, rendered his companionship very congenial to me; and my work as an Assistant Examiner in Physical Geography and Physiography, which occupied me for three weeks or more every summer, almost continually for twenty-seven years. In order to make this record of my life more complete, I have added a supplementary chapter devoted to these two subjects.

My first excursion with Mr. Mitten was in August, 1867, to North Wales, his first visit to that beautiful district. We stayed a few days at Corwen, and our first walk on Sunday morning was along the road to the west up the valley of the Alwen. In about five miles we reached Pont-y-glyn, where a farm-road crossed a very deep ravine. This we descended and found the bottom full of curious hollows, with vertical rocks damp or dripping, overshadowed by trees and shrubs. Here the yellow Welsh poppy grew luxuriantly, as well as the globe-flowers and the subalpine *Rubus saxatilis*. But what delighted Mr. Mitten on this his first walk in Wales was the abundance of mosses and hepaticæ, and for a full

hour he explored every nook and cranny, and every few
minutes cried out to me, "I've got another species I never
gathered before," till I thought he would never tear himself
away ; and during several other visits to Wales in the
Snowdon and Cader Idris districts, and in the Vale of Neath,
I do not think he ever came upon a richer spot for his
favourite group of plants.

The next day we took a much longer ramble along
country lanes, which gradually led us on to the ridge of the
Berwyn mountains, from 2500 to 2700 feet above sea-level.
At the highest summit, where a circle of precipices descends
to a little tarn—Llyn-llyne-caws—my friend crept out along
the face of the rocks to get some rare mosses, till he made
me quite afraid for his safety. But he was very active and
sure-footed, and always managed to get what he wanted.
On the high peaty moors we found the creeping cloud-berry,
and as we had strolled slowly, searching everywhere for
plants and enjoying the scenery and the mountain air, it was
late in the afternoon before we came to a deep valley where
there were some houses, and as we had walked about eight or
nine miles over high mountains since we last saw a house, we
determined to go down and try to find a night's lodging. We
were attracted by glimpses of a waterfall up this valley, and
therefore made for the highest house we could see, a rather
neat small farmhouse. By the time we had reached it the
sun had set, and when we asked if we could have supper
and lodging there, we were told it was impossible, as some
titled person—I forget the name—was coming next day with
some friends to shoot there, and everything was got ready for
him. However, we told them we had walked over the
mountains from Corwen and were very tired, and if we went
down to the village we should have so much to walk back in
the morning, that at last they agreed. I quite forget what
kind of accommodation we had, but I rather think we slept
on the floor. We had, however, a good supper, and breakfast next morning, when, after getting a view of the waterfall, which Mr. Mitten sketched, we walked about three
miles westward over a mountain ridge to a good but very

LLANRHAIDWR WATERFALL.

[*To face p.* 403, VOL. II.

wild road, which led us back through the village of Llandrillo
to Corwen, a distance of about seventeen miles, forming
altogether one of the wildest mountain walks I have ever
taken in our own country.

The waterfall we thus accidentally came upon is called
Pistill Rhaiadwr, and is little known to tourists, as it is a long
way from any beaten track, but it is undoubtedly the finest
in Wales, and has a peculiar feature which is, I think, unique
in the British Isles. Between the upper and the lower part
of the fall the water passes under a natural arch of rock,
along which it is possible to crawl, though when there is
much water the arch is drenched with spray. The photograph
here copied shows this remarkable feature, as well as the
double fall, the upper one being about 150 feet high, the
total height being 240 feet. George Borrow, in his " Wild
Wales," considers this curious bridge to be a blemish, and
remarks, " This unsightly object has stood where it now
stands since the day of creation, and will probably remain
there to the day of judgment. It would be a desecration of
nature to remove it by art, but no one could regret if Nature
herself, in one of her floods, were to sweep it away." The
ancient geology and theology of this passage are very
characteristic.

Two years later we had another excursion together,
accompanied by my friend Geach, going first to Beddgelert,
and then on to Pen-y-gwryd, where we found the little inn
crowded, and had difficulty in finding the roughest accommodation. Next morning we started at five, and had a most
delightful walk up Snowdon by this very picturesque route.
Reaching the summit with excellent appetites, we enjoyed our
breakfast of coffee and bacon in the little hut on the top, and
then, as it was a glorious day with floating clouds whose
shadows below us were a delight, we spent an hour or more
in the enjoyment of the splendid views, with the numerous
lakes in almost all the surrounding *cwms* and valleys which
render this mountain especially interesting to the glacial
geologist. Numbers of swifts were flying about over and
around the peak, and when Mr. Mitten climbed out on some

crags in search of rare mosses, they dashed about so close to
his head as to cause him to retreat. After returning to Beddgelert we went up a small valley to find a very rare water-
moss, which Mr. William Borrer, the well-known botanist,
had told Mr. Mitten was to be found there; and after a long
search in every rock-hole that seemed a likely place, he, at
last, found the treasure, as he almost always did when he
went in search of any rarity. While stopping at a cottage
during a shower, and noticing some large birds of prey
screaming on a mountain near, he asked the woman of the
house what birds they were. To which she replied, "Harpies,"
which made us wonder what remote part of the world we had
got to. We afterwards went to Dolgelly and Cader Idris,
where, in a small lake, we found the uncommon *Lobelia
Dortmanna.*

In 1875 we went again to Snowdon, and afterwards to the
curious ravine called Twll-dû, or the "Devil's Kitchen," near
which I found an umbrella, and Mrs. Mitten, who accompanied
us, found somebody's lunch, consisting of a baked trout and
grapes; while Mr. Mitten revelled as usual in the rare mosses,
and later at the Swallow Falls, on the way to Bettws-y-Coed,
he found a moss quite new to him.

Our next excursion was to South Wales, when my wife
and Mrs. Mitten accompanied us, as I wished to show them
the beautiful scenery of my favourite Vale of Neath. We
stayed a few days at a cottage at Pont-nedd Fychan and
visited the beautiful waterfalls, the rocking-stone, the subterranean river, and the fine Dinas rock. While here one day
we passed a labourer at work on the roadside, and Mrs. Mitten,
thinking to gratify the patriotism of a Welshman, remarked
on the beauty of the scenery, and asked him if he did not
think it a privilege to live in such a fine country? Rather to
our amusement, he told us that he did not think much of the
country, it was all hills and stones, and there was no good
land, and he much preferred his own country, which was
Lincolnshire!

Another year I and Mr. Mitten went to Glen Clova in the
Highlands in search of the many rare plants for which it is

celebrated. But we had little success because we had no guide to the exact localities of the rarities. But we much enjoyed the excursion and the wild scenery, though we had some difficulty in getting the keepers to allow us to enter the glen. Being at the inn on Sunday a number of farmers and their wives came in after church to meet their friends and drink whisky, and on listening to their very voluble talk I could not understand a word that they were saying. I concluded, therefore, that they were speaking Gaelic, and was much pleased to have heard it. But the landlord's daughter told me afterwards that no one spoke Gaelic there, and that all the people I had heard were speaking English! I could not have believed that pronunciation and accent could have produced such complete unintelligibility. On passing through Edinburgh we called on the late Professor Balfour at the Botanical Gardens, and he much regretted that he had not accompanied us, as he could have shown us all the rarities of that botanical treasure-house.

In the spring of 1877 I accompanied Mr. Mitten to Spa in Belgium, where he was taking his youngest daughter to a school to acquire French conversation. We stayed a few days there, botanizing on the moors and hills around, and were interested in noticing some peculiarities of the vegetation as compared with our own. Nowhere did we see a single primrose, but its place was taken by the true oxlip (*Primula elatior*), so local with us. Our rare little fern, *Asplenium septentrionale*, was common by the roadsides. Our Swiss tour has been noticed in Chapter XXXIII. Even during Mr. Mitten's occasional visits to us in Dorsetshire, he had found several plants new to the district or to the county. The most notable of these were the crowberry (*Empetrum nigrum*), never before noticed in Dorsetshire, a quite large bush of which was found on Studland Heath, a well-searched botanical locality. Even more interesting was his discovery of the rare aquatic grass, *Leersia oryzoides*, which he thought should grow in the ditches near Wareham, and knowing its flowering season, he went there and found it, though the very ditch had often been searched by other botanists!

MY EXPERIENCES AS AN EXAMINER.

It was, I think, in 1870, that I heard from Bates of the examinations in Physical Geography under the Science and Art Department, for which he was one of the Assistant Examiners, and he advised me to apply to Professor Ansted, the examiner-in-chief, if I wished to obtain the post of an assistant. I did so; and began the work in 1871, and continued yearly till 1877. In 1871 I also had the examinership in Physical Geography and Geology for the Indian Civil Engineering College, and in 1870 and 1871 for the Royal Geographical Society.

The work under Professor Ansted was hard while it lasted, but was interesting, and often quite amusing, and it was very well paid. The assistant examiners had each over a thousand papers to examine. The work occupied about three weeks more or less, and the remuneration amounted to from £50 to £60, or occasionally even more. In 1878 Professor Judd and Sir Norman Lockyer were appointed joint examiners, the syllabus being altered to include geology and physical astronomy, while the subject of examination was now changed from Physical Geography to Physiography, and I continued to be an assistant examiner till 1897, with the exception of one year during my American tour.

During the earlier period a considerable number of well-known scientific men, mostly geologists or biologists, were among the assistant examiners, such as H. W. Bates, William Carruthers, the botanist, J. F. Collingwood, Major Cooper-King, Professor J. Morris, Professor T. Rupert Jones, Dr. Henry Woodward of the Natural History Museum, Professor H. G. Seeley, and a few others less well known to me. There were three meetings in London to compare results and secure an equal rate of marking, and these afforded an opportunity for a little conversation between persons who rarely met elsewhere, and we also for some years had an annual dinner, which was latterly discontinued when a considerable proportion of the examiners lived in the country.

Although the drudgery and strain of reading through a thousand papers, with replies to the same set of questions, exhibiting every possible degree of ignorance of the subject and often extremely diffuse, was very great, yet a little relief was given by the highly amusing character of some of the answers, of the more curious of which I, as well as several others of the examiners, made notes. During intervals of our more serious work, we often communicated some of these to our fellow-sufferers, and thus contributed a little hilarity to our otherwise strictly business meetings.

On looking over my notes of these examinations extending over more than a quarter of a century, I think it will be both amusing and instructive to give a few examples of these replies, of which I have a rather large collection, as they have an important bearing on the whole question of the utility of such examinations, on which I may, perhaps, afterwards say a few words. The first I will quote are from a rather long series that occurred in 1873. It must be remembered that in Professor Ansted's time sixteen questions were asked, ranging over most of the subjects included in Physical Geography, but only eight were to be answered, so that the candidates need only attempt to answer those about which they knew something. Further, they were all supposed to have had some special teaching in the subject, and were sent up by their masters in the hope of getting the allowance granted by the Government for each one who passed.

The first question was, "Show why the longest day in Edinburgh is longer than the longest day in London." Out of a large number of answers, showing more or less complete ignorance of the cause of this interesting phenomenon which must be known to every one who has spent a winter at any two places in the north and the south of our islands, I have preserved five.

(1) Because it possesses a maritime climate.
(2) Because the manufactures in London produce a smoky atmosphere.
(3) Because it is not in such a warm place as London.
(4) Because London is on a meridian and Edinburgh is not.
(5) Because the first meridian shades the sun from London, while it is shining in Edinburgh.

Now, these answers, and scores of others equally wide of the mark but not so short or so amusing, show that no attempt had ever been made to teach these boys to understand the commonest facts connected with the motions of the earth—such as the seasons, varying lengths of day and night, change of position of the sun at rising and setting, and its altitude at noon, etc.—in the only way in which they can be taught to the majority of people, that is, by simple experiments with a globe and a lamp in a darkened room. In this way the reason of all the changes is *seen* to follow inevitably from the form, position, and motions of the earth, while no amount of *verbal* explanation, even with the help of diagrams, can make it intelligible to any but those who have the special geometrical faculty. By such experiments any intelligent children from eight or ten upwards may be easily made to understand these facts, as well as the apparent motions of all the heavenly bodies. Yet probably to this day not one school in a hundred teaches such things, and not one teacher in a hundred knows how to teach them.

Another question was, " Mention the natural habitat of the horse, the elephant, the hippopotamus, and the rhinoceros," and the following answers were given :—

(1) The horse is used for drawing anything, such as carts, plows, or anything he is taken to do ; the hippopotamus is a. very disagreeable beast and runs about very wild.

(2) The habit of the horse is plowing, the elephant goes to shows.

(3) The principle habitat of the elephant is the fauna, the rhinoceros, the buffalo, and the hippopotamus is the white bear.

The above replies show gross ignorance of the facts of animal distribution or of the terms used in regard to it ; and the following show equal ignorance of common geographical or meteorological phenomena. The answers show sufficiently what were the questions :—

Q. 11. The principal Atlantic icebergs come from the Alleghanies on the east of America ; when they reach the valley below they melt and form small straits, which in time spread out into rivers. They enrich the climate through which they pass.

Q. 11. Iceberg is a mass of ice formed in the polar regions and generally connected with volcanoes.

Q. 11. Icebergs are formed by geysers shooting up in the air out of the sea and frozen there.

In reply to a question as to deep dredging in the Atlantic the following answers were given :—

Q. 15. The depth of the water of the Atlantic is measured by large things called ravines. The depth is 90,000,000 miles. Gold is found at the bottom.

Q. 15. The matter found at the bottom of the Atlantic is copper, pearls, and diamonds.

Q. 15. The material found by deep dredging in the Atlantic is—the Atlantic canal or cable.

The question being, " What is meant by the distribution of plants and animals in vertical and horizontal space, and what do you understand by representative forms? "—I have notes of the four following answers :—

(1) Horizontal distribution is when they grow near the horizon; vertical distribution is when they grow in vertical space, as wheat, or anything on the same level.

(2) Plants grow in gardens, animals live on the earth.

(3) By distribution of plants and animals in vertical and horizontal space, we mean, the plants and animals in the distance between pointed and curved lines.

(4) Representative forms of animals and plants is, how they are represented in books.

In 1878 I had some good examples of the kind of answer in which the candidate evidently has a very high opinion of his own attainments and his mode of explaining the whole matter. The question was, " In what respects do a volcano and a geyser resemble each other, and in what respects do they differ ? " The answer is rather a long one :—

A volcano is a raised piece of land in about a thousand years, then in another thousand years it has become larger and larger till it becomes as high as would be called a volcano. But a geyser is a raised piece of land done all in a night.

Difference. The volcano takes a long long time to be at the point of saturation, but the geyser is done all in one night.

Agreement. They are both raised-up pieces of land. Sometimes a volcano goes on fire and makes a creator, and then it bursts. When it bursts you will always observe that down at the bottom of the volcano and about ten miles round and round about it there lies cinders as large as bricks, and as you proceed to the top of the volcano it always becomes smaller, till at the mouth of it it is all dross, like very small coal.

This last sentence is so precise and clear in its statements that one might suppose it to be the result of personal observation !

Another of the same class occurred in 1879, when in answer to the question, " What evidence have we that lions and tigers once lived in this country ? " the reply was—

We have only this evidence that lions and tigers once lived in this country, that when a man, or even any man or men, have been digging for minerals, wells, or anything else, they have found the fossils, and it has at last after a good long consideration and perseverance it has turned out to be the skeleton of a lion or tiger.

The same paper explains thunder as follows :—

The cause of the noise made during thunderstorms is the meeting of the electric and other gases. It is said that a gentleman caught a glimpse of one of these collisions by means of a kite. It was thus found out what was the cause of thunderstorms, and also what made the flash of lightning.

In 1880 we had the following answers to a question about the causes of the extinction of animals, and as to any which have become extinct since the appearance of man on the earth:—

(1) Giants, and the great fish which swallowed Jonah.

(2) Extinct volcanoes not having erupted for a length of time is one cause which has brought about the extinction of animals.

(3) Animals which lived before the flood no longer exist except their fossilized remains. Iothoraics, Pleathorus, Mammoth, Dothorium, Adam and Eve never saw, having become extinct.

(4) Animals which have become extinct since man has been on the earth are Ammonites, Belemnites, Mammals and Productus horridus.

(5) The unicorn is extinct.

(6) Extinct means that they have gone away, but may become active again. Some of the causes that they have become extinct are that they have been caged up, etc. The animals that have become extinct since the appearance of man are the jaguars.

Many other answers showed a similar absence of knowledge upon this most interesting branch of natural history, and one which may be made easily intelligible even to children.

The equally simple and interesting question as to what geographical range of animals or plants means, is thus answered :—

(1) What is meant by geographical range is, that they are arranged according to their shape and size.

(2) The geographical range of a species of animal or plant is that part of a country in which no species of animal or plant will live, only the species which first originated there.

In 1882 we had a question analogous to that so badly answered in 1873 : "What is the cause of the long days and nights of the Polar regions?" and the answers showed little improvement in the teaching. Here are a few of them :—

(1) The reason why they have long days and nights is because at the poles they have only six hours sun, and the sun does not rise at 6 o'clock a.m. at the poles as it does here, but does not rise till nine and ten o'clock a.m.

(2) Because the sun only visits the polar regions a particular part of the year. When the sun is gone the day only lasts a few hours.

(3) The poles being so far from the equator. That is, it takes the light a certain time to travel that distance.

(4) At the N. Pole the Aurora Borealis ; at the South Pole the South Australis sheds its light upon the Polar regions, the long nights are owing to the Aurora disappearing. Long days may be also owing to the Colures ; long nights to the moon not affecting the Polar regions.

(5) In summer Europe, Asia, Africa, and America, being the bulk of the land of the world, require a great deal of heat from the sun. Again, when it is winter in Europe, etc., it is summer in Australia. Now Australia being a very small part of the earth it will not require as much heat as the other continents did. Consequently more heat can be given by the sun to the Polar regions than in our summer.

(6) The cause of the long days is due to the slowness with which the moon sets, or, more correctly, the long nights, and when the moon does set it remains a long time forming the long days.

(7) The reason they have long days and nights is that the people always catch the sun or the moon ; another reason is that they are nearest the sun.

(8) The cause of the long days and nights of the Polar regions is that the days and nights are just the *opposite* to what it is stated in the

question, namely short days and long nights, it being one continuous winter from one years end to the other, summer being only for a few weeks at a time, and then the days are comparatively short compared with ours.

(9) The long days and nights are caused by the quantity of snow that falls at the poles.

(10) The cause of the long days in the Polar regions is this: when the sun is observed there (which it seldom is) the rays are reflected as it were, and it forms day. The cause of the long nights in the Polar regions is this : the sun only makes his appearance for a very short time, during this time it is day, but after the sun disappears it is night, which by that means is very long. It is to be understood that it is a certain part of the year during which the days are long, and the other part during which the nights are long.

It seems to me a very sad thing that under a vast Government organization at a very great cost, it should be possible for such results as these to be produced. Many of these candidates have evidently good capacity, but are sent up to be examined on subjects of which they are disgracefully ignorant, either from want of any teaching whatever, or through their teachers being themselves disgracefully ignorant —and there are clear indications that the latter is very often the case.

Six years later (1888), we find equal ignorance on another subject of great interest, and as to which knowledge was easy to obtain even without special training. The question was, " How is the depth of the ocean determined ? "

(1) The depth of the ocean is determined by the water carrying the sediment to the mouth of the ocean and depositing it again.

(2) The depth of the ocean is determined by discharging a wire or rope from a cannon, the wire being long with a point fixed, which when it touches something hard an electric current passes immediately to the ship ; they thus go on till they find the lowest sounding.

(3) The depth of the ocean is determined by means of the barometer, an instrument invented for measuring the heights of sea-levels, etc. The barometer is placed by the side of some mountain, and in this manner they calculate taking the readings from the barometer.

(4) The ocean contains poles, insects live at the bottom of the ocean and bore holes in the poles, when the poles are reached they reach the bottom of the ocean.

(5) The depth of the ocean is determined because it i always movin and wearing away the bottom.

(6) The depth of the ocean is determined by fixing a piece of rope to a heavy piece of metal which is lowered into the water, and as soon as it touches the bottom the weight is no longer felt and the rope is cut off at the surface of the water ; the rope is then measured. It is brought up by a diver.

(7) The depth of the ocean is determined by sounding or pianoforte wire which is let down until it reaches the bottom of the ocean ; great care must be taken to catch the sound.

Equally gross ignorance is shown as to the mariner's compass, the question being whether it always points due north ; if not, why not ?

(1) The mariner's compass do not always point due north because if it did on board a ship, the captain of the ship would want to go south and it would guide him the wrong way, instead of south it would guide him north, so it is made to turn N.S.E.W. The mariner's compass is made to turn round in any way in which the captain wishes to turn it, so as to guide him which way he wants to go. If he wants to go to the south he puts the point to the south, etc. They are used by men who want to go to different parts of the world. Say if a man is lost in travelling to Germany he looks at his compass, and if it is north he puts the point north, or if it is south, etc.

(2) The mariner's compass does not point due north because the wind affects it. If the wind is blowing hard the dial points slightly to the north, and when it is a heavy storm the dial points nowhere, but just swings backwards and forwards.

Another subject of the greatest interest and one that can be very easily taught to even young children by a number of simple and easy experiments, is that of the weight and density of the atmosphere, and the construction of the barometer. Some knowledge of these subjects is essential to a clear understanding of a great number of natural phenomena. Yet this is how, so late as 1889, some of these students replied to easy questions about it :—

(1) The weight of the air can be determined by the law of gravitation. For example, take an apple from a tree and let it go. What happens? It falls to the ground. This shows that the air is heavier and attracts the apple at the ground. Therefore we can say the apple does not fall, but it is the ground that attracts it. By that process we could discover or determine the weight of the air. We are able to move about because the earth attracts us, and so we are able to move about in this dense mass of air under us.

(2) To a person who has not studied the question air has no weight. If air has weight, why do we not get tired of bearing that weight? To prove to that person that air has weight, ask—How do you take headaches? We take headaches because the air gets light and some of the usual weight is taken off the head, and we get giddy.

These two young men write with an air of authority, as if they were teachers rather than learners, yet it is hard to say which of the two is the more profoundly ignorant. The other four, while equally ignorant, are more modest in their style.

(3) We are able to move under the pressure of the atmosphere by impurities and other bodies displacing the air. If there were no impurities in the air we could not move about. For example, water-vapour gets into the air, and displaces it making the air lighter.

(4) We are able to bear a certain amount of the weight of the atmosphere and a very little more would kill us.

(5) We are able to move about on the earth's surface because, although the atmosphere is pressing us down we have the sun attracting us.

(6) The reason that we are able to move about under the weight of the atmosphere is that the atmosphere is two hundred miles away from the surface of the earth.

Passing on to 1891, such a common instrument as the barometer, which can be so easily explained by simple experiments, is thus hopelessly blundered :—

(1) Air occupies the space above the mercury. If a hole were bored through the glass above the mercury the air would escape and probably the tube would burst.

(2) The air would escape and the mercury would remain dormant.

(3) The principle on which the action of the mercurial barometer depends is, that it must be enclosed in a strong case and must not be touched in any way.

(4) A water barometer is longer than a mercurial barometer because it has to go down to the bottom of the sea to see how deep it is. A mercurial barometer has to see how high a thing is, and no hill is higher than the depth of the ocean except a few high mountains which nobody can get to the top of. Oxygen occupies the space above the mercury, and if a hole were bored the oxygen would flow out and the mercury rise to the top and flow out also.

In 1893, in order to correct some popular errors, the

following questions were asked : "Point out the errors in the following statements :—

"(*a*) Earthquakes have raised to heaven the ocean bed."

"(*b*) Volcanoes are burning mountains that vomit fire and smoke." To which the following replies were given :—

(1) Earthquakes swallow the ocean bed.

(2) In ancient times volcanoes were called burning mountains, but we do not call them by that name now, because we have a new name for them derived from the Latin words *volca* to burn and *noe* mountain, and the two put together "volcanoe."

In the same year, in reply to the very elementary question, "How is angular space measured?"—without a clear conception of which no knowledge of mechanics or any comprehension of many of the simplest facts of nature is possible—such replies as the following were given :—

(1) By multiplying the number of seconds a body is falling by 32.

(2) Angular space is measured by a delicate instrument which brings the rays to one position on a stand or anything you like to put in the way, and they take the angle and measure it and keep on like this at all times of the year and then find the average.

(3) You take a pair of compasses and put a point on one star and a point on the other, and then you look between your legs where they join and judge the distance between them thus.

In 1895 we again had a simple question as to a very common instrument, the construction and use of which can be taught to any child—"Describe the mariner's compass and its chief uses;"—and we had a set of answers as bad as those seven years earlier :—

(1) The Mariner's Compass is a thin bit of steel cut into 32 points.

(2) The Mariner's Compass is a box with a card and a lot of needles.

(3) The Mariner's Compass is a brass box with 24 circular cards hinged on, no matter which way it rolls it carries these around with it.

(4) The Mariner's Compass is a box and a card with 32 points.

(5) If a sailor was shipwrecked on a desert island he could find a north and south line if he had a Nautical Almanack.

(6) The Mariner's Compass is a circular bit of wood with a nail put through it, and into this is a pivot which is very easily shook about, and the Captain brings this to sea with him. Of course it has the Cardinal points on it, N.E., S.W. etc., and he knows where he is.

(7) To repel the other great magnet, the earth, and to prevent the ship (because of the iron) being attracted to the earth.

Of course it will be said that the examples here given are all extreme cases, and that a majority of the papers show a considerable amount of knowledge. But this is altogether beside the question. I never had time or inclination to interrupt my work in order to copy all the very ignorant answers, but only a few here and there which specially struck me. For each one thus copied there were at least a dozen equally bad, but often so wordy and involved as to take too much time to preserve, while a far greater number exhibited a little knowledge so intermingled with gross ignorance, as for any useful purpose would be equally bad.

But the point I wish to insist upon is, the utter failure of a system which, at the end of twenty years, allows of *any* such candidates as these taking part in an examination. The failure is twofold. First, in the notion that any good can result from the teaching of such a large and complex subject to youths who come to it without any preliminary training whatever, and who are crammed with it by means of a lesson a week for perhaps one year ; and, in the second place, the attempting to teach such a subject at all before a sufficiently capable body of teachers have been found who know the whole range of subjects included in it, both theoretically and practically, and who also know how to communicate to others the knowledge they themselves possess.

In these examinations scores and sometimes hundreds of papers come from single large schools, and it is a familiar thing to examiners to find the same absurd error, often stated in the very same words, running through a whole school, except, perhaps, in the case of one or two exceptionally clever lads who have, by reading or experiment, educated themselves upon the point in question. Now, the absurdity of the system is, that the ignorant teacher never has his ignorance pointed out to him, and imputes the failure of a number of his pupils to *their* stupidity or carelessness, whereas it is really all due to *his own* ignorance.

Another evil result of these examinations under a Government department is, that in order to justify their existence, it is necessary to show a certain considerable amount of success. Hence the "passes" are brought up to good general average, however bad the bulk of the papers may be ; and people are deluded by the idea that because a person has passed in Physiography he has a good general knowledge of the whole subject, whereas many pass who are quite unfit to teach any portion of it to the smallest child. My own conclusion is that all these examinations are an enormous waste of public money, with no useful result whatever. Nature-knowledge of the kind referred to is the most important, the most interesting, and therefore the most useful of all knowledge. But to be thus useful it must be taught properly throughout the whole period of instruction from the kinder-garten onwards, always by means of facts, experiments, and outdoor observation, supplemented, where necessary, by fuller exposition of difficult points in the classroom.

The whole status of the teacher is degraded by the present system, which assumes that any fairly educated person can, by means of a few courses of lectures and a short period of cramming, be qualified to teach these subjects to the young. The real fact is that none can teach them properly who have not a natural taste for them, and have largely taught themselves by personal observation and study. They alone know the difficulties felt by beginners ; they alone are able to go to the fundamental principles that underlie the most familiar phenomena, and are thus able to make everything clear to their pupils. Such men are comparatively rare, but they should be carefully sought for and given the highest rank in the teacher's profession. When that is done, no examinations will be advisable or necessary.

Before quitting the examination question, I wish to say a word in favour of the late Professor Ansted as an Examiner in Physical Geography. On looking over many of the papers set by him from 1871 to 1877, I am greatly impressed by his broad grasp of the whole subject, and the admirable manner in which he dealt in turn with all the natural phenomena

embraced in it, from the simplest to the most complex. He
usually set fifteen to sixteen questions, in both the Elementary
and Advanced stages, only eight of which were to be answered;
and they always comprised a considerable portion of the
whole field embraced in the study. I feel sure that the
questions set by him during any four or five years of the period
named, would serve as an admirable guide to a student who
wished to make himself master of the fascinating study of
earth-knowledge or " physiography."

INDEX

A

A., Mr., anecdotes of, i. 108, 129

Aar, exploring the gorge of the, ii. 214

Abbé Paris, miracles at the tomb of, ii. 309

Aberhonddu, i. 161

Abbey-Cwm-Hir, i. 150, 161

Aberystwith, i. 161

Abyssinia, plants of, ii. 13, 21 ; effects of Christianity in, ii. 53

" Acclimatization," article on, in the " Encyclopædia Britannica," by A. R. Wallace, ii. 98

Academy, The, review by A. R. Wallace, of " The Descent of Man " in, ii. 10

Aden, i. 335

Adelboden, Switzerland, ii. 220

Adirondacks, ii. 188

Adshead, Mr., his interest in spiritualism, ii. 322

"Adventures of Mrs. Leck and Mrs. Aleshine, The," read by A. R. Wallace, ii. 135

Africa, plants of, ii. 13, 21

Agassiz, Louis, on the glacial epoch, i. 134

Agassiz, Mr. Alexander, A. R. Wallace meets, ii. 110

Agassiz Museum of Zoology, ii. 110

" Age of Bronze, The," i. 113

" Age of Reason," Thomas Paine's, i. 87

Aiguilles, view of, i. 325

Ainsworth, W. Harrison, " Rookwood " by, i. 75

Airy, Sir G. B., lecture on Halley's Comet, i. 247

Aksakoff, Hon. Alexander, visits at Grays, ii. 93

Alabama, Fanny Wallace goes to, i. 223 ; returns from, i. 256

Albany Street, London, residence of Mr. and Mrs. Sims at, i. 263

Albury, St. George Mivart builds a house near, ii. 44

d'Alembert, quoted, ii. 284

Alexandria, described in letter to George Silk, i. 332-335

Alexandria Bay, St. Lawrence river, ii. 188

Ali, Malay servant, described, i. 382

Aliven, North Wales, ii. 401

Alleghanies, crossing the, ii. 138

Allen, Mr. Charles, his search for birds of paradise, i. 387-394

Allen, Grant, on " Colour Sense," ii. 71 ; " In Magdalen Tower," by, ii. 121 ; A. R. Wallace's admiration for, ii. 187 ; A. R. Wallace on, ii. 209 ; R. Le Gallienne on, ii. 218 ; on English rule in India, ii. 262, 263 ; A. R. Wallace urges him to write socialistic novel, i. 272, 273

Allen, Rev. J. A., A. R. Wallace's friendship with, ii. 121 ; visit to the House of Representatives, ii. 124, 125 ; A. R. Wallace stays with, ii. 187, 188

Allen, William, shareholder in the New Lanark Mills, i. 98

Allingham, William, introduces A. R. Wallace to Tennyson, ii. 298

Allman, Professor, his sufferings from asthma, ii. 229

All Saints' churchyard at Hertford, i. 49

N

"Origin of Species," Darwin's, i. 255, 358, 372 ; reviews of, ii. 2 ; reference to, ii. 84

Orinoko, A. R. Wallace's expedition to, i. 283, 284 ; Count Stradelli's expedition up, i. 318

Osgood, Mr. Samuel, described, i. 186–189

"Our Destiny," by Gronlund, ii. 267

Ouse, i. 121, 132

Ouzel, i. 132

Owen, Miss, gives plants to A. R. Wallace for his Parkstone garden, ii. 206

Owen, Mr., of San Francisco, ii. 346, 348

Owen, Professor Richard, his review of "Origin of Species" in The Edinburgh Review, ii. 2 ; controversy with Huxley, ii. 46 ; his visits to Augustus Mongredien, ii. 61

Owen, Robert, his influence on A. R. Wallace, i. 87 ; his principles, i. 83 ; sketch of his life and work, i. 91–105, 128 ; reference to, ii. 237, 270

Owen, Robert Dale, his tract on "Consistency," i. 88 ; "The Debatable Land between this World and the Next," by, ii. 294

"Owen, Robert, and his Social Philosophy," by W. L. Sargant, i. 95

Oxford University confers degree of D.C.L. on A. R. Wallace, ii. 201

P

Paine's, Thomas, "Age of Reason,' i. 87

Palembang, i. 376, 379

Pall Mall Gazette, The, accusations against mediums in, ii. 282, 291

"Palms of the Amazon and Rio Negro," by A. R. Wallace, i. 321

Pangenesis, differences of opinion between Darwin and A. R. Wallace on, ii. 21

Pangerango mountain, i. 376

Panshanger, fine oak in the park of, i. 35

Para, Herbert Wallace died at, i. 15 ; A. R. Wallace and H. W. Bates make voyages to, i. 264 ; the city and environments described, i. 268–275

"Paradise Lost," i. 75 ; Sunday readings of, i. 227

Paris, Fanny, John, and Alfred Wallace visit, i. 256 ; A. R. Wallace and George Silk stay at, i. 325

Parkstone, A. R. Wallace goes to live at, ii. 203 ; the garden at, ii. 204 ; orchid growing at, ii. 206, 207 ; A. R. Wallace leaves, ii. 227

Parrots, A. R. Wallace on, i. 397-400

Peabody Museum of Archæology, ii. 110

Pears, Messrs., offer prize for essay on "Depression of Trade," ii. 104

Pearson, Professor Karl, as socialist, ii. 272

Pengelly, William, geniality of, ii. 49 ; his personal experience of seeing a double, ii. 332–334

Pen-y-gwryd, Wales, ii. 403

"Peregrine Pickle," read by A. R. Wallace, i. 75

Perkins, Mr., i. 3

Pernambuco, i. 264 ; butterflies from, i. 266

"Personal Narrative of Travels in South America," by Humboldt, its influence on A. R. Wallace, i. 232, 256

Peru, Dr. Richard Spruce leaves, i. 411

Pestalozzi, educational system of, i. 99

Philippines, A. R. Wallace writes for Daily Chronicle on, ii. 220

Phillips, Colonel, A. R. Wallace stays with, ii. 153, 154

Phrenology, A. R. Wallace's interest in, i. 234 ; his character delineated by, i. 257-262

Physical geography, A. R. Wallace on, i. 404

"Physical History of Man," by Pritchard, comment on, i. 255

"Physician's Problems," by Dr. Elam, ii. 65

"Physiological Selection," paper by Romanes, ii. 316

goes to Batchian, i. 365 ; letter to
George Silk, i. 365 ; letter to Thomas
Sims, i. 367 ; goes to Menado,
Amboyna, and Ceram, i. 369 ; voyage
back to Ternate, i. 370 ; letter to
George Silk, i. 371 ; letter to Bates,
i. 373 ; goes to Timor and Cajeli in
Bouru, i. 375 ; Sourabaya and
Sumatra, i. 376 ; letter to H. W.
Bates, i. 377 ; letter to George Silk,
i. 379 ; journey home, i. 383 ; back
in London, i. 384 ; at work on collections, i. 385 ; "Narrative of Search
after Birds of Paradise," i. 387-
394 ; at work on various papers
dealing with birds, etc., of the Malay
Archipelago, i. 394 ; on parrots and
pigeons, i. 397 ; on butterflies, i. 400 ;
exhibition of birds and butterflies,
i. 404 ; at work on "Malay Archipelago," i. 405 ; on natural selection,
i. 406 ; presidential address to the
Entomological Society, i. 408 ; home
life in London, i. 409 ; goes to Hurstpierpoint, i. 411 ; marriage to Miss
Mitten, i. 412 ; excursions in Wales
and Switzerland, i. 413 ; returns to
Hurstpierpoint, 414 ; tries for the
assistant secretaryship of the Royal
Geographical Society, i. 415 ; applies
for directorship of the Bethnal Green
Museum, i. 415 ; applies for superin-
tendentship of Epping Forest, i. 416 ;
reminiscences of Sir Charles Lyell,
i. 417 ; on the rate of change in
insects, i. 418 ; on the antiquity of
man, i. 419 ; on distribution, i. 420 ;
on dispersal of races of man, i. 421 ;
on pangenesis, i. 422 ; letter from
Sir Charles Lyell on glacial epoch,
on lake basins, i. 426 ; on origin of
solar system, i. 427 ; letter to Sir
Charles Lyell on " Fuel of the Sun,"
i. 429 ; letter to Sir Charles Lyell on
freedom of thought as essential to
intellectual progress, i. 430 ; on dis-
establishment, i. 432 ; friendship with
Sir Charles Lyell, i. 433.

FRIENDS AND ACQUAINTANCES, ii. 1-
89—
reminiscences of Darwin, ii. 1 ; corre-

spondence, ii. 2 ; on colour of caterpillars, ii. 3 ; letter to *The Field* on
" Caterpillars and Birds," ii. 4 ; discussion on natural selection, ii. 7 ;
on meaning of origin of species,
ii. 8 ; Darwin's opinion of " Island
Life," ii. 12 ; discussion on influence
of glacial epoch, ii. 12 ; letters from
Darwin, ii. 13, 14, 15 ; differences of
opinion between Darwin and, ii. 16-
22 ; first meeting with Herbert
Spencer, ii. 23 ; discussion on flight
of birds, ii. 25 ; address on origin
of insects, ii. 26 ; letter from Herbert
Spencer on Land Nationalization
Society just formed, ii. 27 ; letter
from H. Spencer on " Progress and
Poverty," ii. 29 ; Herbert Spencer
on " Bad Times," ii. 31 ; letters from
H. Spencer on Lord Salisbury, ii. 32 ;
scientific friends, ii. 33 ; reminiscences
of Huxley, ii. 34 ; meets Dr. Miklucho Maklay, ii. 34 ; misunderstanding with Huxley, ii. 36 ; on Arthur
J. Bell's works, ii. 37 ; feeling of
inferiority to Huxley, ii. 39 ; on
degrees of latitude, ii. 40 ; reminiscences of Dr. Carpenter, ii. 42 ;
reminiscences of Mivart, ii. 43-45 ;
reminiscences of the meetings of the
British Association, ii. 45-50 ; friendship with Sir James Brooke, ii. 51 ;
his letter to Professor Rolleston on
Christianity, ii. 52 ; on " Government Aid to Science," ii. 55-59 ; his
connection with Mr. Augustus Mongredien, ii. 60 ; W. Wilson's letter
to, ii. 62 ; letter from Dr. Spruce on
the modifications in plant structure
produced by the agency of ants,
ii. 64 ; letter from Dr. Spruce on
aromatic leaves, ii. 65 ; on leaf-
cutting ants, ii. 69 ; letter from Dr.
Spruce on coloration of edible fruits,
ii. 71 ; reminiscences of Dr. Purland,
ii. 75-83 ; his connection with Mr.
Samuel Butler, ii. 83-86 ; his criticism
of Mr. Houghton, ii. 87-89 ; goes to
live at Barking, ii. 90 ; the purchase
of land at Grays, ii. 91 ; building the
house and laying out garden, ii. 92 ;

THE END

PRINTED BY WILLIAM CLOWES AND SONS, LIMITED, LONDON AND BECCLES.

Printed in the United States
By Bookmasters